EXPERIMENTS IN PRINCIPLES OF ELECTRONIC DEVICES AND CIRCUITS

by
David E. LaLond
and
John A. Ross

Delmar Publishers Inc.™

I(T)P™

NOTICE TO THE READER

Publisher does not warrant or guarantee any of the products described herein or perform any independent analysis in connection with any of the product information contained herein. Publisher does not assume, and expressly disclaims, any obligation to obtain and include information other than that provided to it by the manufacturer.

The reader is expressly warned to consider and adopt all safety precautions that might be indicated by the activities described herein and to avoid all potential hazards. By following the instructions contained herein, the reader willingly assumes all risks in connection with such instructions.

The publisher makes no representations or warranties of any kind, including but not limited to, the warranties of fitness for particular purpose or merchantability, nor are any such representations implied with respect to the material set forth herein, and the publisher takes no responsibility with respect to such material. The publisher shall not be liable for any special, consequential or exemplary damages resulting, in whole or in part, from the readers' use of, or reliance upon, this material.

For information, address

Delmar Publishers Inc.
3 Columbia Circle
Box 15015
Albany, New York 12212-5015

Printed in the United States of America
Published simultaneously in Canada
by Nelson Canada,
a division of The Thompson Corporation

1 2 3 4 5 6 7 8 9 10 XXX 00 99 98 97 96 95 94

ISBN: 0-8273-4664-6
Library of Congress Catalog Card Number: 92-36004

CONTENTS

Preface vi

Suggested Experiment Usage vi

EXPERIMENTS

1 VOLTAGE DIVIDERS 1

2 THEVENIN'S THEOREM 5

3 SUPERPOSITION THEOREM 3

4 TEST EQUIPMENT LIMITATIONS 4

5 THE PN JUNCTION DIODE 17

6 RECTIFIER FORMS 21

7 CAPACITOR INPUT FILTERS 27

8 ZENER REGULATION OF POWER SUPPLIES 32

9 VOLTAGE MULTIPLIERS 36

10 SIGNAL CLIPPERS AND CLAMPERS 40

11 BASIC BJT CHARACTERISTICS 47

12 BETA EFFECTS IN THE BJT 51

13 EMITTER BIAS OF THE BJT 55

14 COLLECTOR FEEDBACK BIAS 59

15 VOLTAGE DIVIDER BIAS FOR BJTS 63

16 BIASING PNP TRANSISTORS 67

17 COMMON EMITTER AMPLIFIER 71

18 COLLECTOR FEEDBACK BIASED AMPLIFIER 76

19 COMMON COLLECTOR AMPLIFIER 81

20 COMMON BASE AMPLIFIER 86

21 MULTISTAGE AMPLIFIERS 90

22 CLASS A POWER AMPLIFIERS 94

23 CLASS B PUSH-PULL AMPLIFIERS 98

24 DARLINGTON AMPLIFIERS 103

25 CLASS C AMPLIFIERS 107

26 DIRECT-COUPLED AMPLIFIERS 112

27 JFET CHARACTERISTICS 116

28 SELF-BIASED JFET 120

29 VOLTAGE DIVIDER BIAS 124

30 CURRENT SOURCE BIASED JFET 128

31 COMMON-SOURCE JFET AMPLIFIERS 132

32 COMMMON-DRAIN JFET AMPLIFIERS 136

33 BIASING D MOSFETS 140

34 MOSFET AMPLIFIERS 144

35 BJT SWITCHES 148

36 BJT SCHMITT TRIGGER 153

37 JFET SWITCHES 157

38 UJT RELAXATION OSCILLATORS 160

39 BJT RAMP GENERATORS 164

40 FREQUENCY EFFECTS IN BJT AMPLIFIERS 167

41 FREQUENCY EFFECTS IN JFET AMPLIFIERS 171

42 BJT DIFFERENTIAL AMPLIFIERS 175

43 BASIC OP-AMP PARAMETERS 179

44 OP-AMP SLEW RATE AND CMRR 183

45 NONINVERTING VOLTAGE AMPLIFIERS 187

46 INVERTING VOLTAGE AMPLIFIERS 191

47 OP-AMP CURRENT AMPLIFIERS 195

48 VOLTAGE-TO-CURRENT CONVERTERS 199

49 SUMMING AMPLIFIERS 202

50 RC OSCILLATORS 206

51 COLPITTS AND CLAPP OSCILLATORS 210

52 HARTLEY OSCILLATORS 214

53 RELAXATION OSCILLATORS 217

54 DIFFERENTIATORS AND INTEGRATORS 221

55 OP-AMP DIODE CIRCUITS 226
56 SCHMITT TRIGGER CIRCUITS 231
57 WINDOW COMPARATORS 234
58 ACTIVE LOW-PASS FILTERS 241
59 ACTIVE HIGH-PASS FILTERS 242
60 ACTIVE BAND-PASS FILTERS 246
61 ACTIVE BAND-REJECT FILTERS 250
62 VCVS ACTIVE FILTERS 254
63 INSTRUMENTATION AMPLIFIERS 258
64 DIGITAL-TO-ANALOG CONVERSION 261
65 555 TIMER CIRCUITS 265
66 SILICON-CONTROLLED RECTIFIERS 270
67 TRIACS 274
68 FULL-WAVE PHASE CONTROL 278
69 SERIES PASS REGULATORS 282
70 IC REGULATORS 285
71 SIGNAL MODULATION AND DEMODULATION 288
72 PHASE-LOCKED LOOPS 291
73 VARACTOR DIODES 295

APPENDICES

A COMPOSITE EQUIPMENT AND MATERIALS LIST . . 296
B DATA SHEETS 300

PREFACE

Experiments in *Electronic Devices and Circuits* is designed as a learning companion to the text, *Principles of Electronic Devices and Circuits*. This manual is uniquely structured. The differences are the number of experiments, the extent of the troubleshooting sections, and the use of a format that enhances the learning experience.

The unusually large number of experiments provides a strong base from which instructors can select laboratory activities that support the emphasis of their specific program. The troubleshooting portions of the experiments are located in separate sections that permit assignment as desired. The troubleshooting was designed to simulate realistic circuit faults and, to a major extent, avoid fault simulation by means of a missing element–a questionable technique which tends to produce awkward measurements.

The experiment format is organized to allow the learner to build a circuit, then make functional measurements to see the operating characteristics of the circuit. Coupled with this approach is a discussion section which extends the learner's thinking into further considerations of circuit characteristics and applications.

In summary, these experiments reflect an approach adopted after years of watching students compare measured values with calculated values, and then attempt to describe what they learned from the process.

Several individuals deserve credit and praise for their contributions to this undertaking. Special thanks to Charles A. Heskett for his time and talents. The following people had a part in creating this work: Robert Doyle, Arnie Garcia, David Leigon, Daniel Lookadoo, Daniel Presson, and Pat Thomason. The authors wish to thank all of them and express gratitude for their efforts.

SUGGESTED EXPERIMENT USAGE

It is recognized that, even in a two-quarter program, it would be difficult to assign (and expect every student to complete) all of the experiments available in this manual. The tables on the following pages list a nominal assignment rate that is based on an assumed laboratory time of approximately eight lab hours per week. In this list, some experiments are marked in italics. These specially marked experiments could be deleted to accommodate shorter laboratory times or slower work rates. In addition, the tables list supplementary experiments that might be assigned for a particular program emphases or extra work for advanced learners.

Suggested Experiments Two-Quarter Program		
Text Chapter(s)	Main Experiments	Supplementary Experiments
1 1 1 2	1 Voltage Dividers 2 Thevenin's Theorem 4 *Test Equipment Limitations* 5 The PN Junction Diode	3 Superposition Theorem
3 3 3	6 Rectifier Forms 7 Capacitor Input Filters 8 Zener Regulation Of Power Supplies	
3 3 4	9 *Voltage Multipliers* 10 Signal Clippers and Clamper 11 Basic BJT Characteristics	
4 4 4	12 *Beta Effects in the BJT* 13 Emitter Bias of the BJT 15 Voltage Divider Bias for BJTs	14 Collector Feedback Bias
4 5 5	16 *Biasing PNP Transistors* 17 Common Emitter Amplifier 19 Common Collector	18 Collector Feedback Biased Amplifier
5 5	20 Common Base Amplifier 21 Multistage Amplifiers	
6 6 6	22 Class A Power Amplifiers 23 Class B Push-Pull Amplifiers 24 *Darlington Amplifiers*	25 Class C Amplifiers 26 Direct-Coupled Amplifiers
7 7 7	27 *JFET Characteristics* 28 Self-Biased JFET 29 Voltage Divider Bias	30 Current Source Biased JFET
7 7 8	31 Common-Source JFET Amplifiers 32 Commmon-Drain JFET Amplifiers 33 *Biasing D MOSFETS*	34 MOSFET Amplifiers
9 9 9	35 BJT Switches 36 *BJT Schmitt Trigger* 37 JFET Switches	38 UJT Relaxation Oscillators 39 BJT Ramp Generators
10 10	40 Frequency Effects in BJT Amplifiers 41 Frequency Effects in JFET Amplifiers	
10 11 11	42 BJT Differential Amplifiers 43 Basic Op-Amp Parameters 44 Op-Amp Slew Rate and CMRR	

Suggested Experiments Two-Quarter Program (Cont'd)		
Text Chapter(s)	Main Experiments	Supplementary Experiments
12 12 12	45 Noninverting Voltage Amplifiers 46 Inverting Voltage Amplifiers 48 *Voltage-to-Current Converters*	47 Op-Amp Current Amplifiers
12 13	49 Summing Amplifiers 50 RC OScillators	
13 13 13	51 Colpitts and Clapp Oscillators 52 *Hartley Oscillators* 54 Differentiators and Integrators	53 Relaxation Oscillators
14 14	55 Op-Amp Diode Circuits 56 Schmitt Trigger Circuits	57 Window Comparators
15 15 15	58 Active Low-Pass Filters 59 Active High-Pass Filters 60 Active Band-Pass Filters	
16 16	61 Active Band-Reject Filters 62 VCVS Active Filters	
16 16	63 Instrumentation Amplifiers 64 Digital-to-Analog Conversion	
17	65 555 Timer Circuits	
18 18	66 Silicon-Controlled Rectifiers 67 Triacs	68 Full-Wave Phase Control
19 19	69 Series Pass Regulators 70 IC Regulators	
20 20 20		71 Signal Modulation and Demodulation 72 Phase-Locked Loops 73 Varactor Diodes

Suggested Experiments One-Semester Program		
Text Chapter(s)	**Main Experiments**	**Supplementary Experiments**
1 1 2	1 Voltage Dividers 4 Test Equipment Limitations 5 The PN Junction Diode	2 Thevenin's Theorem 3 Superposition Theorem
3 3 3	7 Capacitor Input Filters 8 Zener Regulation Of Power Supplies 9 Voltage Multipliers	6 Rectifier Forms
4 4 4 4 4	10 Signal Clippers and Clamper 12 Beta Effects in the BJT 13 Emitter Bias of the BJT	11 Basic BJT Characteristics 14 Collector Feedback Bias 15 Voltage Divider Bias for BJTs 16 Biasing PNP Transistors
5 5 5 5	17 Common Emitter Amplifier 19 Common Collector Amplifier 21 Multistage Amplifiers	18 Collector Feedback Biased Amplifier 20 Common Base Amplifier
6 6 6 6	22 Class A Power Amplifiers 23 Class B Push-Pull Amplifiers 26 Direct-Coupled Amplifiers	24 Darlington Amplifiers 25 Class C Amplifiers
7 7 7	28 Self-Biased JFET 29 Voltage Divider Bias 31 Common-Source JFET Amplifiers	27 JFET Characteristics 30 Current Source Biased JFET 32 Commmon-Drain JFET Amplifiers
8 8	33 Biasing D MOSFETS 34 MOSFET Amplifiers	
9 9 9	35 BJT Switches 37 JFET Switches	36 BJT Schmitt Trigger 38 UJT Relaxation Oscillators 39 BJT Ramp Generators
10 10	40 Frequency Effects in BJT Amplifiers 42 BJT Differential Amplifiers	41 Frequency Effects in JFET Amplifiers
11 11	43 Basic Op-Amp Parameters 44 Op-Amp Slew Rate and CMRR	
12 12 12	45 Noninverting Voltage Amplifiers 46 Inverting Voltage Amplifiers	47 Op-Amp Current Amplifiers 48 Voltage-to-Current Converters 49 Summing Amplifiers
13 13 13 13	50 RC OScillators 51 Colpitts and Clapp Oscillators 54 Differentiators and Integrators	52 Hartley Oscillators 53 Relaxation Oscillators

Suggested Experiments One-Semester Program (Cont'd)		
Text Chapter(s)	Main Experiments	Supplementary Experiments
14 14 14	55 Op-Amp Diode Circuits 56 Schmitt Trigger Circuits 57 Window Comparators	
15 15 15	58 Active Low-Pass Filters 59 Active High-Pass Filters 60 Active Band-Pass Filters	
16 16	61 Active Band-Reject Filters 63 Instrumentation Amplifiers	62 VCVS Active Filters
16 17	 65 555 Timer Circuits	64 Digital-to-Analog Conversion
18 18	66 Silicon-Controlled Rectifiers 68 Full-Wave Phase Control	67 Triacs
19 19 20 20 20	69 Series Pass Regulators 70 IC Regulators	 71 Signal Modulation and Demodulation 72 Phase-Locked Loops 73 Varactor Diodes

1

VOLTAGE DIVIDERS

INTRODUCTION

The theory of voltage dividers is an interesting and important tool when used to analyze series electronic circuits. The first part of this laboratory experiment will demonstrate to you three facts about voltage dividers:

1. The sum of all voltage drops is equal to voltage applied (V_A).

2. Current flow is the same at any point in the circuit.

3. The voltage dropped across any one resistor in a series circuit is "equal to the ratio of that resistance value to the total resistance (R_X / R_T) times the applied voltage (V_A)."

You will see that the voltage divider principle is also used in series-parallel circuits. This occurs when you place a "load" on one of the resistors in the voltage divider. The troubleshooting part of the experiment permits you to see the effect of a resistor failure in your voltage divider circuit. You will also be able to relate measurement values to circuit fault.

REFERENCE

Principles of Electronic Devices and Circuits - Chapter 1, Section 1.2

OBJECTIVES

In this experiment you will:

✓ Prove that multiple voltages are available with the use of only one power supply

✓ Study the effect of a load on a voltage divider circuit

✓ Be able to relate measured values to circuit faults

EQUIPMENT AND MATERIALS

DC power supply
Digital multimeter
Circuit protoboard
Resistors: 1 kΩ, 2 kΩ, 3 kΩ, 5 kΩ, 1 MΩ

The circuit shown below in Figure 1.1 is the same one explained in your textbook. Since all calculations for the circuit are done, you are ready to construct this circuit on your breadboard.

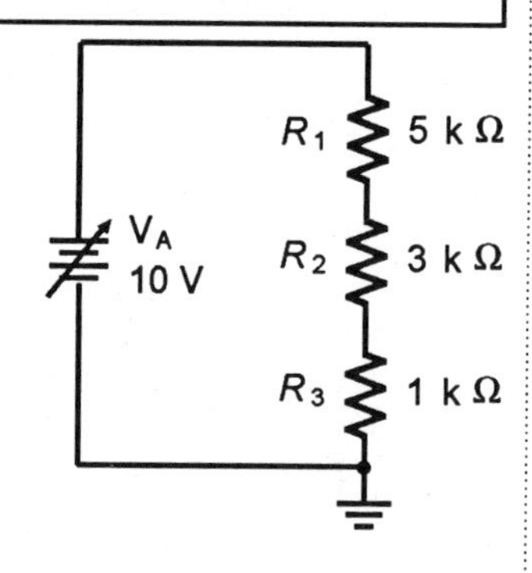

Figure 1.1

SECTION I FUNCTIONAL EXPERIMENT

1. Adjust the variable power supply to 10 V. With the power supply turned off, connect power to the circuit.

2. Since you already know the expected voltage drops on each resistor, it is now time to verify these calculated results. Use your multimeter to measure each voltage drop. Record your readings in the chart below.

   V_{R1} = __________ V_{R2} = __________ V_{R3} = __________

3. To prove that the sum of these three voltage drops is equal to the voltage applied (V_A) to the circuit, sum the individual voltage drops data from step 2 above to calculate V_A.

   V_A = ________________

Your calculated value of V_A should agree (within the voltmeter accuracy). If there is a significant difference, recheck your measurements of steps 1, 2, and 3.

4. Using Ohm's law, you can also prove that the current flowing through this circuit is the same at any point in that path. This is accomplished by taking each voltage drop and dividing it by its respective resistance value. Record your results below.

   I_{R1} = __________ I_{R2} = __________ I_{R3} = __________

5. Construct the circuit shown in Figure 1.2 and adjust the power supply to 10 V.

The portion of this lab starting with step 5 will enable you to study the effect of a "load" on a voltage divider. The circuit used in Figure 1.1 will be utilized so that we can have a good reference point from which to make comparisons

The addition of a 1-kΩ load across R_1 alters the makeup of the circuit in Figure 1.2. This is due to the fact that the new load has provided another path for current to flow. Another item to consider will be the effect this load will have on total resistance and, therefore, total current.

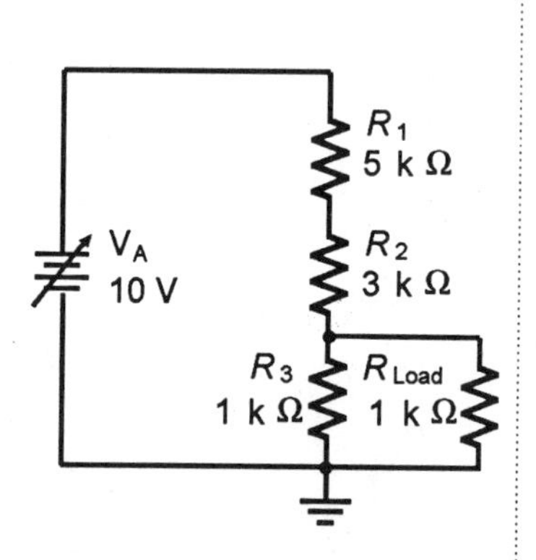

Figure 1.2

6. Recalculate the new circuit values and record the data in the *Calculated* column in Table 1.1. Measure the same circuit values and record them in the *Measured* column in Table 1.1

	Calculated	Measured
I_T		
R_T		
I_{R1}		
I_{R2}		
I_{R3}		
V_{R1}		
V_{R2}		
V_{R3}		
V_{RLoad}		
I_{RLoad}		

Table 1.1

SECTION II TROUBLESHOOTING

Fault 1 - Changing resistance in the voltage divider

1. If one resistor in your original circuit fails, then the voltage drops in the circuit would be expected to change also. This can be simulated easily by merely replacing one resistor with a large resistance to simulate a failed (open) resistor. With the power supply turned off, remove R_3 (5 kΩ) from the circuit and replace it with a 1-MΩ resistor.

2. Using your digital multimeter, measure the voltage drop on each resistor and record this reading below.

   V_{R1} = __________ V_{R2} = ____________ V_{R3} = ____________

Your measurements in Step 2 should clearly indicate that the circuit fault is associated with R_3.

3. Although the the voltage measurements for Figure 1.2 are different from the ones for Figure 1.1, their sum should still be equal to the total voltage applied. Verify this concept below.

   $V_A = V_{R1} + V_{R2} + V_{R3}$ = ________________________________

Fault 2 - Load resistor open

1. Using Figure 1.2, let's assume that the 1-kΩ load resistor fails, or opens. The parallel combination of R_1 (1 kΩ) and R_{Load} is different from before. Construct the circuit of Figure 1.2. Modify the circuit by replacing the 1-kΩ R_L with a 1-MΩ resistor.

   R_T = ________________ I_T = ______________

   $V_{R1} = V_{RLoad}$ = ____________

   V_{R2} = ________________ V_{R3} = ______________

2. Apply 10 V to the circuit and measure the values listed below.

   $V_{R1} = V_{RLoad}$ = ____________

   V_{R2} = ________________ V_{R3} = ______________

3. From your measured data it should be clear that the circuit fault lies with R_3 or R_L. List the measurement you would make to isolate the failure to the specific resistor that failed. ______________________________

DISCUSSION

Section I

1. In Section I you measured voltage drops and compared those to the calculated values in your textbook. Discuss the relationship of each resistor's ratio to the measured voltage drop.

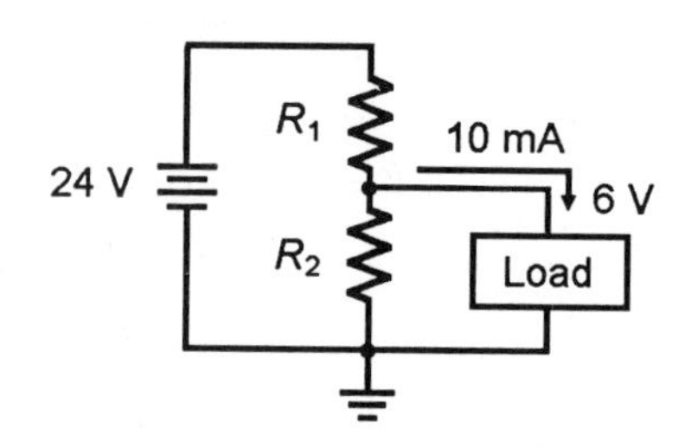

Figure 1.3

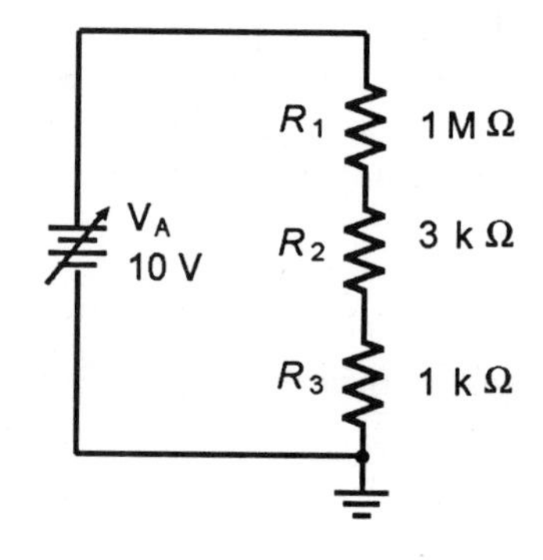

Figure 1.4

2. Discuss how a voltage divider could be used to supply a certain amount of voltage to a circuit or to a component.

3. Discuss how you would connect a load that operates with a certain voltage/current demand to a voltage divider circuit (see Figure 1.3).

Section II

1. If the circuit of Figure 1.4 were mounted in a printed circuit board, describe the measurements you would make to be certain that the failure was an open resistor rather than a break in the solder connection on the board connecting one leg of R_3 to the circuit board.

2. Referring to the loaded voltage divider circuit failure, could you with just one voltage measurement determine if the failure was R_L instead of R_3? Explain why you chose your answer.

Quick Check

1. Solve for the voltage drops of the voltage divider shown in Figure 1.5.

2. If R_2 of Figure 1.5 was removed and replaced with an 8 kΩ resistor, what would be the new value of current flow? What effect, if any, does this have on V_{R1} and V_{R3}?

3. Figure 1.6 shows a loaded voltage divider. With the information given, determine the following values:

V_{R3} = ____________ I_{R1}= ____________

I_{R2} = ____________ R_2 = ____________

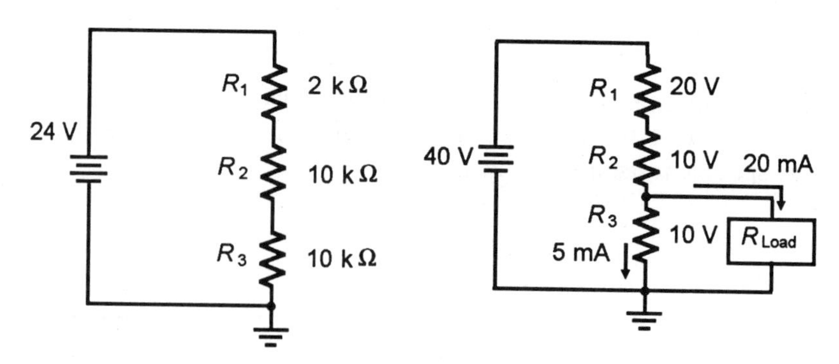

Figure 1.5 Figure 1.6

2

THEVENIN'S THEOREM

INTRODUCTION

Thevenin's theorem provides a way to take a complex circuit and reduce it to a simple Thevenin voltage (V_{TH}) source in series with a Thevenin resistance (R_{TH}). Thevenin's theorem is also used for simplifying circuits that involve more than one power source. This experiment provides a functional review of the application of Thevenin's theorem and an experimental application.

REFERENCE

Principles of Electronic Devices and Circuits - Chapter 1, Section 1.4

OBJECTIVES

In this experiment you will:

✓ Reduce a complex resistive circuit to a single resistance (R_{TH}) in series with a single voltage source (V_{TH})

✓ Experimentally verify Thevenin's theorem through voltage measurements

EQUIPMENT AND MATERIALS

DC power supply
Digital multimeter
Circuit protoboard
Resistors: 2.7 kΩ, 82 kΩ
Potentiometers: 5 kΩ [2], 10 kΩ

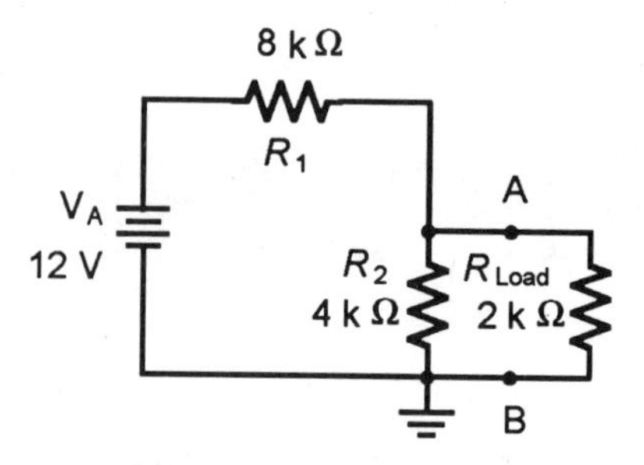

Figure 2.1

1. Construct the circuit of Figure 2.1 on your circuit breadboard. Use the potentiometers adjusted to the required resistance value.

2. The total current (I_T) and load current (I_L) have been calculated in the textbook on page 28. With that information, determine the expected value of voltage on the load.

   $V_{Load\ (calc.)}$ = ____________

3. Adjust the power supply for 12 V and connect this voltage to the circuit. Measure load voltage and record below.

   $V_{Load\ (measured)}$ = ____________

4. Using the circuit of Figure 2.1, Thevenize and reconstruct the circuit on your breadboard.

   NOTE: The calculations for this particular circuit have been completed on pages 30 and 31 of the textbook. Figure 2.2 shows the equivalent (Thevenized) circuit after your calculations.

   A standard 2.7-kΩ resistor is used instead of the calculated 2.67-kΩ resistance shown on page 31 of the text. If you want to be more accurate, use a potentiometer that is adjusted to 2.67 kΩ in place of the 2.7-kΩ resistor.

   $V_{Load(calc.)}$ = ____________

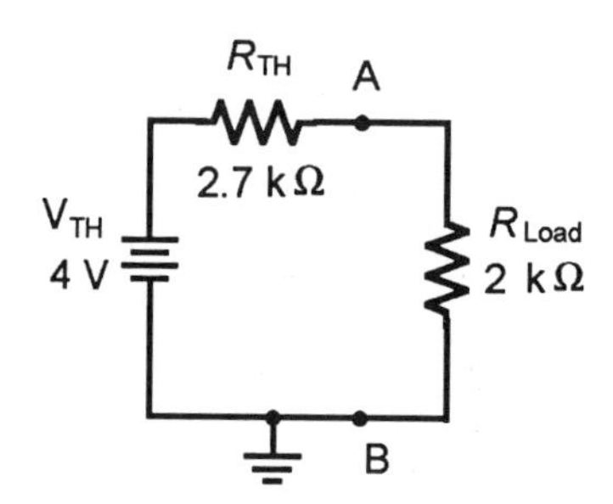

Figure 2.2

5. Apply 4 V to the Thevenized circuit and measure the voltage drop across the load.

   $V_{Load(measured)}$ = ____________

   How does this measurement compare to the measurement across the load in step 3?

   ____

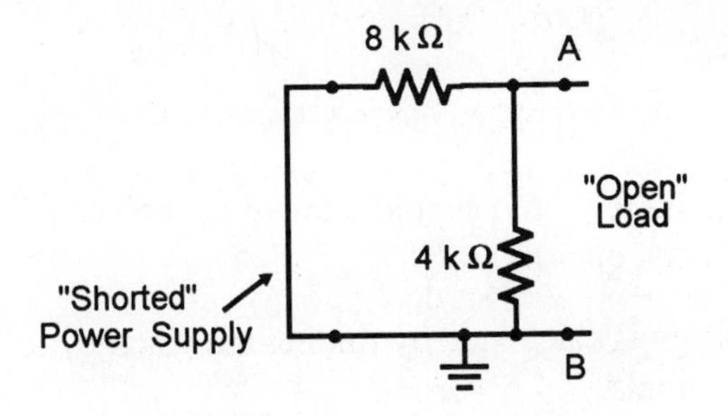

Figure 2.3

6. To verify R_{TH}, construct the circuit of Figure 2.3 and measure the resistance at points A and B.

7. Turn off the circuit power and restore your circuit to that of Figure 2.1, except use an 82-kΩ load resistance. Apply 12 VDC to the circuit and measure the value of V_L.

   V_L = ______________

8. Calculate for your circuit the value of I_T, I_L, and V_L.

   I_T = __________ I_L = ___________ V_L = __________

9. Modify your circuit to the Thevenin equivalent (Figure 2.2 where $R_L = 82$ kΩ).

10. Apply 4 VDC and measure the value of V_L.

    V_L = ______________

11. Using the Thevenin equivalent circuit form, calculate the value of I_T and V_L.

    I_T = ______________ V_L = ______________

    Can you agree that calculations of circuit values are much easier using the "Thevenized" circuit form?

This completes the measurements of this experiment.

DISCUSSION

1. Discuss the reasons for any differences you might have encountered between your calculated values and your measured values.

2. Discuss how Thevenin's theorem could be useful in the "real world."

Quick Check

1. When calculating R_{TH}, you should remove the power supply and replace it with a/an (short, open).

2. Thevenin voltage is calculated by finding the voltage drop at the (loaded, unloaded) terminals.

3. Determine the Thevenin voltage and resistance for Figure 2.4.

   V_{TH} = ______________ R_{TH} = ______________

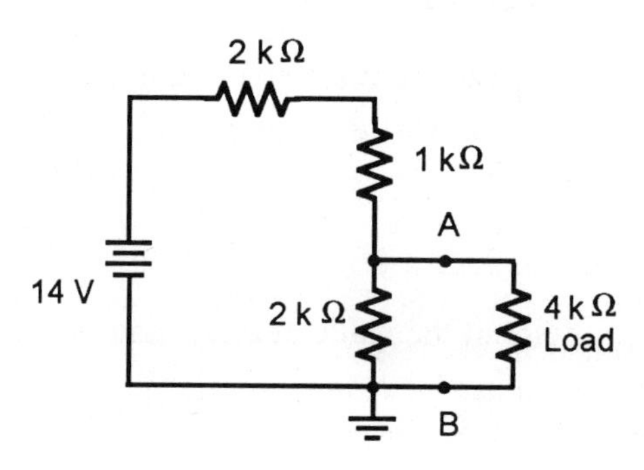

Figure 2.4

3

SUPERPOSITION THEOREM

INTRODUCTION

The superposition theorem is helpful when analyzing circuits that have more than one source. The superposition theorem can, with some restrictions, be used in both AC and DC and in circuits where both sources are used. In this experiment you will construct and analyze two circuits, one with both DC sources and the other with an AC and a DC source.

REFERENCE

Principles of Electronic Devices and Circuits - Chapter 1, Section 1.6

OBJECTIVES

In this experiment you will:

✓ Demonstrate the superposition theorem in DC circuits

✓ Demonstrate the superposition theorem in AC- and DC-sourced circuits

EQUIPEMENT AND MATERIALS

DC power supply [2]
Function generator
Digital multimeter [2]
Oscilloscope
Circuit protoboard
Resistors: 220 Ω, 330 Ω, 1 kΩ

SECTION I FUNCTIONAL EXPERIMENT

NOTICE
When a step in this experiment instructs you to "short" a power supply, you should disconnect the designated power supply and replace its connections with a jumper wire.

Two DC Sources

1. Construct the circuit in Figure 3.1.

2. Calculate R_T, I_3, and V_3 for V_B shorted. Record as V_B *Shorted* in the *Calc.* columns of Table 3.1.

3. Short power supply V_B. Disconnect V_A and measure R_T. Record as V_B *Shorted* in the R_T column of Table 3.1.

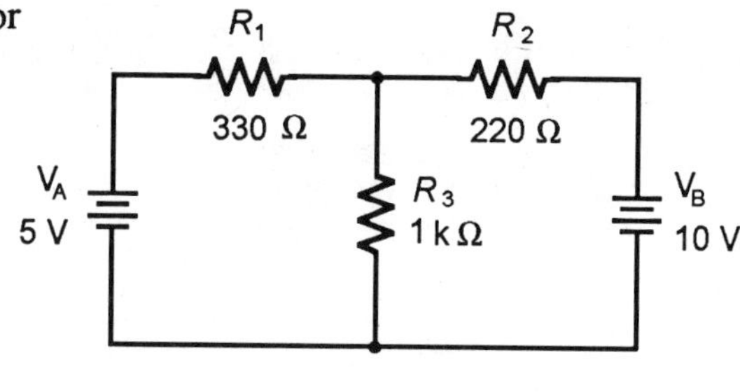

Figure 3.1

4. Reconnect the V_A supply, but *do not turn it on at this time.*

5. Connect an ammeter in the R_3 branch.

6. Turn on the V_A supply. Measure I_3 and V_3, and record the values as V_B *Shorted* in the *Meas.* columns of Table 3.1.

7. Turn off the power.

8. Calculate R_T, I_3, and V_3 for V_A shorted. Record as V_A *Shorted* in the *Calc.* columns of Table 3.1.

9. Short V_A. Disconnect V_B and measure R_T. Record as V_B *Shorted* in the R_T column of Table 3.1

10. Repeat steps 4, 5, 6, and 7, using V_B as the active supply.

11. Use the superposition theorem to calculate the values of I_3 and V_3 in the full circuit. Record as *Full Circuit* items in the *Calc.* columns of Table 3.1.

12. Reconnect V_A so that both supplies are in the circuit, and insert an ammeter in the R_3 branch.

13. Turn on the power and measure I_3 and V_3.

Do your measured values in step 13 agree with the calculated values for I_3 and V_3 with two supplies? If they do not, recheck your calculations and procedural steps.

	R_T		I_3		V_3	
	Calc.	Meas.	Calc.	Meas.	Calc.	Meas.
V_B Shorted						
V_A Shorted						
Full Circuit	---					

Table 3.1

Mixed DC and AC Sources

1. Construct the circuit in Figure 3.2.

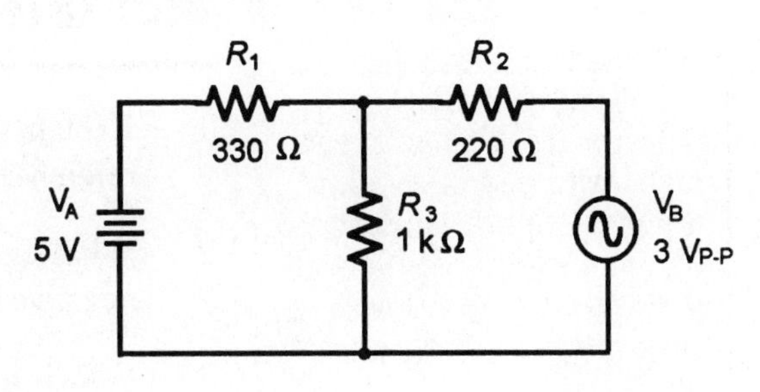

Figure 3.2

2. Calculate the values for R_T, I_3, and V_3 for V_A and V_B shorted. Record the values in the *Calc.* columns of Table 3.2.
3. Short supply V_B. Disconnect V_A and measure R_T. Record as *V_B Shorted* in the *R_T Meas.* column of Table 3.2.
4. Reconnect the V_A supply, but *do not turn it on at this time*.
5. Connect an ammeter in the R_3 branch.
6. Turn on the power. Measure I_3 and V_3, and record as *V_B Shorted* in the *Meas.* columns of Table 3.1.
7. Turn off power.
8. Short supply V_A. Disconnect V_B and measure R_T. Record as *V_B Shorted* in the *R_T Calc.* column of Table 3.2.
9. Repeat steps 4, 5, 6, and 7, using V_B as the active supply.
10. Use the superposition theorem to calculate the values of I_3 and V_3 in the full circuit. Record as *Full Circuit* items in the *Calc.* columns of Table 3.2.
11. Reconnect the circuit so that both supplies are active.
12. Insert an ammeter in the R_3 branch.
13. Turn on the power. Measure I_3 and V_3, and enter the values as *Full Circuit* items in the *Meas.* columns of Table 3.2.

    Do your measured values in step 13 agree with the calculated values for I_3 and V_3 with two supplies? If they do not, recheck your calculations and procedural steps.

14. Connect your oscilloscope, DC coupled, across R_3. Sketch the waveform in the space provided as Graph 3.1.

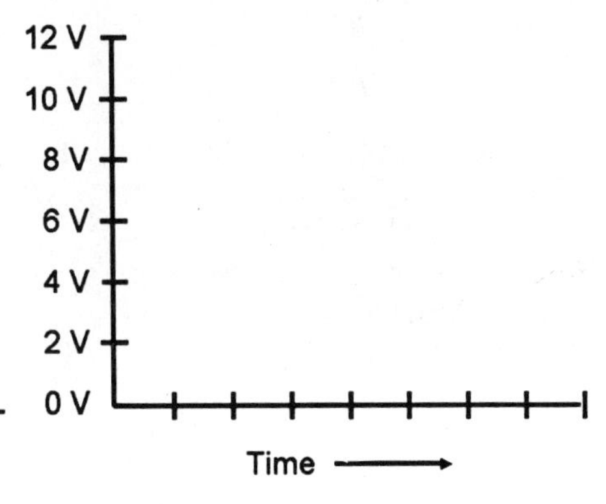

Graph 3.1

	R_T		I_3		V_3	
	Calc.	Meas.	Calc.	Meas.	Calc.	Meas.
V_B Shorted						
V_A Shorted						
Full Circuit	---					

Table 3.2

DISCUSSION

1. Explain the signal you drew in step 13. Indicate for example, the reference voltage. Was it zero? Why?

2. What signal would you expect to see if the DC and AC levels were negative?

3. Why is it necessary to remove the source from the circuit before measuring the total resistance of the circuit?

Quick Check

1. In circuit of Figure 3.1, your calculated R_T for V_B shorted was approximately:

   (a) 150 Ω (b) 510 Ω
   (c) 330 Ω (d) 1330 Ω

2. The equation for finding R_T in question 1 is $R_T = R_1 + \frac{R_2 R_3}{R_2 + R_3}$.

   True False

3. In the superposition theorem, currents and voltages are analyzed with one source applied at a time.

   True False

4. The superposition theorem can be used to analyze circuits with both DC and AC sources.

   True False

5. It is not possible to determine the direction of current flow using the superposition theorem.

   True False

4

TEST EQUIPMENT LIMITATIONS

INTRODUCTION

Measuring instruments are very important in technology. Whether you are troubleshooting or gathering data for an engineering project, having good measurement skills and knowing the limitations of your test equipment are imperative. When using a DMM, VOM, or an oscilloscope, several things you should remember.

Each piece of measurement equipment, oscilloscope, CMM, or VOM has a finite input impedance. This impedance, in parallel with the circuit element where the measurement is being made, can alter the circuit, and thus the measurement. Secondly, when you are measuring AC voltages, frequency limitations of the measuring equipment can result in misleading data being obtained.

In this experiment you will observe the effect of meter loading on a circuit and measure the input impedance of a meter. You will also examine the frequency limitations of the AC voltmeter.

REFERENCE

Principles of Electronic Devices and Circuits - Chapter 1, Section 1.7

OBJECTIVES

In this experiment you will:

✓ Demonstrate the effect of meter loading

✓ Learn a technique to determine the input impedance of a meter

✓ Learn the frequency limitations of the DMM (or VOM)

EQUIPMENT AND MATERIALS

DC power supply
Function generator
Oscilloscope
Circuit protoboard
VOM and DMM (Also the meters' specifications)
Resistors: 10 kΩ [2], 1 MΩ
Potentionmeter: 2-MΩ or 5-MΩ ten-turn trimpot

SECTION I FUNCTIONAL EXPERIMENT

Input Impedance

1. Determine the input impedance from you meters' specification for two voltage scales (ranges). Record below.

   Scale: ______________ Z_{in} ______________

   Scale: ______________ Z_{in} ______________

2. Adjust the DC supply voltage to equal the first voltage scale of step 1.

3. Connect the circuit of Figure 4.1. Do not readjust the power supply value set in step 2. While measuring the voltage across R_1, adjust the potentiometer until the meter displays one-half of the initial voltage set in step 2. Since this is a series circuit and each impedance-potentiometer and meter has one-half the total supply voltage, their impedances are equal.

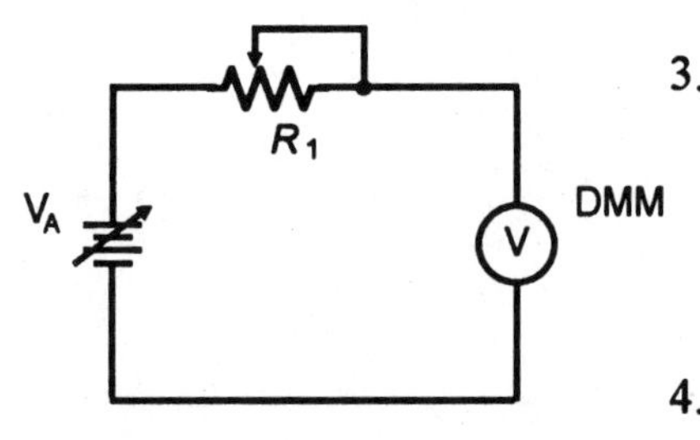

Figure 4.1

4. Turn off the power supply and remove the potentiometer.

5. Measure the resistance of the potentiometer. Compare this measurement to the rated Z_{in} of step 1 and record below.

   $Z_{in\ (meter)}$ = ____________

6. Repeat steps 3, 4, and 5 for each of the voltage ranges of step 1.

Effects of Input Impedance

In this part of the experiment, you will observe the effect of meter loading. It will be necessary to measure the resistance of R_2 in Figure 4.2.

1. Using an ohmmeter, measure the resistance values of the 1-MΩ resistor. Measure and adjust your potentiometer to the same value.

2. Construct the circuit in Figure 4.2. Adjust the DC supply to provide 10 V, and record the supply value. Measure and record the voltage across R_2 using your DMM.

   V_{DC} = ______________

   $V_{R2\ (measured)}$ = ____________

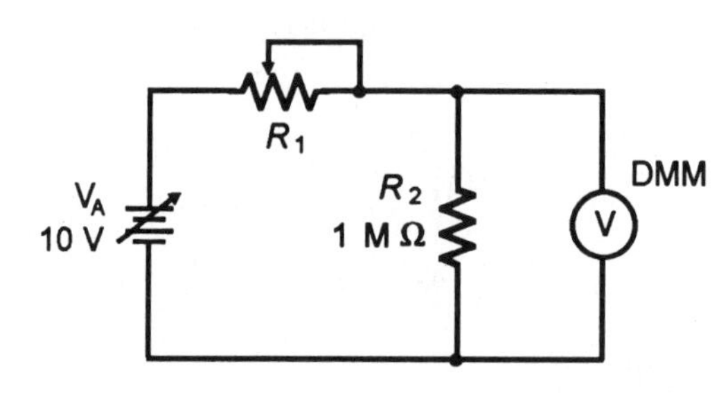

Figure 4.2

3. Repeat the measurement of step 2 using a VOM, if available. Record your meter reading below.

   $V_{R2(measured)} =$ ____________

4. The expected voltage reading across R_2 is one-half the DC supply voltage. Calculate and record the expected meter reading.

   $V_{R2\ (expected)} =$ ____________

   Calculate the error in the actual voltage reading using the formula below.

$$\%\ \text{Error} = \frac{V_{Expected} - V_{Measured}}{V_{Expected}}$$

   Record the % Error for the readings of Procedure steps 2 and 3.

   step 2 %Error = ____________

   step 4 % Error = ____________

These errors, while they include meter accuracy, are due to loading of the circuit by the voltmeter.

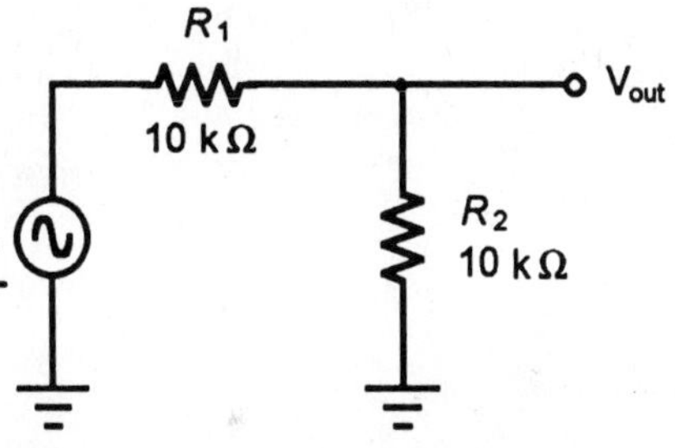

Voltmeter Frequency Limitations

1. Construct the circuit of Figure 4.3. Using your oscilloscope, set the function generator to provide a sinewave signal of 3 $v_{p\text{-}p}$ at 1 kHz.

2. Using your oscilloscope, measure and record the voltage across R_2.

   $V_{R2} =$ ____________ (p-p)

   Calculate the RMS value of the R_z value recorded above.

   $V_{R2} =$ ____________ (RMS)

   Use your DMM to measure and record the voltage across R_2.

   V_{R2} (DMM) = ____________

3. Increase the input frequency to 20 kHz. Repeat the measurements of steps 1 and 2.

   $V_{R2} =$ ____________ (p-p)

   $V_{R2} =$ ____________ (RMS)

   $V_{R2\ (DMM)} =$ ____________

You should have found that at the higher frequency, your DMM measured a lower value than it did at 1 kHz. This is due to frequency response characteristics of the DMM.

Figure 4.3

DISCUSSION

1. Describe the effect the meters had on the voltage measurements in the Procedure for **Effects of Input Impedance**. Was there a difference when using the DMM versus the VOM?

2. Explain why it is important that you understand circuit loading by your measurement equipment.

3. Given an oscilloscope with a bandwidth of 20 MHz and your DMM, discuss which one you would use in making measurements of a circuit operating at 50 kHz. Also, why would you select the one you did?

Quick Check

1. To obtain an accurate measurement, the ohmmeter must have an input impedance of at least 10 times greater than that of the component being measured.

   True False

2. It is okay to use an ohmmeter in a circuit where power is applied.

   True False

3. Voltage is always measured across the component.

   True False

4. An oscilloscope is used only to observe waveforms.

   True False

5

THE PN JUNCTION DIODE

INTRODUCTION

The PN junction diode in the simplest sense is a device that will conduct current in one direction and block current in the opposite direction. When forward biased to overcome the internal barrier potential, the diode will conduct. Since its forward-biased resistance is low, current must be limited by external resistance of the circuit. When the diode is reverse biased, the diode current is very small, typically in the nano amp range, thus approximating an open circuit.

In this experiment, you will perform measurements to let you see the characteristics of the PN junction diode. Also, from your measured data, you will plot a typical diode characteristic curve.

REFERENCE

Principles of Electronic Devices and Circuits - Chapter 2, Sections 2.6 and 2.7

OBJECTIVES

In this experiment you will:

✓ Determine forward and reverse resistance of the diode

✓ Measure the forward voltage and current of a diode and plot the result

EQUIPMENT AND MATERIALS

DC power supply
Digital multimeter [2]
Circuit protoboard
Small-signal diode, 1N914 or similar
Resistors: 220 Ω, 220 kΩ

SECTION I FUNCTIONAL EXPERIMENT

The first measurement of the diode is a simple resistance test that will give you an idea of the diode forward and reverse resistances. This test is also a good quick to verify a good (or failed) diode. Set your DMM to ohms and the 1-kΩ range. If your meter has a diode test range use that range.

CAUTION

When making resistance checks of a diode, do not use low meter ranges. Some ohmmeters can supply sufficient voltage with minimum resistance to damage a low-current diode.

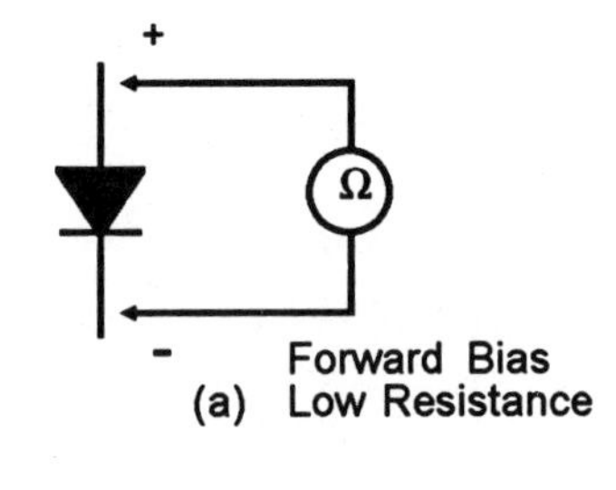

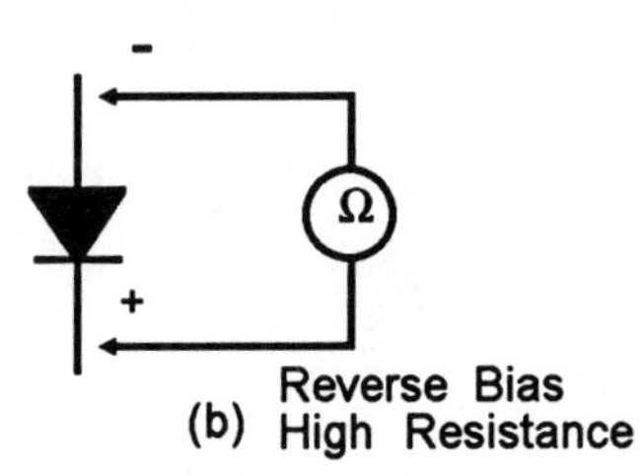

Figure 5.1

1. Connect the diode to the ohmmeter as shown in Figure 5.1a. Record the resistance reading.

   R_F = ______________

2. Reverse the ohmmeter leads connection to that of Figure 5.1b. In the reverse bias connection you may want to increase your ohmmeter range setting. Record the reverse bias resistance reading.

   R_R = ______________

In the next procedure steps you will be measuring the forward-bias characteristics of the diode. To make this measurement you will adjust the source to obtain the required current reading of Table 5.1, and at each current value step you will measure and record the diode forward voltage drop.

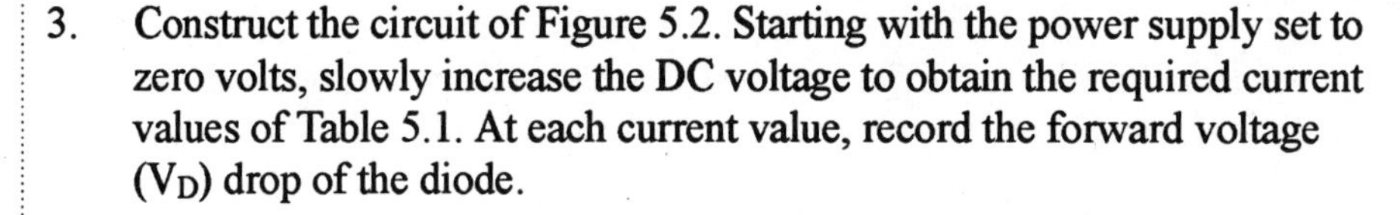

3. Construct the circuit of Figure 5.2. Starting with the power supply set to zero volts, slowly increase the DC voltage to obtain the required current values of Table 5.1. At each current value, record the forward voltage (V_D) drop of the diode.

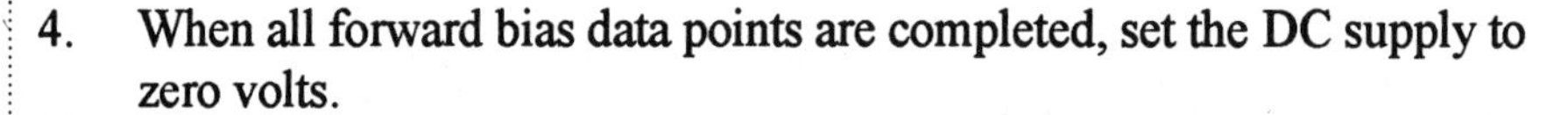

4. When all forward bias data points are completed, set the DC supply to zero volts.

In the next procedure steps, you will be measuring the reverse bias values for the diode. Since reverse current is too low to read directly on your ammeter, your values will be derived by the IR drop across a 220-kΩ resistor.

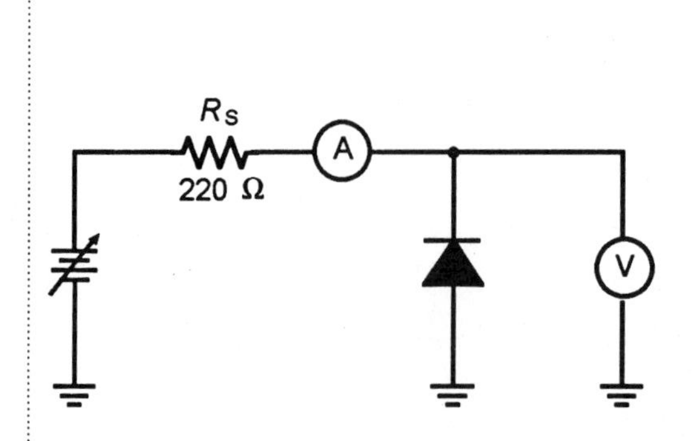

Figure 5.2

I_F	V_D
0.25 mA	
0.5 mA	
1.0 mA	
2.0 mA	
5.0 mA	
10.0 mA *	
20.0 mA *	
30.0 mA	

Table 5.1

5. Construct the circuit of Figure 5.3. Starting with the power supply set to zero volts, slowly increase the supply while reading the diode reverse voltage. At each diode reverse voltage step of Table 5.2, measure the voltage drop across R_S (220 kΩ) and calculate the current to record in Table 5.2.

6. When data measurements for Table 5.2 are complete, turn off the DC supply.

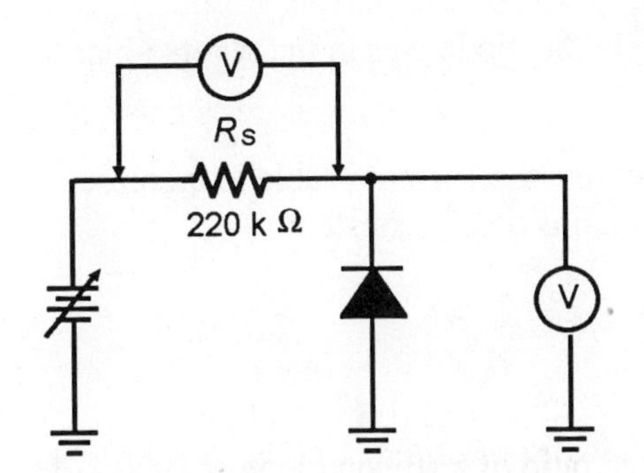

Figure 5.3

V_D	I_R
1.0 V	
2.5 V	
5.0 V ×	
10.0 V ×	
15.0 V	

Table 5.2

7. Plot the data of Tables 5.1 and 5.2 in Graph 5.1. Your plot should resemble that of your text Figure 2.19.

8. Calculate the diode dynamic forward-bias resistance using the formula below and your data of Table 5.1. Use the data points × of I_R of 10.0 mA and 20.0 mA from your table. Record your calculated forward resistance.

R_F = ________________

9. Calculate the diode reverse-biased resistance, using the formula below and your data from Table 5.2. Use the × data points of 5.0 V and 10.0 V. Record your calculated value.

R_R = ________________

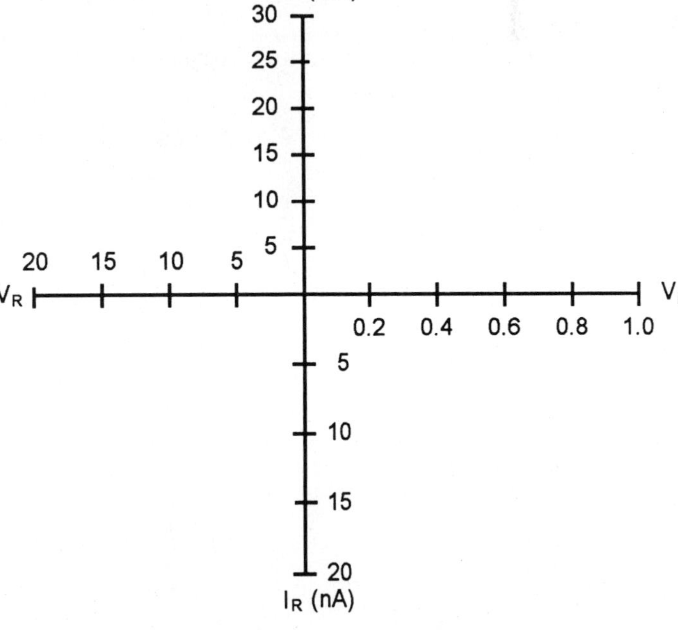

Graph 5.1

DISCUSSION

1. Describe how to determine the anode and cathode of a diode by using the diode test of the DMM. Assume that the diode has no markings.

2. When forward-biased, at what diode voltage did you notice the diode begin to conduct?

3. Did you observe any conduction while the diode was in the reverse-biasd mode?

4. For a forward-biased diode, describe how the current is able to increase while the voltage across the diode remains nearly constant.

Quick Check

1. The typical reverse/forward resistance ratio of a silicon diode is 1000:1.

   True False

2. The diode has high resistance in the forward-biased mode.

   True False

3. One characteristic of a silicon diode is that as the current increases, the resistance decreases.

   True False

4. A diode in reverse bias will conduct only at low current levels.

   True False

6

RECTIFIER FORMS

INTRODUCTION

The purpose of a rectifier circuit is to convert AC power line voltage to DC. Essentially every piece of electronic equipment that operates on AC line power must use a rectifier circuit.

You will work with three different types of rectifier power supply circuits in this experiment: the half-wave rectifier, the full-wave, and full-wave bridge rectifier. You will contrast the advantages and disadvantages of each rectifier as you observe their differences.

In the troubleshooting section, you will observe the effects on the output voltage and ripple frequency of the bridge rectifier if a diode opens, if the secondary opens, or if half of the secondary shorts.

REFERENCE

Principles of Electronic Devices and Circuits - Chapter 3, Section 3.2

OBJECTIVES

When you complete this experiment, you will:

✓ Understand the operation of half-wave, full-wave, and full-wave bridge rectifiers

✓ Be able to contrast the differences in each rectifier circuit

✓ Be able to relate the measured values of a failed circuit to the circuit fault

EQUIPMENT AND MATERIALS

115:12.6 V center-tapped transformer, equipped with AC power cord, fused primary, and power switch (see **Materials Note**)

Circuit protoboard
Digital multimeter
Oscilloscope
Circuit protoboard
Rectifier diode [4], 1N4001 or similar
Resistor, 1 kΩ (1/2 W)

Materials Note

If only a 115:12.6 V transformer is available, the following additional items are required for safe connection of the transformer:

AC power line cord
In-line fuse holder
1/2-A fuse
SPST toggle switch
3 ea. wire nuts

BE CAREFUL!

There will be 120 VAC on the primary side of the transformer. This is sufficient voltage to be a hazard.
USE CAUTION!

SECTION I FUNCTIONAL EXPERIMENT

If your transformer is equipped with an AC line cord, fuse, and switch, begin this procedure from step 2.

1. Assemble the power line cord, fuse holder, and transformer as follows (refer to Figure 6.1):

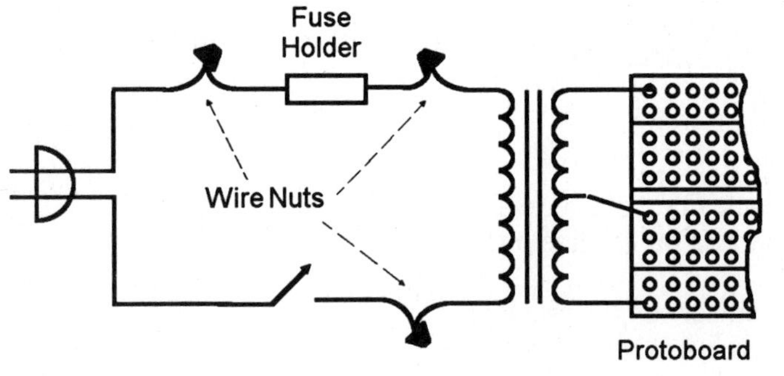

Figure 6.1

   A. Install the in-line fuse holder in series with one of the two leads on the line cord. Install the glass fuse in the fuse holder.

   B. Install the toggle switch in series with the other line cord lead. Solder the switch in place.

   C. Install the line cord to the transformer primary. Twist the wires together in a clockwise direction for these connections. This will allow the wire nut to tighten the connection.

Note: It is a good idea to place the two secondary and center-tap transformer leads in the protoboard as shown. The connections can then be made as needed and the center tap will not be loose to cause trouble.

Half-Wave Rectifier

2. Build the rectifier circuit of Figure 6.2. Apply AC power.

3. Use your digital voltmeter to measure the transformer secondary V_{rms} and DC output of the rectifier circuit. Record the data in the *Half-Wave* column of Table 6.1.

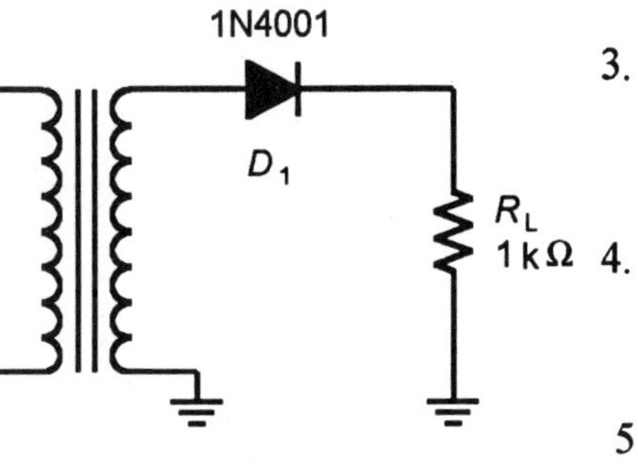

Figure 6.2

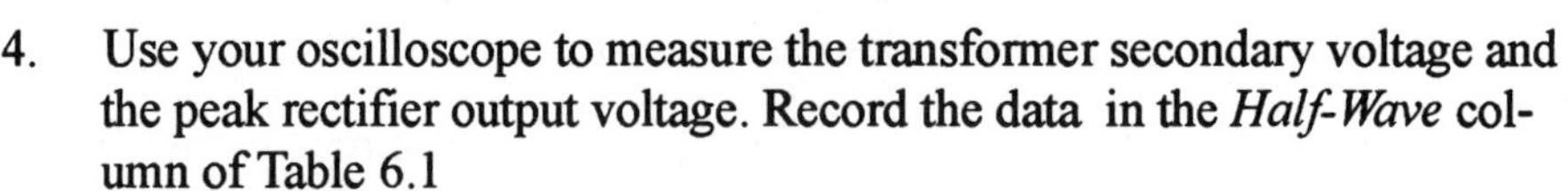

4. Use your oscilloscope to measure the transformer secondary voltage and the peak rectifier output voltage. Record the data in the *Half-Wave* column of Table 6.1

5. Connect your oscilloscope to the rectifier circuit output. Sketch the output waveform on the scale provided in Graph 6.1. Determine the frequency of the output waveform and record this value in the *Half-Wave* column of Table 6.1.

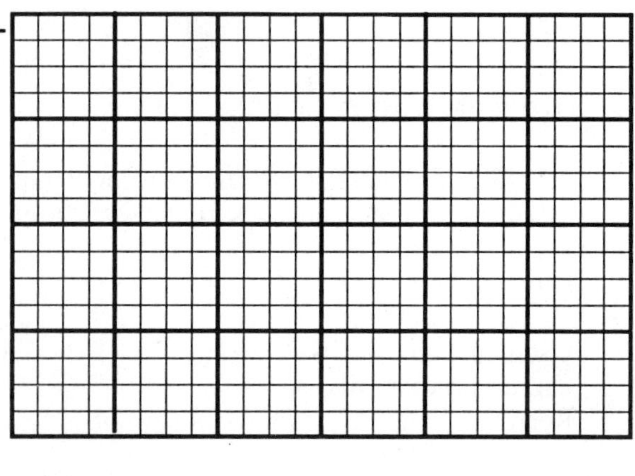

Graph 6.1

Full-Wave Rectifier

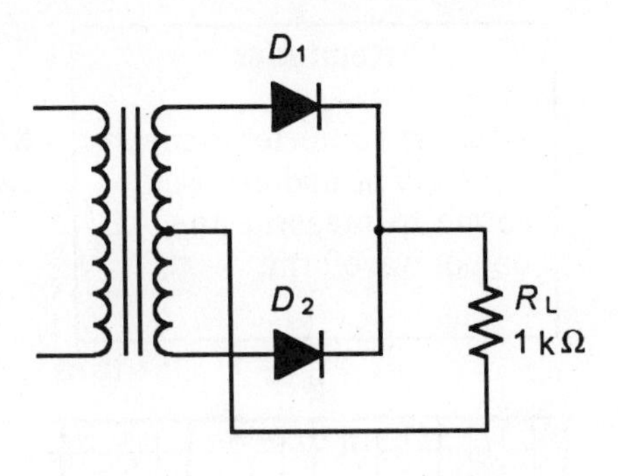

Figure 6.3

6. Build the rectifier circuit of Figure 6.3. Apply AC power.

7. Use your digital voltmeter to measure the transformer secondary V_{RMS} and DC output of the rectifier circuit. Record the data in the *Full-Wave* column of Table 6.1.

8. Use your oscilloscope to measure the transformer secondary voltage and
the peak rectifier output voltage. Record the data in the *Full-Wave* column of Table 6.1

9. Connect your oscilloscope to the rectifier circuit output. Sketch the output waveform on the scale in Graph 6.2. Determine the frequency of the output waveform and record this value in the *Full-Wave* column of Table 6.1.

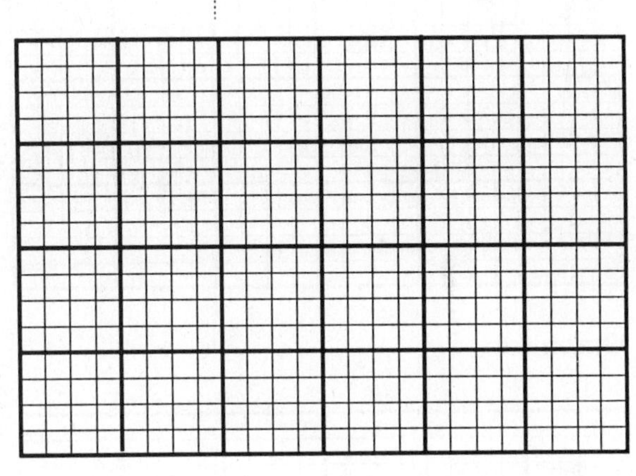

Graph 6.2

Bridge Rectifier

10. Build the circuit of Figure 6.4. Following construction, take a minute and check your circuit to ensure that the diodes are installed correctly.

11. Apply AC power. Measure and record in Table 6.1 the required circuit
values.

12. With your oscilloscope connected to the rectifier circuit output, observe and sketch the output waveform on the scale in Graph 6.3. Measure the frequency of the output waveform and record the value in Table 6.1.

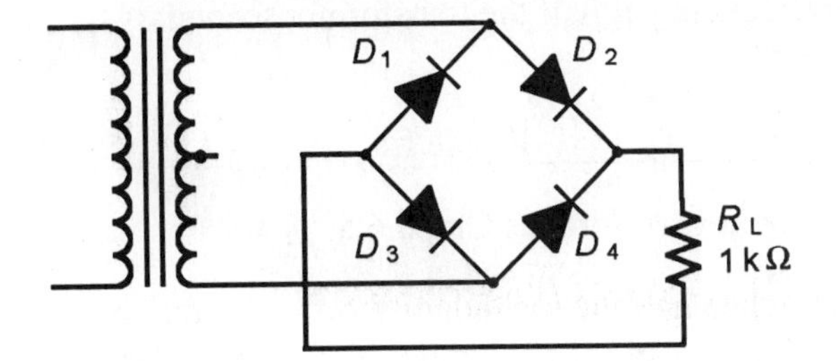

Figure 6.4

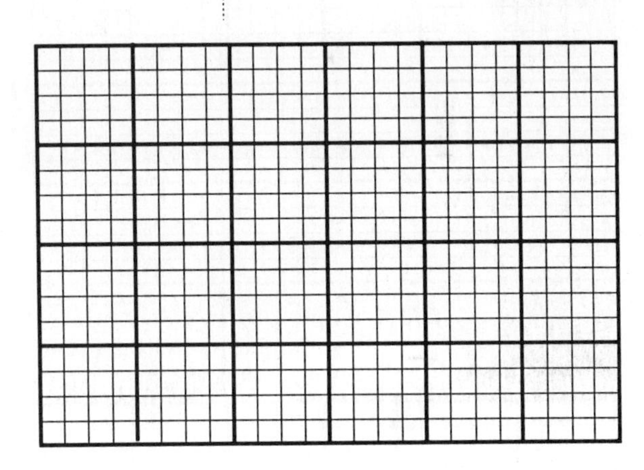

Graph 6.3

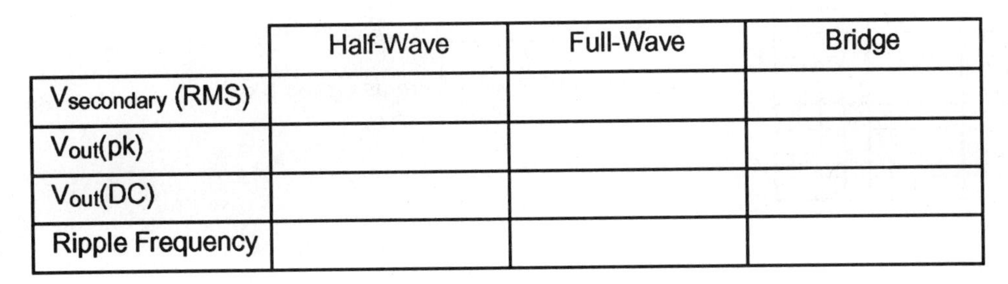

	Half-Wave	Full-Wave	Bridge
$V_{secondary}$ (RMS)			
V_{out}(pk)			
V_{out}(DC)			
Ripple Frequency			

Table 6.1

Reminder

User the voltmeter to measure DCV_{out} and the oscilloscope to measrue the AC output waveform.

SECTION II TROUBLESHOOTING

Fault 1 - D_1 open

1. For the circuit of Figure 6.4 with diode D_1 open, calculate the DC V_{out} and the ripple frequency.

   DC V_{out} = ____________ Ripple freq. = ____________

2. Lift one end of D_1 of the circuit of Figure 6.4. Apply AC power; then measure and record the values below.

   DC V_{out} = ____________ Ripple freq. = ____________

Fault 2 - Transformer secondary open

1. Turn off the AC power, reconnect diode D_1, and calculate the values of DC V_{out} and ripple frequency if the transformer secondary were open.

   DC V_{out} = ____________ Ripple freq. = ____________

2. Disconnect one lead of the transformer secondary. Apply AC power. Measure and record the following values:

   DC V_{out} = ____________ Ripple freq. = ____________

   Draw the output waveform as Graph 6.4.

Graph 6.4

Fault 3 - Half the transformer secondary shorted

1. Turn off the AC power. Calculate the values of DC V_{out} and ripple frequency if half the transformer secondary were shorted.

   DC V_{out} = ____________ Ripple freq. = ____________

   Draw the output waveform as Graph 6.5.

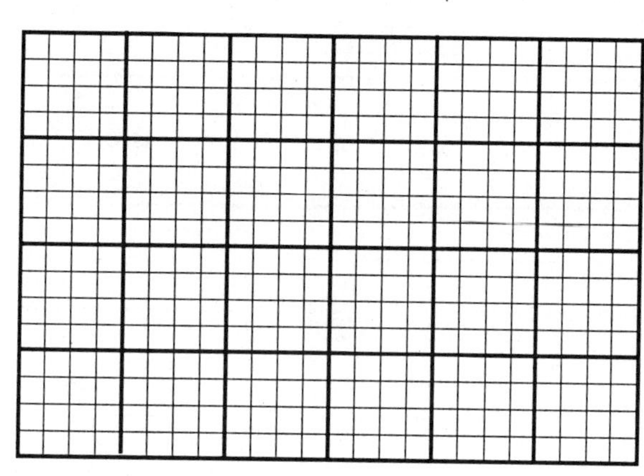

Graph 6.5

2. You can simulate half the secondary shorting by removing one secondary lead and replacing it with the center tap. Make the connection changes and apply AC power.

3. Measure and record the values of DC Vout and ripple frequency.

   DC V_{out} = ____________ Ripple freq. = ____________

   Draw the output waveform as Graph 6.6.

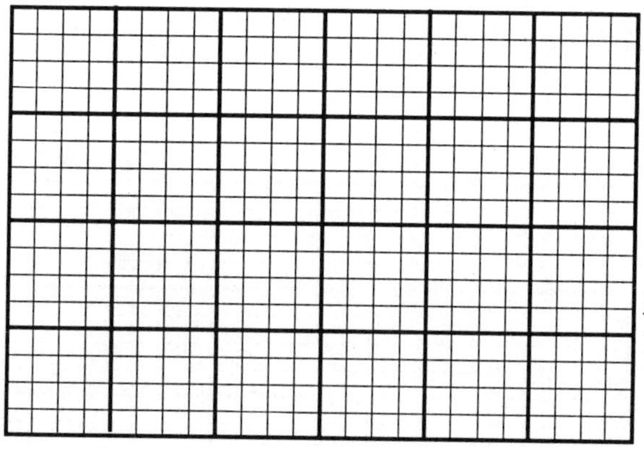

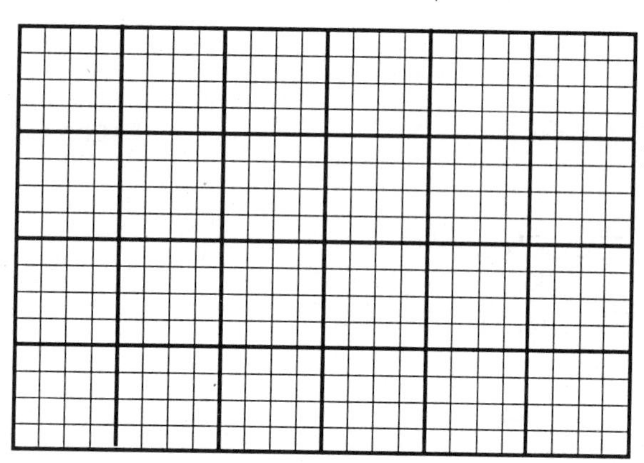

Graph 6.6

DISCUSSION

Section I

1. All three rectifier circuits had the full transformer secondary connected, yet the DC output of the bridge rectifier was larger than the others (Refer to your data of Table 6.1). Briefly, for each rectifier circuit explain why you obtained the DC output voltage values measured.

2. Refer to your data of Table 6.1 and the output waveform drawings made for all three rectifier circuits. Two of the rectifier circuits had a higher ripple frequency. Discuss the rectifier circuits, describing why the ripple frequency differences.

3. The bridge rectifier circuit is considered to be the most efficient of the three rectifier forms. It also has another advantage over the other two rectifier circuits. Can you identify this advantage? Hint: Look at Figure 6.5.

Section II

Fault 1 - Diode opens

1. Referring to your fault measurements in Step 2, you should have found a low DC V_{out} and a lower than expected ripple frequency. Explain how the oscilloscope helps you diagnose this problem easier than the DVM.

Fault 2 - Transformer secondary open

1. Fault 2 causes the V_{out} to disappear completely. One of the most obvious causes for this is an open secondary winding. The fastest way to troubleshoot this problem would be to measure the secondary voltage with a DVM or an oscilloscope.

2. How would you verify that the secondary is open as opposed to a primary failure or the rectifier circuit shorting the secondary?

Fault 3 - Half the transformer secondary shorted

1. You should have found that Vout measured about half of the expected V_{out}. Notice that using a DVM would not show you that the V_{out} waveform is still full wave. Using the oscilloscope quite often speed up troubleshooting by allowing the waveform to be observed. You will use the oscilloscope much more often as you progress in electronics for this reason. Why do you think you would not always use the oscilloscope to troubleshoot?

Quick Check

1. How many diodes does the bridge rectifier use ? ____________

2. When a diode of the half wave rectifier is biased off, the current through the load is ____________ amps.

3. The fuse is usually located in the ____________ side of the transformer.

4. The output frequency of the half wave rectifier is _____.

   (a) 120 Hz
   (b) 60 Hz
   (c) 0 Hz

5. In a bridge rectifier is there one or two diodes conducting at one time ?

   ____

7

CAPACITIVE INPUT FILTERS

INTRODUCTION

The capacitive filter is used to smooth out the pulsating DC voltage of the rectifier circuit. The capacitor changes to the AC peak value, thus providing an output larger than the average value. This gives a DC output voltage of a much higher value than the unfiltered output voltage.

In Section I of this experiment, you will observe the effects on your bridge rectifier output voltage when you add an output filter capacitor. You will then add a series resistance and a second capacitor to improve the output ripple voltage even further. You will also explore how different size load resistors Affect the ripple output voltage. In Section II you will observe the effect of an open diode and open filter capacitor on the supply ripple voltage. Learning to recognize the effects of these common faults will make troubleshooting power supplies much simpler and faster.

REFERENCE

Principles of Electronic Devices and Circuits - Chapter 3, Section 3.3

OBJECTIVES

In this experiment you will:

✓ Learn the effects of capacitive filters on the output voltage of rectifier circuits

✓ Understand the effects of the load resistance on the capacitive filter

✓ Recognize common problems with capacitive filtered power supplies

EQUIPMENT AND MATERIALS

The bridge rectifier circuit of Experiment 6
Oscilloscope
Digital multimeter
470 μF capacitor [2]
Resistors: 200 Ω (2 watt), 470 Ω (1/2 watt), 33 kΩ

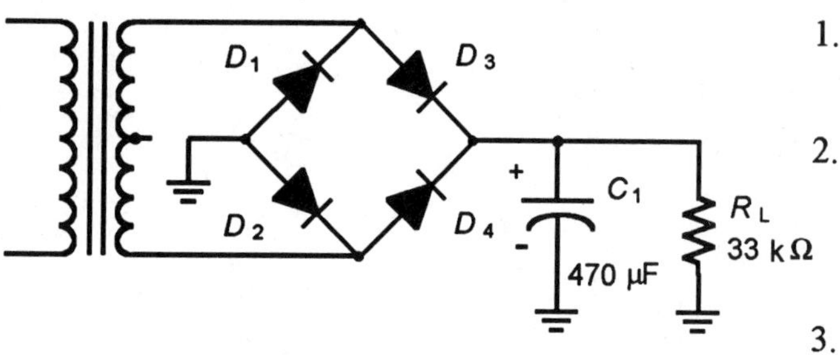

Figure 7.1

1. Build the circuit of Figure 7.1, omitting C_1 at this time.

2. From Table 6.1 or your calculation, determine the average DC output of the bridge rectifier.

   DC V_{avg} = ____________

3. Apply AC power to your circuit. Using your DC voltmeter, measure the output voltage across the load resistor (R_L).

   Measured DC V_{avg} = ____________

4. Using your oscilloscope to observe the output voltage, sketch the waveform on the scale provided in Graph 7.1. Also note the peak output voltage value.

   V_p = ____________

5. Turn off the rectifier power and install C_1, a 470 μF capacitor connected in parallel with the load resistor. Ensure that the negative side of the capacitor is at ground or the most negative part of the circuit.

6. Turn on the rectifier and measure the output voltage with your DC voltmeter. With your oscilloscope coupling set to DC, observe the supply output. You should find a DC level that is essentially a horizontal straight line.

   V_{out} = ____________

7. Now switch the oscilloscope coupling to AC and adjust the vertical range selector until the ripple is at least 1 division in height. Measure and record the peak-peak ripple voltage (V_{rip}). Sketch this waveform on the scale provided in Graph 7.2.

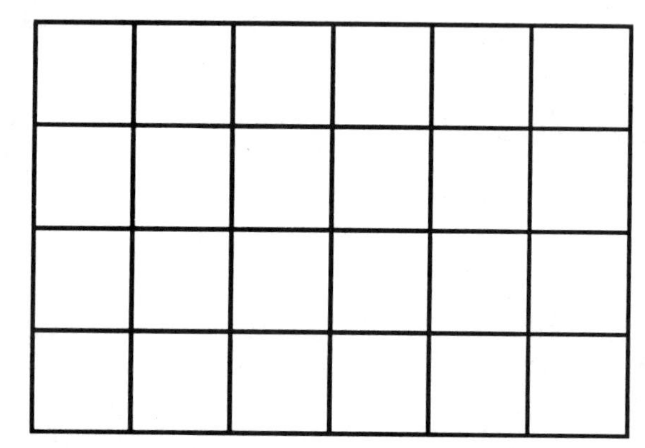

Graph 7.1

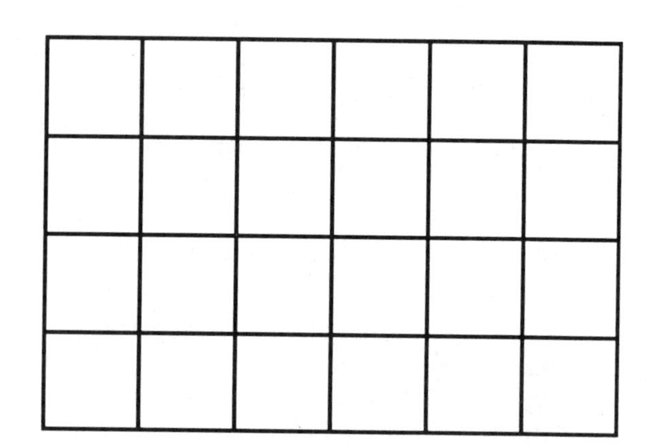

Graph 7.2

V_{rip} = ________________

8. Calculate the capacitor charge and discharge time constant for your circuit. Use 2 Ω for the diode resistance (TC = RC).

   Charge TC = __________

   Discharge TC = ________

9. Turn off the AC power and exchange the 33-kΩ load resistor for the 200-Ω resistor.

10. Measure and record both the DC V_{out} and peak-peak ripple voltage, and draw the ripple voltage waveform in Graph 7.3.

   DC V_{out} = ____________ V_{rip} = ____________

Graph 7.3

11. Calculate the time constant for charge and discharge again.

   Charge TC = __________ Discharge TC = __________

12. Build the circuit of Figure 7.2.

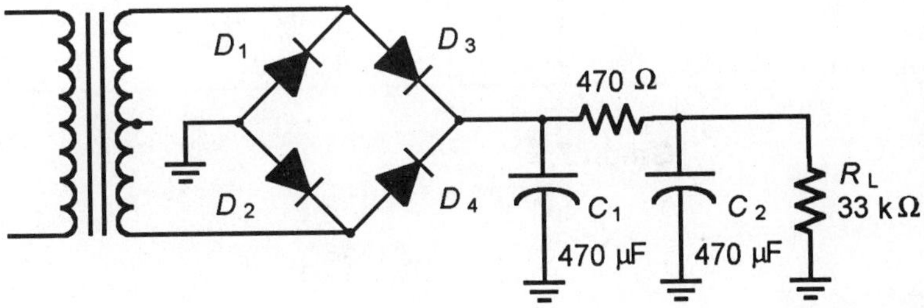

D1 D3 470 Ω D2 D4 C1 470 µF C2 470 µF RL 33 kΩ

Figure 7.2

Graph 7.4

13. Apply AC power. Measure the DC output voltage with your voltmeter, and use the oscilloscope to measure the peak-peak ripple voltage. Sketch the output voltage waveform in Graph 7.4.

   DC V_{out} = ____________ V_{rip} = ____________

14. Turn off the AC power and remove the 33-kΩ R_L and the 470-µF capacitor C_2. Notice that the peak-peak ripple voltage increases slightly. While it appears that the addition of C_2 did not have much effect, if you look closely, you will see that as a percentage it helped quite a bit.

SECTION II TROUBLESHOOTING

Fault 1 - Diode open

1. With the circuit of Figure 7.1, disconnect one leg of D_1.

2. Turn on the rectifier. Measure and record DC V_{out} and V_{rip}, and draw the output waveform voltage in Graph 7.5. You should see an increase in the ripple.

   DC V_{out} = ____________ V_{rip} = ____________

3. Contrast this reading with the reading taken in Procedure Step 4.

Fault 2 - Filter capacitor open

1. With AC power off, reconnect diode D_1. Disconnect one end of capacitor C_1.

2. Apply AC power. Measure and record DC V_{out} and V_{rip}, and sketch the output waveform in Graph 7.6.

   DC V_{out} = ____________ V_{rip} = ____________

Graph 7.5

Graph 7.6

DISCUSSION

Section I

1. You should have found that the average DC voltage decreased while the ripple increased as the load resistor got smaller in value. Why do you think the output voltage dropped as the load resistance decreased?

2. Explain in your own words why the shorter time constant for charge and the longer time constant for discharge provided this increased output voltage magnitude.

Section II

Fault 1 - Bridge rectifier diode open

Most power supply problems show up indirectly in other parts of an electronic circuit. An audio amplifier, for example, may be found to be operating improperly because of a low power supply voltage. When you check V_{CC} with your DVM you find that the voltage is reading below normal. At this point you would suspect the power supply and troubleshoot accordingly.

You also found that the load resistance can effect the power supply Vout. Can you think of a way to determine if the power supply or the load is the problem?

Fault 2 - Filter capacitor open

You found that when the filter capacitor opens, the V_{out} as measured on the DVM goes down. What does the oscilloscope waveform tell you about the condition of the power supply that the DVM does not?

Quick Check

1. The filter capacitors charge slowly and discharge quickly.

   True False

2. The output voltage, after being filtered, is called the voltage.

   True False

3. A half-wave rectifier is easier to filter than a full-wave rectifier.

   True False

4. The most-used filter today is a _____.

   (a) pi filter (b) capacitive input filter
   (c) coil and capacitor filter

8

ZENER REGULATION OF POWER SUPPLIES

INTRODUCTION

The zener diode can act as a regulator for power supplies since its voltage drop (V_Z) is nearly constant for a wide range of zener current. This characteristic also functions to reduce power supply ripple variations.

In Section I of this experiment you will measure the current through the zener and the output voltage for three different values of load resistors. You will also measure ripple voltage and see that the zener improves V_{out} ripple as well.

In Section II you will observe the effect on the zener regulated voltage if either a rectifier diode or a filter capacitor fails open.

REFERENCE

Principles of Electronic Devices and Circuits - Chapter 3, Section 3.4

OBJECTIVES

In this experiment, you will:

✓ Better understand the zener as a voltage regulator

✓ Understand zener operation under different load conditions

✓ Be able to relate circuit measurements to fault conditions

EQUIPMENT AND MATERIALS

Bridge power supply of Experiment 7
9.1 V zener diode, 1N5239B or equivalent
Circuit protoboard
Digital multimeter
Oscilloscope
Resistors: 330 Ω, 470 Ω, 560 Ω, 1.2 kΩ

SECTION I FUNCTIONAL EXPERIMENT

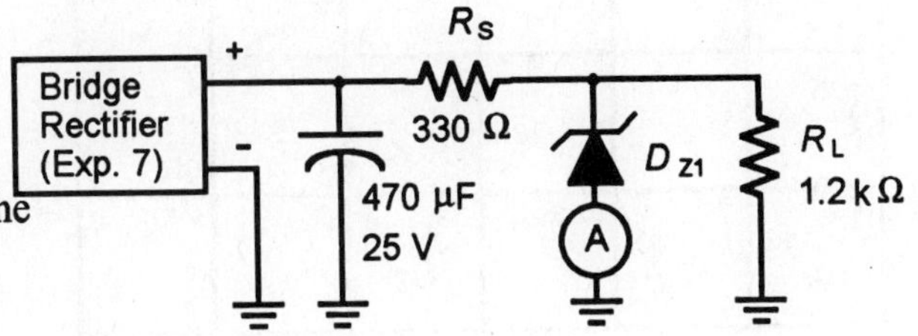

1. Build the circuit of Figure 8.1, leaving the zener diode out of the circuit.

2. Measure the output peak-to-peak ripple voltage using the oscilloscope.

   V_{rip} = ______________

Bridge Rectifier (Exp. 7)
+
-
R_S
330 Ω
470 µF
25 V
A
D_{Z1}
R_L
1.2 kΩ

Figure 8.1

3. Turn off the power and add the zener diode. (Remember to reverse bias the zener diode). Insert your ammeter in series with the zener to monitor zener current.

4. Turn on the power supply and measure the output peak-to-peak ripple voltage again. Also measure and record the zener current (I_Z).

   V_{rip} = ______________ I_Z = ______________

5. With circuit power off, disconnect your zener current ammeter. Reapply AC power. Use your DC voltmeter and measure and record DC V_{out}.

   V_{out} = ______________

6. Turn off the power supply and replace the 1.2-kΩ load resistor (R_L) with a 560-Ω resistor.

7. Turn on the power supply and measure the output peak-to-peak ripple, I_Z, and DC V_{out}.

   V_{rip} = __________ I_Z = ______________ V_{out} = ______________

8. Turn off the power supply and replace the 560 Ω R_L with a 470 Ω resistor.

9. Turn on the power supply and measure output pk-pk ripple, I_Z, and DC V_{out}.

   Vrip = __________ I_Z = ______________ V_{out} = ______________

SECTION II TROUBLESHOOTING

PROCEDURE

Fault 1 - D_1 open

With circuit power off, disconnect one leg of D_1 in the circuit of Figure 8.1. Measure the peak-to-peak output ripple, I_Z, and DC V_{out}.

Graph 8.1

V_{rip} = ________ I_Z = ____________

V_{out} = ________

Additionally, sketch the output waveform in Graph 8.1, and note on your sketch the peak values.

Fault 2 - C_1 open

With circuit power off, reconnect diode D_1. Disconnect one leg of C_1 in the circuit of Figure 8. 1. Measure the peak-to-peak output ripple, I_Z, and DC V_{out}.

V_{rip} = ________ Iz = __________ V_{out} = __________

DISCUSSION

Section I

1. From your measured data, you should have found that with changing load resistors, DC V_{out} remained nearly constant and zener current was different for each value of R_L. Looking at the relationship of I_Z value vs. R_L value, briefly discuss this relationship and the zener regulation.

2. You should have found that the magnitude of ripple voltage was significantly less with the zener diode than without it. Discuss why this occurred.

Section II

Fault 1 - Diode D_1 open

Of the measurements you made, at least one is a very strong indicator of a diode failure in the bridge rectifier. Identify the measurement(s) indicative of diode failure, and explain for each measurement you selected why it indicates a diode failure.

Fault 2 - C_1 open

Refer to your output waveform sketch:

(a) Can you verify that the zener diode is operating? Why do you believe this is so?

(b) Is there any indication in the output waveform that suggests that the filter capacitor failed? Describe the indication in the waveform that supports your answer.

Quick Check

1. The zener diode is a current regulator.

   True　　　　　False

2. The zener diode must be ___________ biased to operate as a zener diode.

   (a) forward　　　　　(b) reverse

3. If the zener were installed backwards in the power supply of Figure 8.1, what do you think the result would be?

4. What is the total current through R_S in the circuit of Figure 8.1 when the load resistor is:

   (a) $R_L = 1.2\ k\Omega$, $I_{RS} =$ ___________

   (b) $R_L = 470\ \Omega$, $I_{RS} =$ ___________

9

VOLTAGE MULTIPLIERS

INTRODUCTION

Voltage multipliers are an AC-to-DC voltage converter whose DC output is the product of an integer (2, 3, or greater) and the peak AC input. They cannot, however, supply a large load current. And the larger the multiplication, the smaller the load current capability. They find application where a large DC voltage is required and the load current is small.

In Section I you will build the voltage doubler and tripler circuits and evaluate these circuits. You will substitute a small R_L on the doubler to observe the effects on the DC output.

In Section II you will observe the effect of open capacitors in the doubler and tripler circuits.

REFERENCE

Principles of Electronic Devices and Circuits - Chapter 3, Section 3.5

OBJECTIVES

In this experiment you will:

✓ Experimentally evaluate the operation of voltage multipliers

✓ Understand the relationship between the load and the voltage multiplier

✓ Measure the effects of open capacitors on the operation of voltage multipliers

EQUIPMENT AND MATERIALS

Oscilloscope
Circuit protoboard
Digital multimeter
Diode [3], 1N4001
Capacitor [3], 470 µF (50 V)
12.6 V center-tapped transformer
Resistors: 500 Ω, 100 kΩ

SECTION I FUNCTIONAL EXPERIMENT

1. Build the doubler circuit of Figure 9.1.

   NOTE: You may use the line cord and transformer from Experiment 6. Also notice that you use one secondary and the center tap leads for this experiment. This gives an input AC of 6.3 V_{RMS}.

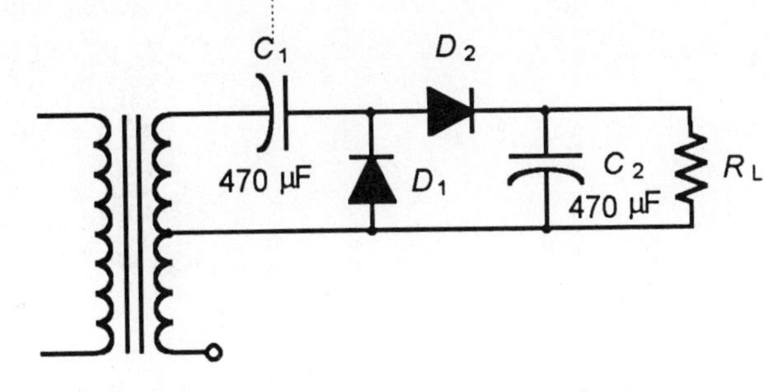

Figure 9.1

2. Apply power to the circuit. Measure and record the peak input voltage and peak-to-peak ripple voltage output of the voltage doubler with the oscilloscope. Use your DC voltmeter to measure the DC output.

   V_{in} pk = ________ V_{rip} = ____________ DC V_{out} = ________

   Notice that if V_{in} were unknown, the DC V_{out} reading would tell you what peak-to-peak value was connected to the input. This is why this circuit is also called a peak-to-peak detector.

3. Turn off the power and replace the 100-kΩ R_L with a 500-Ω R_L. Measure the DC V_{out} and the peak-peak ripple voltage and record below.

   R_L = 500 Ω

   DC V_{out} = ____________ V_{rip} = ____________

4. Build the tripler circuit of Figure 9.2. Measure the peak AC input, DC output and the pk-pk ripple output and record below.

   V_{in} pk = ____________

   DC V_{out} = ____________

   V_{rip} = ____________

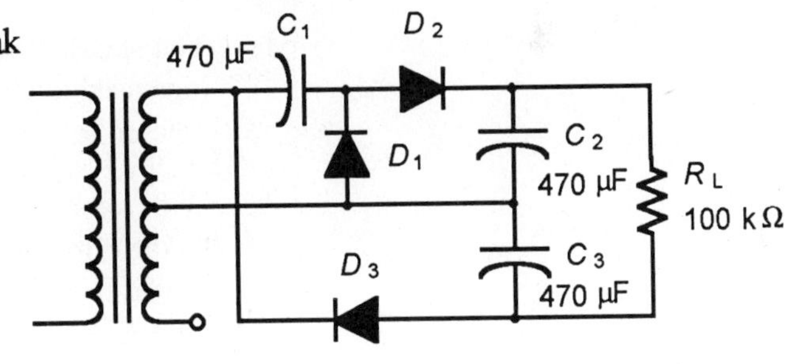

Figure 9.2

5. With the circuit turned off, replace the 100-kΩ R_L with the 500-Ω R_L. Apply AC power. Measure the DC output and peak-peak ripple voltage and record below.

   R_L = 500 Ω

   DC V_{out} = ____________ Vrip = ____________

SECTION II TROUBLESHOOTING

Fault 1 - C_1 open

1. Starting with the circuit of Figure 9.2 using a 100-kΩ load, ensure that the AC power is off. Disconnect one leg of capacitor C_1.

2. Apply AC power. Measure the DC output and the V_{C2} and V_{C3} voltages. Record these values below. Using your oscilloscope, observe the circuit output.

   V_{out} = __________ V_{C2} = ____________ V_{C3} ______________

Fault 2 - D_1 open

1. With circuit power off, reconnect capacitor C_1. Disconnect one leg of diode D_1.

2. Apply AC power. Measure DC output, V_{C2} and V_{C3} voltages. Record these values below. Using your oscilloscope, observe the circuit output.

   V_{out} = __________ V_{C2}= ____________ V_{C3} ______________

DISCUSSION

Section I

1. You should have discovered that the voltage multipliers were sensitive to increased load. Explain why you think these circuit's DC output decreased to a lower value when the R_L went down.

2. With reference to the voltage doubler or tripler, what do you think would be the effect on the DC output if one or more capacitors in the multiplier circuit were leaky?

Section II

Fault 1 - C_1 open

You should have found that if C_1 failed open, the circuit output dropped to that of a doubler. Explain why this should be.

Fault 2 - D_1 open

Briefly explain why you obtained the DC output you measured for this fault. Also indicate what other (if any) component failures could have produced the same result.

Quick Check

1. A voltage doubler is also called a(n) _____.

   (a) rf detector
   (b) peak-to-peak detector
   (c) audio detector

2. The major disadvantage of the voltage multiplier is that it _____.

   (a) uses expensive diodes
   (b) raises the V_{out}
   (c) lowers the available current
   (d) all the above

3. With reference to the circuit of Figure 9.1, the PIV seen by diode D_1 is _____.

   (a) V_{in} pk
   (b) 2 V_{in} pk
   (c) 3 V_{in} pk
   (d) 4 V_{in} pk

4. Refer to the circuit of Figure 9.2. If AC V_{in} were 12.6 V_{RMS}, the DC output would be _____.

   (a) 25.2 V
   (b) 35.6 V
   (c) 37.8 V
   (d) 53.5 V

5. Refer to the circuit of Figure 9.2. The maximum DC voltage of capacitor C_3 is _____.

   (a) 8.9 V
   (b) 17.8 V
   (c) 25.2 V
   (d) 35.6 V

10

SIGNAL CLIPPERS AND CLAMPERS

INTRODUCTION

Clippers are circuits that operate to limit (clip) a signal waveform at a given voltage. Clampers are circuits that change (clamp) the reference voltage level of a signal waveform. In this experiment you will study series, shunt, and biased clippers. You will also see how clamper circuits operate.

In Section I you will build and make measurements on series, shunt, and biased clipper circuits as well as a clamper circuit. In Section II you will examine the effects of several common defects on the operation of clippers and clampers.

REFERENCE

Principles of Electronic Devices and Circuits - Chapter 3, Sections 3.6 and 3.8

OBJECTIVES

In this experiment you will:

✓ Observe and draw the output voltage signals of clippers and clampers

✓ Learn how to clip a signal at a certain level using biased clippers

✓ Observe the effects on a signal voltage when clamped

✓ Through measurement, learn the effect of typical clipper and clamper failures

EQUIPMENT AND MATERIALS

Circuit protoboard
Dual-trace oscilloscope
Function generator
Small-signal diode [2], 1N914 or similar
Resistors: 1 kΩ, 4.7 kΩ, 10 kΩ, 2.2 MΩ
Capacitor, 4.7 μF

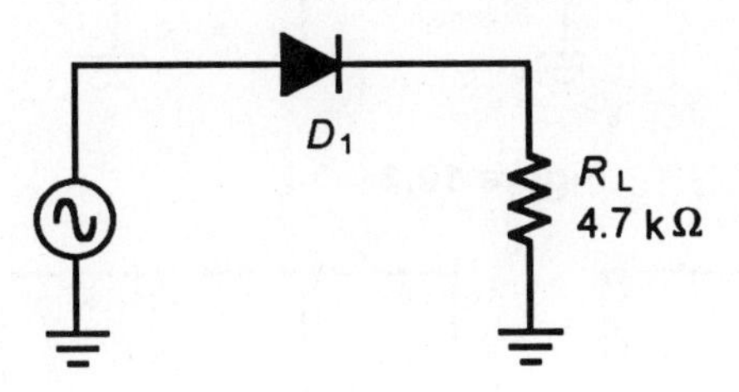

Figure 10.1

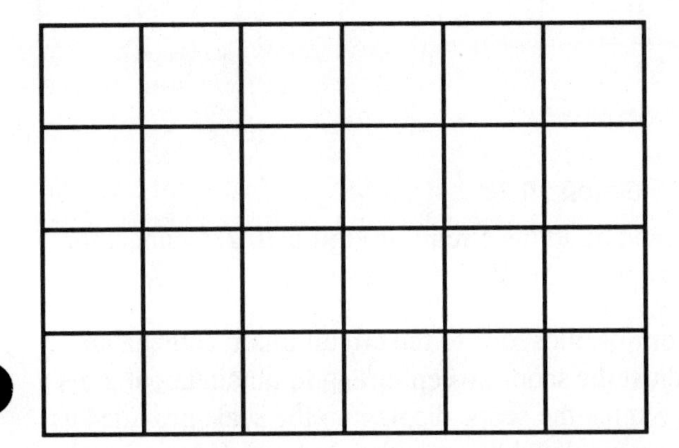

Graph 10.1a

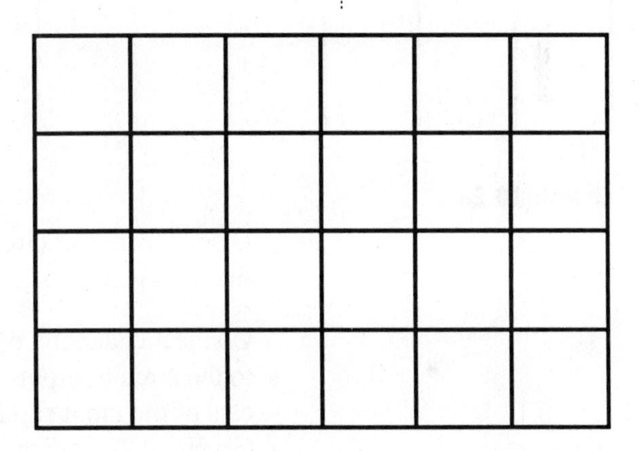

Graph 10.1b

1. Build the circuit of Figure 10.1. Set your function generator to supply a 1-kHz, 10-$V_{p\text{-}p}$ input.

2. Connect your oscilloscope channel 1 to the circuit input, and channel 2 to the output. This will permit you to measure the signal voltage levels and observe the relationship of the output to input signal.

   Adjust the scope sweep timing to obtain about 2 cycles of the input waveform. Sketch the oscilloscope display on the scale in Graph 10.1a. Note on the sketch the values of the positive and negative peak voltage.

3. Switch the input signal off. Reverse the diode in the circuit. Apply the input signal and repeat the measurements of step 2. Use Graph 10.1b for this version of the oscilloscope display.

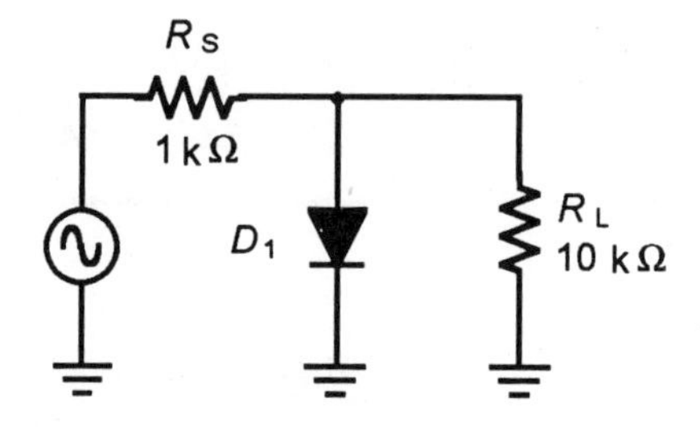

Figure 10.2

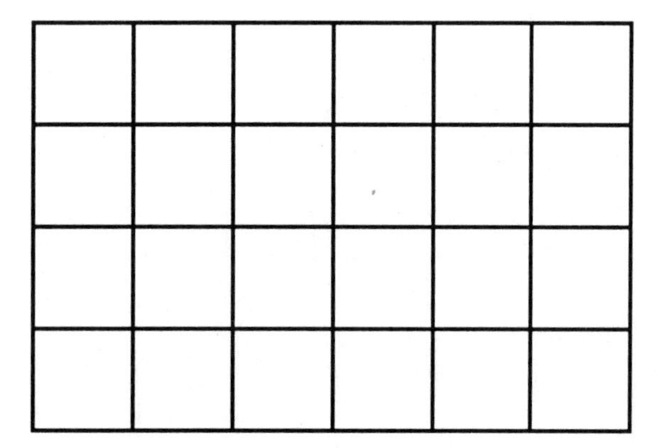

Graph 10.2a

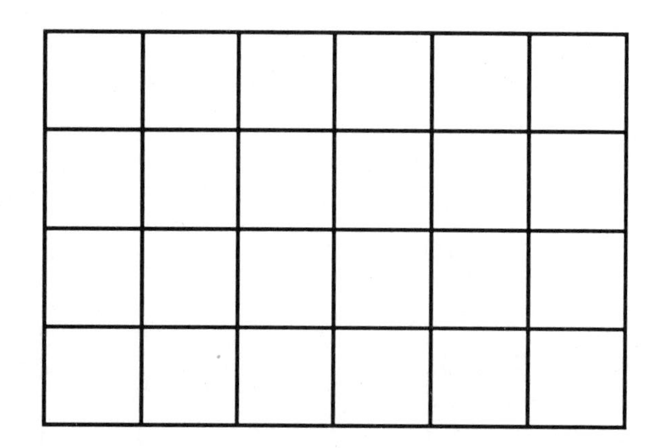

Graph 10.2b

4. Disconnect your circuit and build the circuit of Figure 10.2. Connect the function generator set to apply a 10-V_{p-p}, 1-kHz signal.

5. Connect channel 1 of your oscilloscope to the circuit input, and channel 2 to the circuit output. Adjust the scope sweep timing to obtain about 2 cycles of the input signal. Sketch the scope display on the scale provided as Graph 10.2a. Indicate on your sketch the positive and negative peak voltage values.

6. Turn off the input signal and reverse the diode connections in the circuit. Apply the input signal and repeat the measurements of step 5. Use Graph 10.2b for this version of the oscilloscope display.

7. Build the circuit of Figure 10.3.

8. Connect a DC source of 5 V to the diode (plus to the cathode of the diode, minus to circuit ground). Connect the function generator set to provide a 12-V_{p-p}, 1-kHz signal input.

9. Connect your oscilloscope channel 1 to the circuit input, and channel 2 to the circuit output with DC coupling. Set the scope sweep timing to obtain a display of about 2 cycles of the input signal. Sketch the scope display on the scale in Graph 10.3a. Mark the positive and negative voltage values on your sketch.

10. Adjust the DC source for a value of 3 V, and repeat the measurements of step 9. Use Graph 10.3b for this sketch of the oscilloscope display.

    What effect did reducing the DC voltage level have on your output signal?

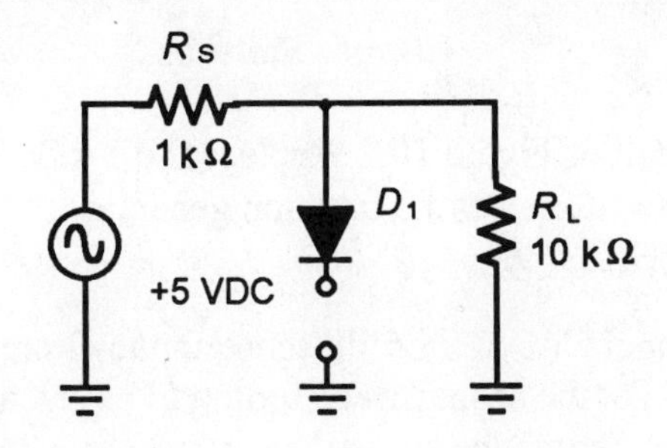

Figure 10.3

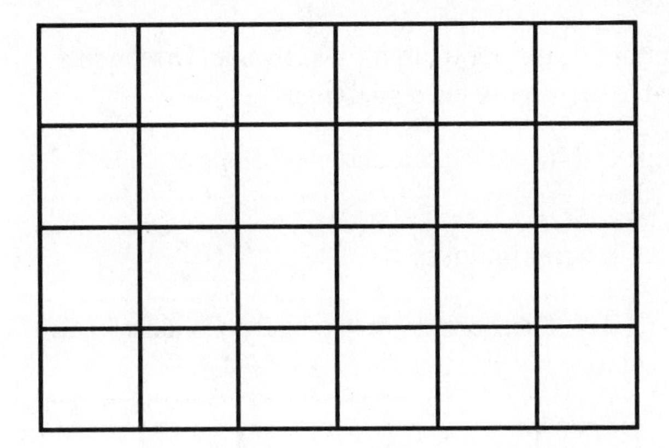

Graph 10.3a

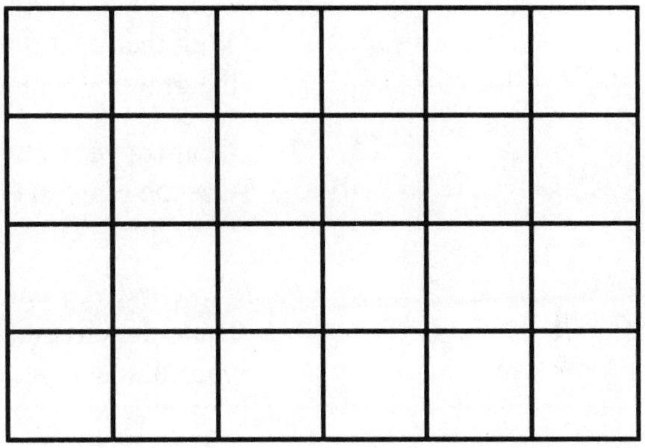

Graph 10.3b

11. Disconnect your circuit and build the clamper circuit of Figure 10.4. Observe the capacitor polarity. Connect the function generator set to provide a 10-$V_{p\text{-}p}$, 1-kHz signal.

12. Connect channel 1 of your oscilloscope to the circuit input, and channel 2 set to DC coupling to the circuit output. Sketch the scope display on the scale provided in Graph 10.4. Mark the signal positive and negative peak voltage values.

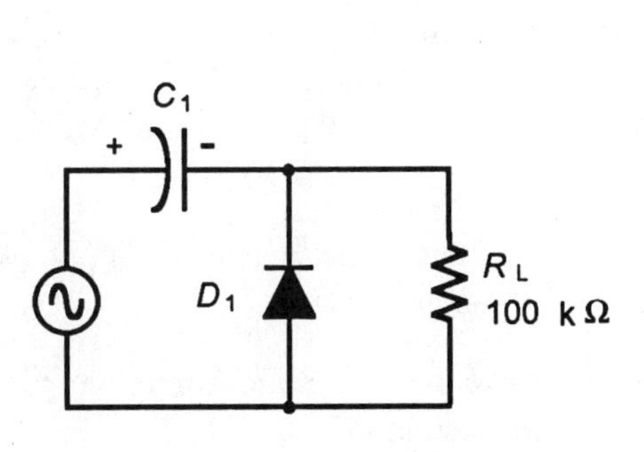

Figure 10.4

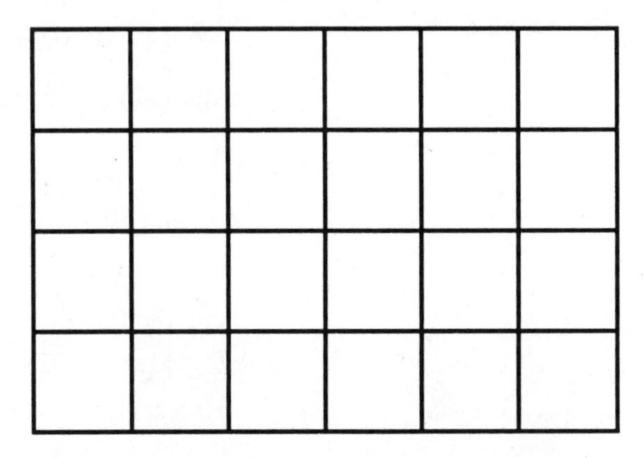

Graph 10.4

Clipper Failures

Fault 1 - R_S open

1. Build the circuit of Figure 10.2. Replace the 1-kΩ resistor of R_S with a 2.2-MΩ resistor. Connect the function generator set to provide 10 V_{p-p} at 1 kHz.

2. Connect channel 1 of your oscilloscope to the circuit input, and channel 2 to the output. Set the scope sweep timing to obtain about 2 cycles of input waveform. Sketch the scope display in Graph 10.5. Mark the positive and negative peak voltage values.

Fault 2 - Diode D_1 open

1. With the function generator off, restore R_S to its 1-kΩ value. Disconnect the ground (cathode) end of D_1. Apply the signal input.

2. Connect your oscilloscope and repeat the measurement steps of Fault 1. Use the scale in Graph 10.6.

Clamper Failures

Fault 1 - Diode D_1 open

1. Build the circuit of Figure 10.4. Disconnect one leg of D_1. Connect your function generator set to apply an input of 12 V_{p-p} at 1 kHz.

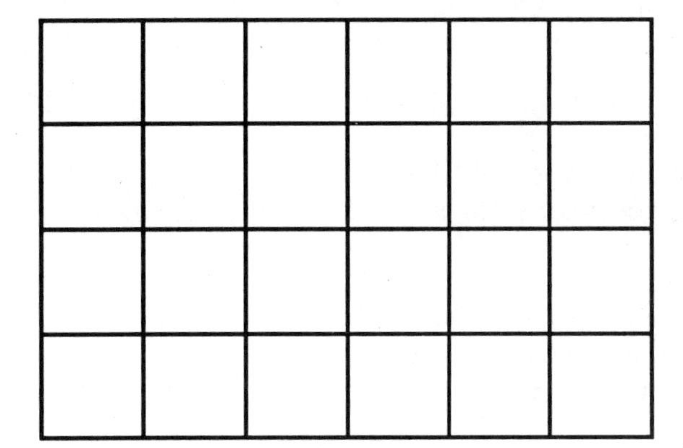

Graph 10.5

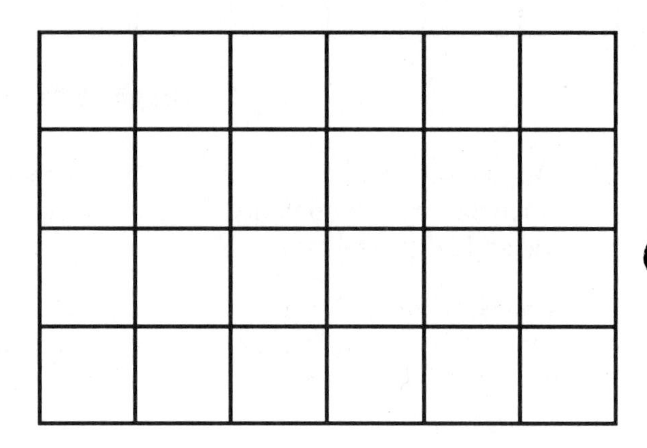

Graph 10.6

2. Connect channel 1 of your oscilloscope to the circuit input and channel 2, set to DC coupling, to the circuit output. Set the sweep timing to obtain about 2 cycles of the input signal. Sketch the scope display on the scale in Graph 10.7. Mark the positive and negative peak voltage values.

Fault 2 - C_1 leaky

1. With the function generator off, connect a 1-kΩ resistor in parallel with C_1. Apply an input signal of 12 Vp-p at 1 kHz.

2. Connect your oscilloscope and make the circuit measurements of Clamper Fault 1, step 2, using the scale in Graph 10.8.

Graph 10.7

Graph 10.8

DISCUSSION

Section I

1. Considering the output waveform of the circuit of Figure 10.1, you should have found that the output is the same shape as that of the half-wave rectifier. Explain why you think we call this circuit a clipper instead of a rectifier.

2. Compare the output waveform and signal levels of the series and shunt clipper. With the understanding that for the shunt clipper R_S and R_L do not have to be a 10:1 ratio, list the advantages and disadvantages for each circuit.

3. In Figure 10.3 you saw that you could pick the amount of voltage you wished to clip. Why do you think we might wish to clip only the top of a waveform?

4. Finally, in the circuit of Figure 10.4, you saw how to place the entire signal on a DC voltage level. These circuits are necessary in some electronic equipment, such as television sets, to force a signal to a certain DC voltage level for proper operation. What is another name for the clamper circuit?

Section II

Fault 1 - R_S open

When R_S is open in the circuit of Figure 10.2, the output goes essentially to 0 V because there is no current flow in the circuit. During your troubleshooting procedures, how would you isolate the fault to R_S?

Fault 2 - D_1 open

When D_1 in Figure 10.2 opens, the output has the complete signal waveform. Briefly explain why you believe (or don't believe) that the output waveform is sufficient data to determine the diode failure.

Fault 3 - D_1 open

In the clamper of Figure 10.4, if the diode D_1 opens, the normal signal voltage appears across R_L. Why does this occur?

Quick Check

1. In our shunt clipper of Figure 10.2, what would happen to the V_{out} if R_S shorted?

2. Clamping circuits are also called ________.

   (a) DC restorers
   (b) DC inserters
   (c) baseline stabilizers
   (d) all the above

3. The capacitor in a clamper circuit is in parallel with the voltage source.

   True
   False

4. If you wish to make a negative shunt clipper out of a positive shunt clipper, what should you do?

11

BASIC BJT CHARACTERISTICS

INTRODUCTION

In a base biased transistor circuit, the quiescent collector current is determined by the base current and the beta of the transistor. Also, the base current may be varied by changing the amount of the base-emitter bias voltage.

In Section I of this experiment, you will see how the current in a transistor varied by changing the amount of forward bias of the base-emitter junction with its corresponding change in base current.

In Section II you will explore two of the most common transistor failures: the C-E open and the C-E short. You will also learn to perform a quick measurement to determine the status of a transistor.

REFERENCE

Principles of Electronic Devices and Circuits - Chapter 4, Section 4.3

OBJECTIVES

In this experiment you will:

✓ Observe the effect of a changing base forward bias voltage on collector current

✓ Know how to recognize a shorted or open transistor using V_{CE} measurements

✓ Learn how to make a quick check of transistor status using an ohmmeter

EQUIPMENT AND MATERIALS

DC power supply
Digital multimeter
Circuit protoboard
NPN transistor, 2N3904 or equivalent
Resistors: 1 kΩ [2] , 22 kΩ, 36 kΩ, 100 kΩ

SECTION I FUNCTIONAL EXPERIMENT

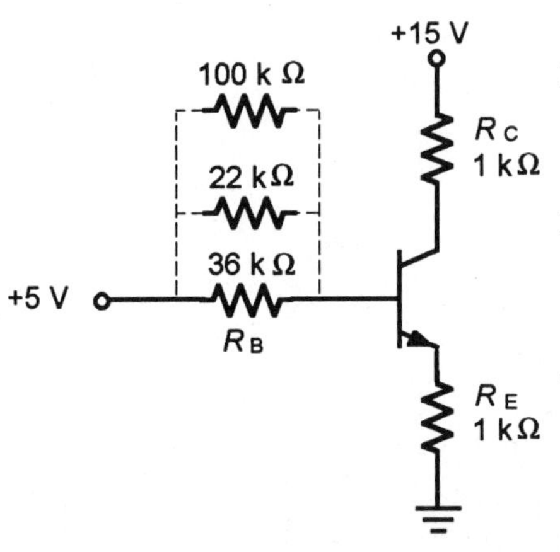

Figure 11.1

1. Build the circuit shown in Figure 11.1.

2. Measure and record the following voltages.

   V_{BE} = ________________ V_{CE} = ________________

3. Measure and record the collector and base currents.

   I_C = ________________ I_B = ________________

4. Calculate the values of emitter current and beta (β); and then measure and record the value of I_E.

   Calculated:

   I_E = ________________ DC β = ________________

   Measured:

   I_E = ________________

5. Replace the 36-kΩ R_B with a 22-kΩ RB, measure the V_{BE} and I_C, and record these values below. *Note:* The value of V_{BE} will be small, in the range of 0.6 to 0.8 volt. Make and record as accurate a value as you can.

   R_B = 22 kΩ:

   V_{BE} = ________________ I_C = ________________

6. Replace the 22-kΩ R_B with a 100-kΩ R_B; measure and record the V_{BE} and I_C.

   R_B = 100 kΩ:

   V_{BE} = ________________ I_C = ________________

SECTION II TROUBLESHOOTING

Fault 1 - Collector-emitter short

1. You will simulate a shorted collector-emitter transistor failure by placing a piece of wire from the collector to base and from the base to emitter.

2. Apply power to the circuit and measure the C-E voltage. You should find this voltage very low. Remember: The voltage drop across a short is ideally zero volts.

   V_{CE} = ____________

3. Remove the transistor and the shorting wires from the circuit and check the V_{CE} circuit point again. You should now find the voltage drop very close to the power supply level. (Remember: The voltage source will be dropped across an open circuit.)

   V_{CE} = ____________

Testing a Transistor with an Ohmmeter

Voltage checks are the easiest and less time-consuming tests for a technician to make. The information gained from this type of testing plus the knowledge of how the circuit and components operate is essential for successful troubleshooting. Once you have made the decision to change a transistor that is inoperative, it is good to make a quick check with an ohmmeter to verify that the transistor is bad.

The following is a quick check procedure to determine the condition of a transistor.

CAUTION

Do not use the very low (R × 1 or R × 10) range of your ohmmeter for testing transistors. On some ohmmeters, these ranges can cause a large current that will damage a diode or transistor. If your ohmmeter has a range identified for diode tests, use that range.

1. Refer to Figure 11.2a. For an NPN transistor connect the positive lead of your ohmmeter to the base, and the negative lead to the collector. A good transistor will show a low resistance reading.
2. Refer to Figure 11.2b. Leaving the positive lead on the base, place the negative lead on the emitter. This should also show a low resistance reading for a good transistor.
3. Refer to Figure 11.2c. Next place the negative lead on the base and the positive lead on the collector. This should show an extremely high resistance reading on the ohmmeter for a good transistor.
4. Leaving the negative lead on the base, place the positive lead on the emitter. This should also show a very large resistance reading for a good transistor.

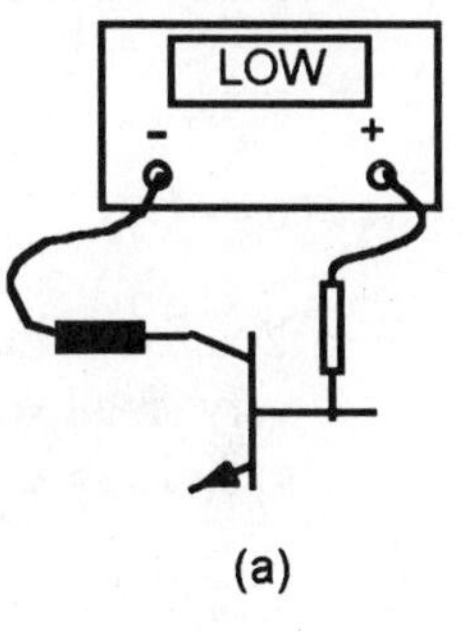

(a)

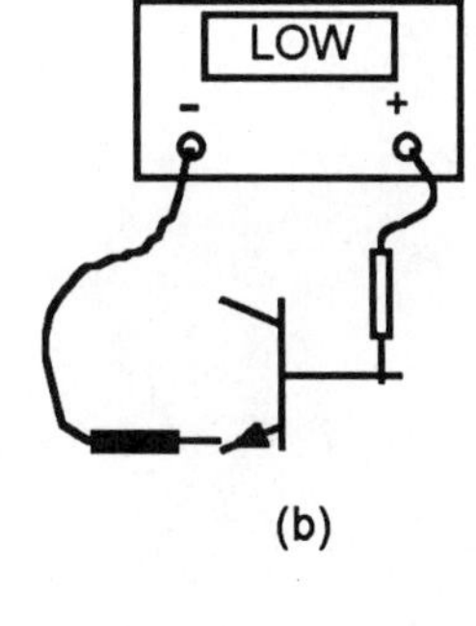

(b)

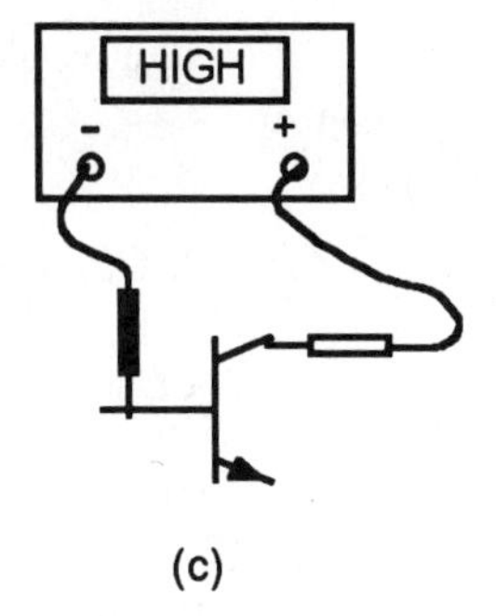

(c)

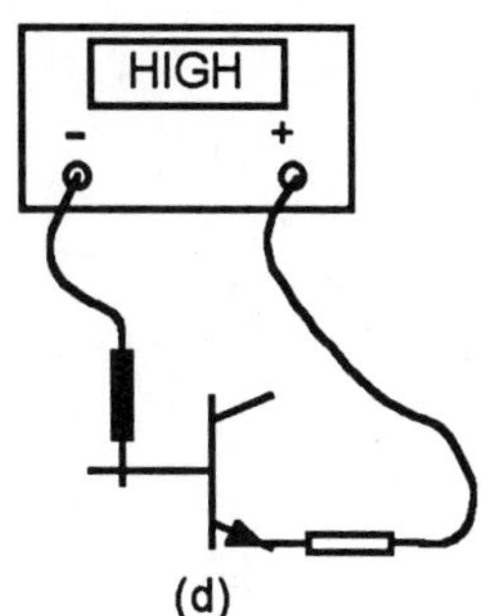

(d)

Figure 11.2

The forward to reverse resistance ratio for a good transistor should be 1000:1.

To check a PNP transistor, simply reverse the above process.

DISCUSSION

Section I

1. You found in Section I that beta is an expression of a BJT current relationship. State in your own words the beta relationship.

2. You also found that the amount of forward bias voltage between the base and emitter controls the amount of collector current flow in the transistor. What use do you think this control of a transistor by the base voltage may be?

3. In step 4 of the experiment, you calculated the beta of the transistor used in the experiment. Using the beta box model for the circuit of Figure 11.1, predict the value of collector current for a transistor with beta 50% larger than your calculated value.

   Based on your measured and predicted collector current values, comment on the ability to know the value of collector current in a base biased circuit.

Section II

1. You found that a transistor that was shorted from collector to emitter has a very small voltage drop across it. How can this knowledge be useful to a technician as a troubleshooting aid?

2. You discovered that when a transistor is open from collector to emitter, V_{CE} would read a very high voltage. Can you think of a way this knowledge would be helpful to a technician during troubleshooting?

Quick Check

1. A transistor will operate correctly even without any base current flow.

   True False

2. The V_{CE} of a shorted transistor will be _____.

   (a) very high (b) very low
   (c) normal

3. A varying forward bias voltage between the base and emitter of a transistor will not affect the collector current.

   True False

12

BETA EFFECTS IN THE BJT

INTRODUCTION

Base biasing, although the simplest of transistor biasing forms, provides the least stable quiescent operating point. This experiment introduces transistor biasing and allows you to observe variations in the "Q" point for different transistors and two circuit forms of base biasing.

In the troubleshooting section you will make measurements to identify and enable you to learn the effects of failure of the base biasing resistor and circuit power supply.

REFERENCE

Principles of Electronic Devices and Circuits - Chapter 4, Sections 4.4 and 4.5

OBJECTIVES

In this experiment you will:

✓ Learn how to use a single biasing supply

✓ Verify experimentally the inherent instability of base biasing

✓ Learn how to relate measured circuit values to specific circuit faults

EQUIPMENT AND MATERIALS

DC power supply
Digital multimeter
Circuit protoboard
NPN transistor [3], 2N3904 or equivalent
Resistors: 1 kΩ, 1.2 kΩ, 2.2 kΩ, 510 kΩ, 680 kΩ, 4.7 MΩ

NOTE

Mark or identify each transistor as A, B, or C so that the transistor effect on each circuit can be compared.

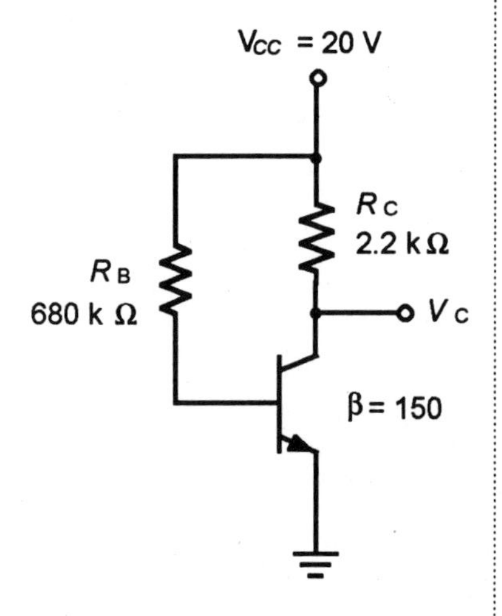

Figure 12.1

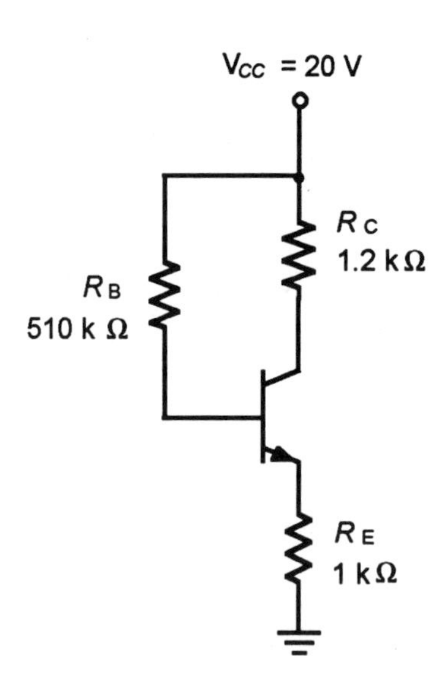

Figure 12.2

SECTION I FUNCTIONAL EXPERIMENT

The resistor values of the circuits in Figures 12.1 and 12.2 are selected to provide a quiescent collector current near midpoint with an assumed transistor beta of 150. In this experiment, you will try three different transistors and measure the quiescent circuit values in the circuit of Figure 12.1.

Then you will use the same three transistors in a modified circuit (Figure 12.2) to permit comparison of the individual transistor beta effect on the two circuits.

1. Using transistor A, build the circuit in Figure 12.1. Measure and record the following circuit values:

   V_{BE} = ________________ V_{RB} = ________________

   I_C = ________________ V_{CE} = ________________

2. Turn off the circuit power. Remove transistor A and insert transistor B. Reapply circuit power. Make and record the following measurements:

   V_{BE} = ________________ V_{RB} = ________________

   I_C = ________________ V_{CE} = ________________

3. Turn off the circuit power. Remove transistor B and insert transistor C. Reapply circuit power. Make and record the following measurements:

   V_{BE} = ________________ V_{RB} = ________________

   I_C = ________________ V_{CE} = ________________

4. Disconnect your circuit, and connect the circuit in Figure 12.2, using transistor A. Reapply circuit power; then measure and record the following circuit values:

   V_{BE} = ________________ V_{RB} = ________________

   I_C = ________________ V_{CE} = ________________

5. Turn off the circuit power. Remove transistor A and install transistor B. Reapply circuit power; then measure and record the following circuit values:

   V_{BE} = ________________ V_{RB} = ________________

   I_C = ________________ V_{CE} = ________________

6. Turn off the circuit power. Remove transistor B and install transistor C. Reapply circuit power, measure and record the following circuit values:

   V_{BE} = ________________ V_{RB} = ________________

   I_C = ________________ V_{CE} = ________________

SECTION II TROUBLESHOOTING

Fault 1 - R_B open

Build the circuit in Figure 12.1. Simulate an open R_B by replacing the 680 kΩ resistor with a 4.7 MΩ resistor. Apply circuit power. Measure the collector current and V_{CE}. Contrast this with your measurement of Section I, step 1.

Fault 2 - DC supply failure

Here you will simulate the DC power supply voltage falling from 20 V to 15 V. Adjust your power supply for 15 V and again measure collector current and V_{CE}. Contrast this measurement with the measurement of Section I, step 1.

DISCUSSION

Section I

1. In your measurements of the two circuits (Figures 12.1 and 12.2), you should have found that one circuit showed less variation in collector current and collector-emitter voltage for the three different transistors used. Which circuit seemed to have the more consistent quiescent operating point, and why would you expect it to be so?
2. As you know from your text and this experiment, you cannot depend on a base-biased transistor to have a stable quiescent operating point. This means that this bias form is not useful in an amplifier circuit. Can you think of a circuit application where a base-biased transistor would be useful? Why?
3. You should have found in Procedure steps 1, 2, and 3 that each transistor most likely resulted in a different value of collector current. Why do you think this was the case?

Section II

1. In Fault 1 you found that there is no collector current flow if R_B becomes open. Why is this true?
2. In Fault 2 you found that collector current was greatly reduced for a reduced V_{CC}. Ohm's law tells us that if the voltage is reduced, the current also is reduced. Thinking back to the chapter on power supplies, what problem can you think of that could cause the DC voltage to fall in value?

Quick Check

1. Base biasing is one of the most stable biasing arrangements.

   True   False

2. Base biasing does not make beta independent in amplifier circuits.

   True   False

3. Why do our circuit calculations differ from our actual measurements?

   (a) Resistor tolerances
   (b) Different transistor betas.
   (c) All of the above
   (d) None of the above

13

EMITTER BIAS OF THE BJT

INTRODUCTION

In Experiment 12 you found that base bias could not provide stable collector current or voltage for transistors having different values of beta. In this experiment you will explore emitter bias. Emitter bias makes the circuit much more independent of beta, thereby stabilizing changes in collector current and voltage better than the base bias circuit. Although emitter bias provides an improvement in circuit stability, it has the disadvantage of requiring two power supplies to operate.

In the troubleshooting section you will examine the effect on your circuit if R_B opens, R_C opens, or the transistor shorts.

REFERENCE

Principles of Electronic Devices and Circuits - Chapter 4, Section 4.6

OBJECTIVES

In this experiment you will:

✓ Learn to construct an emitter bias circuit and perform measurements of that circuit

✓ Understand how emitter bias improves circuit quiescent stability

✓ Learn how to relate the effects of common emitter biased circuit failures with measured circuit values

EQUIPMENT AND MATERIALS

Dual DC power supply
Circuit protoboard
Digital multimeter
NPN transistors [3], 2N3904 or equivalent
Resistors: 470 Ω, 1 kΩ [2]

SECTION I FUNCTIONAL EXPERIMENT

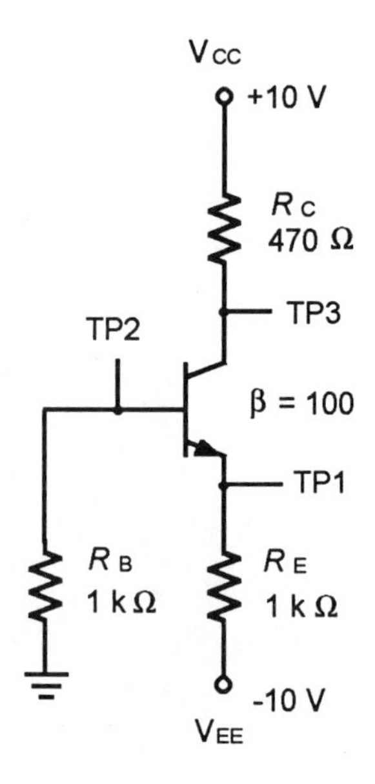

Figure 13.1

1. Using the first transistor, construct the circuit in Figure 13.1. Check your circuit to be sure that your connections are correct. Apply circuit power; then measure and record the following circuit values:

   I_C = ________ V_{RE} = ________ V_E = ________

   V_B = ________ V_C = ________ V_{CE} = ________

2. Turn off the circuit power. Remove the transistor and replace it with transistor #2. Reapply circuit power. Measure and record the following circuit values:

   I_C = ________ V_{RE} = ________ V_E = ________

   V_B = ________ V_C = ________ V_{CE} = ________

3. Turn off the circuit power. Remove transistor #2 and replace it with transistor #3. Reapply circuit power. Measure and record the following circuit values:

   I_C = ________ V_{RE} = ________ V_E = ________

   V_B = ________ V_C = ________ V_{CE} = ________

Step 4 completes the measurement portion of the experiment. If you are proceeding to the troubleshooting portion of the experiment, leave your circuit connected.

4. Using the largest and smallest values of collector current measured for your circuit, calculate the mean value of collector current using the formula below.

$$I_{C(mean)} = \sqrt{I_{C(max)} \times I_{C(min)}}$$

   $I_{C(mean)}$ = ________

5. Using the beta box model with an assumed beta of 173, calculate the following values for the circuit in Figure 13.1.

   I_C = ________ V_E = ________

   V_C = ________ V_{CE} = ________

SECTION II TROUBLESHOOTING

Fault 1 - R_B open

Ensure that circuit power is off. Replace the 1-kΩ R_B with a 2.2-MΩ resistor. Turn on circuit power. Measure and record the following circuit values:

V_B = ________ V_C = ________ V_{RE} = ________

V_{CE} = ________ I_C = ________

Fault 2 - R_C open

Turn off the circuit power. Replace R_B with a 1-kΩ resistor. Exchange the 470-Ω R_C with a 2.2-MΩ resistor. Reapply circuit power. Measure and record the following circuit values.

V_B = __________ V_C = __________ V_{RE} = __________

V_{CE} = __________ I_C = __________

Fault 3 - Transistor shorted

Turn off the circuit power. Replace R_C with a 470-Ω resistor. Place shorting wires from the collector to emitter and base to emitter of the transistor. Reapply circuit power. Measure and record the following circuit values.

V_B = __________ V_C = __________ V_{RE} = __________

V_{CE} = __________ I_C = __________

DISCUSSION

Section I

1. Your measurements for the emitter biased circuit should have shown that the base voltage is almost 0 volts. Explain how it possible to forward bias the emitter-base diode with 0 volts on the base.

2. Considering the values of measured and calculated collector current, do you feel that emitter bias provides a relatively stable quiescent operating point? Explain your answer using the measured data (steps 1, 2, and 3), mean collector current (step 4) and calculated collector current (step 5).

Section II

Fault 1 - R_B open

You should have found, with R_B open, that the base voltage dropped completely to zero and there was no base current. Since base voltage in this circuit form is small (about 0.09 V), can you suggest a better voltage test that would indicate the transistor circuit operation?

Fault 2 - R_C open

According to your measured data, what is the most obvious indication that R_C is open in this circuit form?

Fault 3 - Transistor shorted

In Fault 3 the shorted transistor is indicated by a V_{CE} voltage of 0 volts. What other condition in a transistor circuit could give almost this same indication and yet not have a failure?

Quick Check

1. The emitter bias transistor circuit should provide better stability than the base biased circuit.

   True　　　　　　　　False

2. The base voltage of the emitter bias is typically close to _____.

   (a) 5 V　　　　　　(b) 3.5 V
   (c) 0 V　　　　　　(d) 0.7 V

3. The fact that the emitter bias circuit requires two power supplies is considered _____.

   (a) an advantage　　　　(b) a disadvantage
   (c) no problem

4. The saturated value of I_C for the circuit in Figure 13.1 is _____.

   (a) 9.3 mA　　　　(b) 15.1 mA
   (c) 21.3 mA　　　　(d) 36.2 mA

14

COLLECTOR FEEDBACK BIAS

INTRODUCTION

Collector feedback bias is an older form of bias originally designed to provide a simple circuit independent of variations in beta. In this experiment you will calculate and measure the DC values and then introduce beta changes to observe the effect on circuit stability.

The troubleshooting section will explore the effects of an open R_B and R_C.

REFERENCE

Principles of Electronic Devices and Circuits - Chapter 4, Section 4.6

OBJECTIVES

In this experiment you will:

✓ Observe the effect of feedback to control circuit quiescent point stability

✓ Measure the effects of temperature-induced beta changes on your circuit

✓ Learn to relate measured values to circuit failures

EQUIPMENT AND MATERIALS

DC power supply
Digital multimeter
Circuit protoboard
Heat gun or freeze spray
NPN transistor, 2N3904
Resistors: 2.2 kΩ, 470 kΩ, 2.2 MΩ

SECTION I FUNCTIONAL EXPERIMENT

In this experiment you will first measure the quiescent values at normal room temperature. You will then repeat the measurements with temperature-induced beta changes to observe the effect of feedback in controlling circuit stability.

1. Construct the circuit in Figure 14.1.

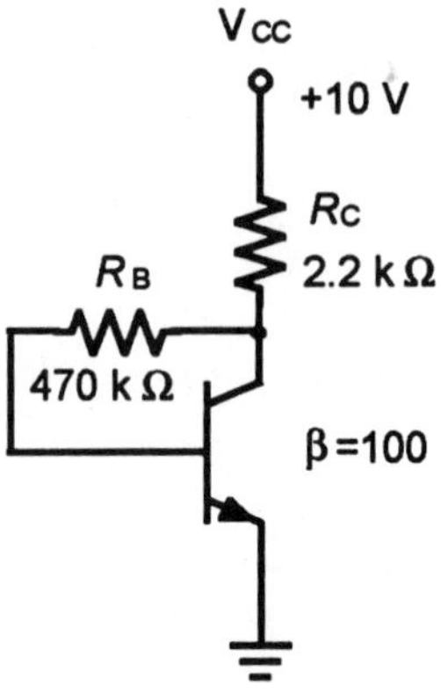

V_{CC}
+10 V
R_C
2.2 kΩ
R_B
470 kΩ
β=100

Figure 14.1

2. Apply power to the circuit. Measure and record the circuit DC values indicated below.

I_E = __________ I_C = __________ V_B = __________

V_E = __________ V_C = __________ V_{CE} = __________

If possible, use two digital voltmeters for Procedure step 3. If you cannot use two DMMs, make the collector current measurement first; then make the collector voltage measurement.

3. Connect one DMM to measure collector current and the other to measure collector voltage. Apply power and record these measurements. They should be approximately as recorded in step 2.

I_C = ______________ V_C = ______________

4. While monitoring these values use the heat gun to warm the transistor or the freeze spray to cool it. Use care to avoid overheating the transistor. Limit heat gun application to approximately 1/2 minute to 1 minute maximum, and let the transistor cool before reheating. As collector current increases, you should notice the collector voltage decreasing; and as collector current decreases, collector voltage increases. If you have access to both a heat gun and freeze spray, you may wish to make the measurements first with the heat gun and then with the freeze spray.

Step 4 completes the measurement portion of the experiment. You may leave the circuit connected if you are going to proceed to the troubleshooting section.

SECTION II TROUBLESHOOTING

Fault 1 - R_C open

Make sure the circuit power is turned off. Leaving the DMMs connected to read collector current and collector voltage from Section I, step 5, replace the 2.2-kΩ R_C with a 2.2-MΩ resistor. Apply power to the circuit; measure and record the values indicated below.

I_C = __________ V_C = __________ V_{RC} = __________

V_{BE} = __________ V_{CE} = __________

Fault 2 - R_B open

Turn off the circuit power. Again leaving the DMM connected as in step 4 of Section I, replace the 2.2-kΩ collector resistor, and exchange the R_B with a 2.2-MΩ resistor. Apply circuit power; measure and record the values below.

I_C = ___________ V_C = ___________ V_{RC} = _________

V_{BE} = _________ V_{CE} = _________

DISCUSSION

Section I

1. While the transistor was heated in Procedure step 4 of the experiment, you should have observed a small increase in the collector current and a decrease in collector voltage. In your own words, explain how this relationship helps to provide beta stabilization in this transistor circuit.

2. If the circuit in Figure 14.1 were modified to include an emitter resistance, how do you think this would affect the beta-independent nature of this circuit? *Hint:* Set up a beta box model for the modified circuit form.

Section II

Fault 1 - R_C open

1. You have found that if R_C opens, collector current and forward base-emitter voltages are zero. V_C also goes to zero volts. Why does the base emitter voltage also disappear when R_C opens?

2. You also observed that when R_C opens, the entire V_{CC} (10 V) is dropped across R_C. Why does this occur?

Fault 2 - R_B open

Explain why you found that with R_B open, collector current was zero and $V_{CE} = V_{CC}$.

Quick Check

1. The feedback used in the collector feedback bias circuit demonstrates the ability of the output current or voltage to regulate a circuit by affecting the input current or voltage.

   True False

2. One advantage of collector feedback bias over base bias is that it uses only one DC power supply.

True False

3. The collector saturation current for the circuit in Figure 14.1 is:

(a) 2.55 mA (b) 4.55 mA
(c) 8.55 mA (d) 9.55 mA

4. At saturation, the collector-emitter voltage will be approximately 0.2 V.

True False

5. Assuming a transistor beta of 150, changing R_B in the circuit in Figure 14.1 to 330 kΩ would put the transistor into saturation.

True False

15

VOLTAGE DIVIDER BIAS FOR BJTS

INTRODUCTION

Voltage divider bias is one of the best and most commonly used bias methods for stabilizing BJT circuits. In Section I of this experiment you will set up a voltage divider-biased circuit, predict circuit values, and make circuit measurements to evaluate the stability of the quiescent circuit values.

Section II of this experiment deals with troubleshooting. In this section you will simulate three different faults and make measurements of the circuit for each fault. This will develop your troubleshooting knowledge by letting you observe the effects of a component's failure and by giving you the measurement data of the fault.

REFERENCE

Principles of Electronic Devices and Circuits - Chapter 4, Sections 4.7 and 4.10

OBJECTIVES

In this experiment you will:

✓ Add to your understanding of voltage divider bias

✓ Verify the stability of a voltage divider-biased circuit

✓ Determine circuit values for different component failures

EQUIPMENT AND MATERIALS

DC power supply
Digital multimeter
Circuit protoboard
NPN transistor [2], 2N3904 or equivalent
Resistors: 270 Ω, 1.2 kΩ, 2.7 kΩ,12 kΩ, 2.2 MΩ

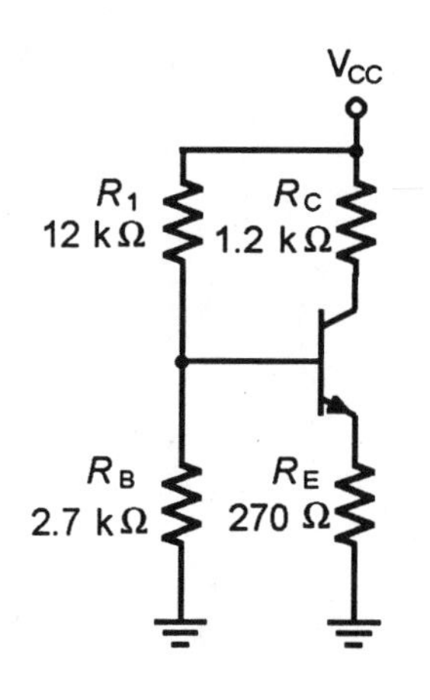

Figure 15.1

Changing transistors in the circuit gave two measurements with two different (although unknown) values of beta. To test the circuit stability further, we can use the fact that beta increases with temperature. The next set of measurements will be made while the transistor is being heated.

> **NOTE**
> An alternative method to heating the transistor is to cool it using component cooler spray. This will decrease beta.

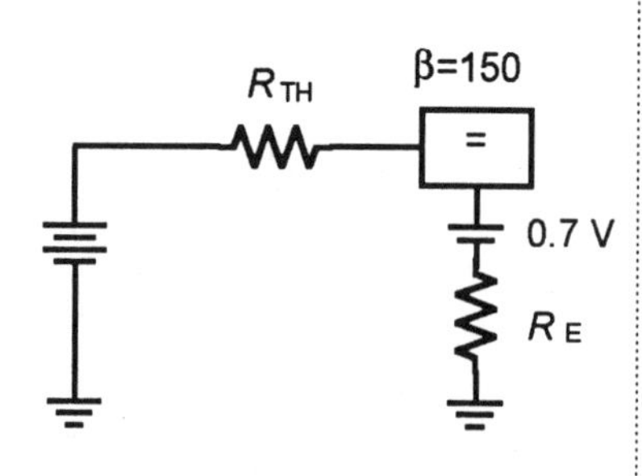

Figure 15.2

SECTION I FUNCTIONAL EXPERIMENT

1. Construct the circuit shown in Figure 15.1. Set the DC power supply to 12 V, and connect the power supply to your circuit.

2. Using the digital multimeter, measure and record the circuit parameters listed below.

   I_E = ____________ V_B = ____________ V_E = ____________

   V_C = ____________ V_{CE} = ____________

3. Turn off the DC power supply and substitute a second 2N3904 transistor. Reapply circuit power; measure and record the circuit parameters again.

   I_E = ____________ V_B = ____________ V_E = ____________

   V_C = ____________ V_{CE} = ____________

4. You can use a heat gun or soldering iron to heat the transistor. If you use a soldering iron, do not actually touch the transistor with the hot tip of the soldering iron—hold the soldering iron close to the transistor. Heat the transistor for about 1 minute, or until the case is very warm to the touch. Make the following measurements quickly following the heating of the transistor before it returns to ambient temperature:

   I_E = ____________ V_B = ____________ V_E = ____________

   V_C = ____________ V_{CE} = ____________

5. Now you have three sets of data from your circuit. Solve the circuit for the same parameters as those measured. Calculate the Thevenin values of the voltage divider—R_{TH} and V_{TH}. Enter these values in the blanks provided below and in the beta box model in Figure 15.2. Assume a beta of 150, and, using the beta box model, calculate emitter current, collector voltage, base voltage, and emitter voltage. Also enter these results in the appropriate blanks provided below.

   V_{TH} = ____________ R_{TH} = ____________ V_B = ____________

   V_E = ____________ V_C = ____________ V_{CE} = ____________

SECTION II TROUBLESHOOTING

Fault 1 - R_B divider resistor fails
When a resistor fails, it becomes a high resistance, or open circuit. In this procedure, you will simulate an open-resistor failure by substituting a large resistance value for resistor R_B.

1. Modify your circuit of Figure 15.1 by replacing the 2.7-kΩ resistor with the 2.2-MΩ resistor.

2. Apply 12 VDC to the circuit. Using the digital multimeter, measure and record the circuit values listed below:

V_B = ______________ V_E = ______________

V_C = ______________ V_{CE} = ______________

3. Compare the circuit values just obtained to those of a normal circuit (Section I, Step 2 or 3). Notice in particular the readings that had the large change.

Fault 2 - R_1 divider resistor fails

1. Turn off the circuit power and replace R_B with the 2.7-kΩ resistor. Exchange the 12-kΩ value of R_1 with the 2.2-MΩ resistor.

2. Reapply 12 VDC to the circuit. Using the multimeter, measure and record the circuit values listed below.

V_B = ______________ V_E = ______________

V_C = ______________ V_{CE} = ______________

You should have found that the measured circuit values for this fault are quite different from those of a normal circuit. Also notice the differences from the values obtained for the R_B fault.

Fault 3 - Emitter resistor fails

1. Turn off the DC power to the circuit and replace resistor R_1 with the correct value of 12-kΩ. Exchange the 270-Ω emitter resistor for the 2.2-MΩ resistor.

2. Reapply 12 VDC to the circuit. Using the multimeter, measure and record the circuit values listed below.

V_B = ______________ V_E = ______________

V_C = ______________ V_{CE} = ______________

Notice that each fault gave a different and unique set of circuit values, and that in each case, the transistor shifted to either a cutoff or a saturation condition.

DISCUSSION

Section I

1. Discuss, from the standpoint of a stable operating point, the voltage divider portion of a voltage divider-biased circuit. Consider the effects of large resistances in the divider, both firm and stiff.

2. Discuss the stability of voltage divider-biased transistors based on the data you have taken. Consider such factors as the use of two different transistors, as well as the forced beta changes of heating or cooling the transis-

tor, and the percentage difference in the values of emitter current and collector voltages.

3. Compare the stability of voltage divider bias and emitter bias (Experiment 13).

4. If you used the quick method of your text for solving the firm or stiff divider for the circuit in Figure 15.1, do you believe this would give a satisfactory value for emitter current? Explain your answer.

Section II

1. For each fault, describe the transistor state (cutoff or saturation). Describe how you would use this knowledge in troubleshooting a circuit.

2. Based on your measured data of circuit failures, what circuit measurements would you need to make in order to identify each fault?

Quick Check

1. The circuit in Figure 15.1 is near mid-point bias.

   True   False

2. The saturation current of the circuit in Figure 15.1 is _____.

   (a) 5.5 mA   (b) 8.16 mA
   (c) 10.23 mA   (d) 11.04 mA

3. One advantage of voltage divider bias compared to emitter bias is _____.

   (a) beta is not a factor   (b) fewer components
   (c) alpha is not a factor   (d) simpler power supply needed

4. Based on your circuit calculations, with an assumed beta of 150, the voltage divider is _____.

   (a) firm   (b) stiff
   (c) neither

16

BIASING PNP TRANSISTORS

INTRODUCTION

The operating voltages for PNP transistors have the opposite polarity from those used to bias NPN transistors.

PNP transistors are used less often than NPN transistors, and when they are, they are often used in circuits with NPN transistors. For convenience, and to avoid two power sources, the PNP transistors are usually connected upside down with the emitter connected to the same $+V_{CC}$ that the collector of NPN transistors are. This is the circuit form you will use in this experiment.

In Section I of this experiment you will calculate and measure the DC values for a PNP voltage divider biased transistor circuit.

In Section II you will observe the effects of open bias resistors R_1, a shorted collector/emitter, and an open collector emitter.

REFERENCE

Principles of Electronic Devices and Circuits - Chapter 4, Section 4.8

OBJECTIVES

In this experiment you will:

✓ Learn to set up an upside down voltage divider biased PNP transistor circuit

✓ Understand the references used to make voltage measurements in a PNP transistor circuit

✓ Recognize fault voltage values for common troubles in a PNP voltage divider biased circuit

EQUIPMENT AND MATERIALS

DC power supply
Digital multimeter
Circuit protoboard
PNP transistor, 2N3906 or equivalent
Resistors: 1.2 kΩ, 2.2 kΩ, 3.3 kΩ, 10 kΩ

SECTION I FUNCTIONAL EXPERIMENT

1. Construct the circuit of Figure 16.1.

In step 2 you will be measuring the DC values of your circuit. As a reminder, where a parameter has a single subscript such as V_C or V_E the measurement is to be made from the element to circuit common (ground).

2. Apply circuit power, measure and record the DC values Table 16.1, paying close attention to Figure 16.1.

3. Turn off circuit power. In the area labeled Figure 16.2, redraw the circuit of Figure 16.1 right side up. Shift the circuit ground to the positive terminal of the power supply.

4. Construct the circuit you drew in Figure 16.2. Be sure to check your power supply connections carefully.

5. Apply circuit power. Measure and record in Table 16.1 the same circuit parameters you made in step 2.

This completes the procedure. Turn off circuit power. Disconnect your circuit.

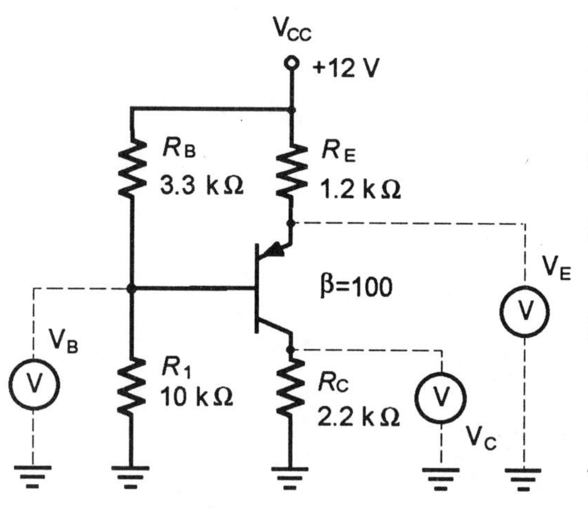

Figure 16.1

Figure 16.2

	Data from step 1	Data from step 2
V_B		
V_E		
V_C		
I_E		
I_B		

Table 16.1

SECTION II TROUBLESHOOTING

Fault 1 - R_1 open

1. Construct the circuit of Figure 16.1. Check your circuit to ensure proper operation. Turn off circuit power. Remove the 10-kΩ R_1 and replace with a 2.2-MΩ resistor.

2. Reapply circuit power. Measure and record the following circuit values:

V_B = __________ V_E = __________ V_C = __________

Fault 2 - Transistor shorted

1. Turn circuit power off. Replace the transistor in the circuit. Place a shorting wire from the emitter to base and the base to collector. Turn on circuit power.

2. Measure and record the following DC values:

   V_B = __________ V_E = __________ V_C = __________

Turn off circuit power and disconnect your circuit.

DISCUSSION

Section I

1. In the measurements of steps 2 and 5, you found that the transistor element voltages, V_B , V_E , and V_C, were quite different yet the transistor operated at the same quiescent point. Why do you think these differences are important to the technician?

2. Because the circuits of Figure 16.1 and your circuit of Figure 16.2 had the same values of I_C and V_{CE}, the values of the base-emitter voltage and the voltage from collector to emitter were the same. Identify other voltages and currents that were the same in the two circuits.

Section II

Fault 1 - R_1 open

In Fault 1 you found that V_B and V_E were at 12 V, while V_C had no voltage. Explain why you think these values were measured.

Fault 2 - Transistor shorted

You discovered that when the transistor was shorted, V_{CC} was divided across R_E and R_C. Explain why you think this occurred.

Quick Check

1. PNP transistor schematics are usually drawn:

   (a) right side up (b) sideways
   (c) upside down

2. The bias voltage polarities for PNP transistors are:

   (a) same as NPN (b) reversed from NPN
   (c) higher voltage than NPN

3. In the circuit of Figure 16.1, the bias is:

(a) firm
(b) stiff
(c) neither

4. The quiescent operating point of the circuit of Figure 16.1 would depend on the transistor beta.

True

False

17

COMMON EMITTER AMPLIFIER

INTRODUCTION

One of the most commonly used small-signal amplifiers is the common emitter configuration. In this experiment you will be testing a single-stage common emitter amplifier. You will be measuring voltage gain for both an unswamped and a swamped configuration, measuring amplifier phase shift, and observing the amplifier response to loading.

In the troubleshooting section, you will insert amplifier faults and observe the failed amplifier AC and DC circuit values.

REFERENCE

Principles of Electronic Devices and Circuits - Chapter 5, Sections 5.3 and 5.4

OBJECTIVES

In this experiment you will:

✓ Learn how to determine voltage through circuit measurements

✓ Gain understanding of the effects of loading on amplifier gain

✓ Be able to relate amplifier AC and DC voltages to component failures

EQUIPMENT AND MATERIALS

DC power supply
Digital multimeter
Dual-trace oscilloscope
Function generator
NPN transistor, 2N3904 or equivalent
Circuit protoboard
Resistors: 1 kΩ, 2.2 kΩ, 3.9 kΩ [2], 10 kΩ, 47 kΩ
Capacitors: 1 μF [2], 470 μF

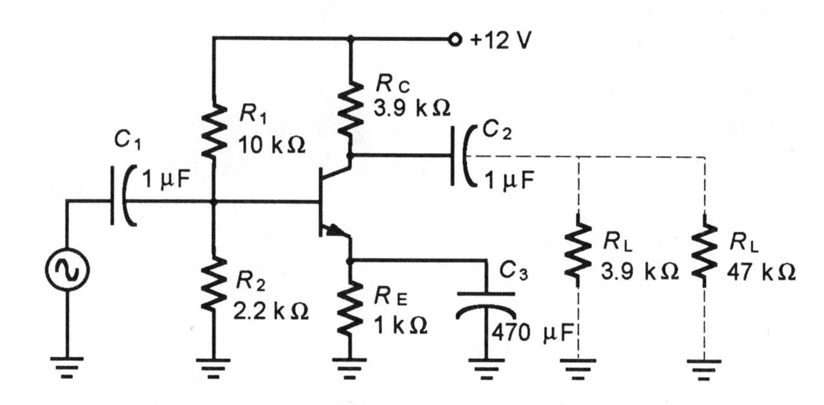

Figure 17.1

1. Build the circuit of Figure 17.1, without connecting a load resistance at this time.

2. Connect the DC power supply set to 12 V. *Do not connect the audio function generator* at this time. Measure and record the following DC voltages:

   V_B = ________________ V_E = ________________

   V_C = ________________ V_{CE} = ________________

   Is the amplifier operating at or close to the load line midpoint?

   ____

3. Connect the function generator, and set it to provide a signal of 20 $mV_{p\text{-}p}$ at the base of the transistor.

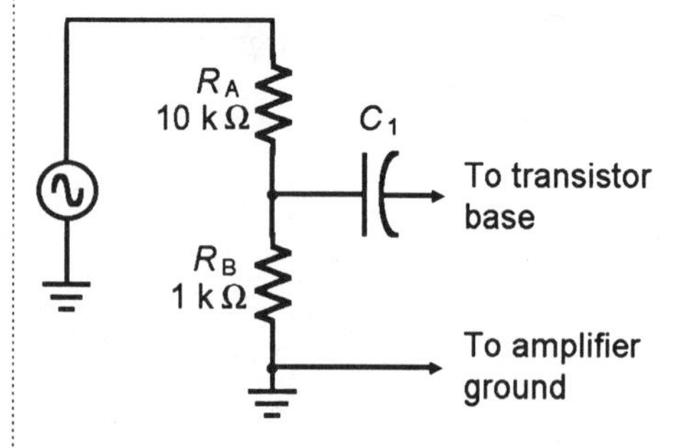

Figure 17.2

If you have difficulty obtaining a low enough output from your function generator in step 3, add the voltage divider network of Figure 17.2 to your circuit.

Do not rely on the divider value to determine the amplifier input signal. Measure the signal at the base of the transistor and adjust the function generator to obtain the required value.

4. Set the oscilloscope coupling to AC and connect the scope to the transistor collector. Adjust the function generator to obtain the maximum output signal possible without clipping. Observe the shape of this output waveform. Reduce the input signal to obtain an output signal of approximately 4 $V_{p\text{-}p}$. (The exact value is not critical.)

5. Set the oscilloscope coupling to DC. Measure and sketch the combined AC and DC signals at the base (V_B), emitter (V_E), and collector (V_C) in Graph 17.1. Record the positive and negative peak values obtained at the collector in the waveform sketch.

6. Using your dual trace oscilloscope, connect channel 1 to the transistor base and channel 2 to the collector. Set the scope coupling to AC. Set the scope to trigger from channel 2. What is the phase angle between the input and output of the amplifier?

   Phase angle = ___________

7. Adjust the function generator to provide a 20 $mV_{p\text{-}p}$ input signal to the
transistor base. Using your oscilloscope, measure and record V_{in} and V_{out} of the amplifier. Calculate and record the value of unloaded voltage gain.

   V_{in} =__________ V_{out} = __________ A_V =__________

8. Turn off the circuit power and connect the 3.9-kΩ load. Measure the input signal and output across the load. Record these values. Calculate the amplifier loaded voltage gain.

   V_{in} =__________ V_{out} = __________ A_{VL} = __________

9. Turn off the circuit and replace the 3.9-kΩ load with 47-kΩ measurements of step 8 to determine the loaded voltage gain.

   A_{VL} = __________

10. Turn off the circuit power. Modify your circuit to the swamped amplifier of Figure 17.3. Apply the circuit power. Set the function generator to apply a 20 $mV_{p\text{-}p}$, 1 kHz input signal, using your oscilloscope. Measure and record the amplifier input and output signal values. Then calculate the amplifier loaded voltage gain.

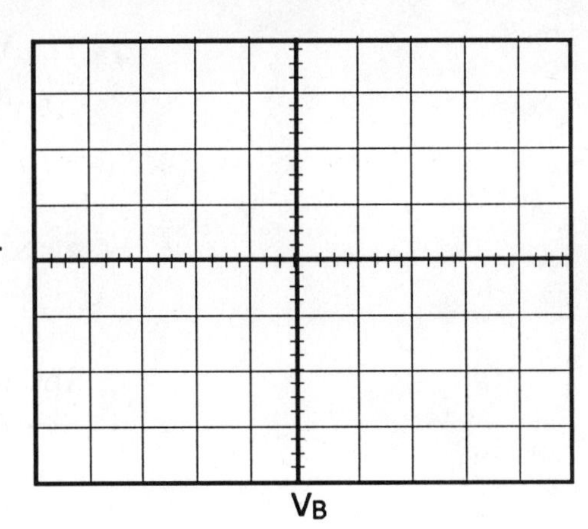

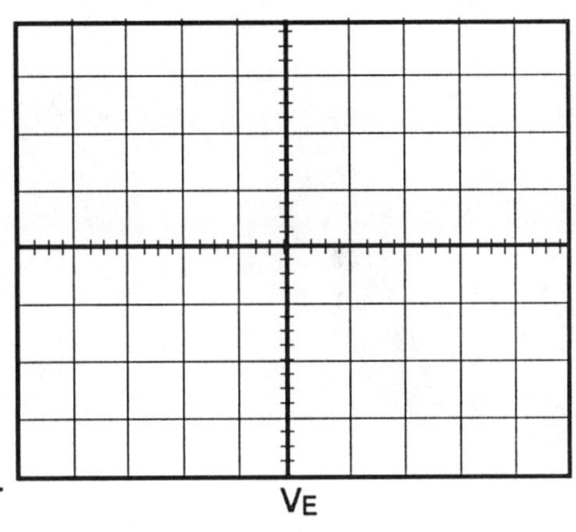

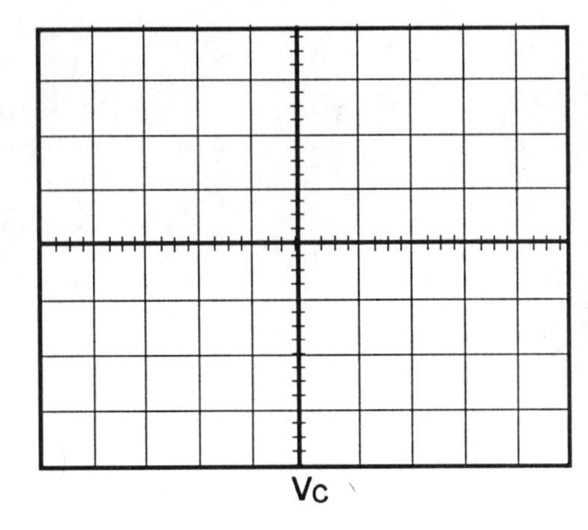

Graph 17.1

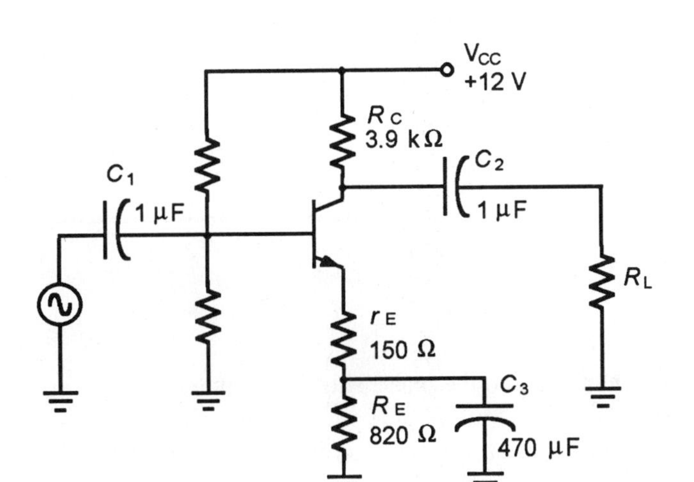

Figure 17.3

V_{in} = __________

V_{out} = __________

A_{VL} = __________

11. While monitoring the output signal at the load, increase the input signal to the point just short of clipping. Observe this waveform. You should observe a clean and undistorted waveform. Contrast this waveform to that obtained in step 4.

Step 11 completes your measurements. You may disconnect your circuit.

SECTION II TROUBLESHOOTING

Fault 1 - Emitter bypass capacitor open

1. Capacitors have three failure modes: They can open, short, or leak. You will now explore what happens to the V_{out} when the Emitter bypass capacitor opens.

2. Apply 12 VDC and the AC signal of 1 kHz at 20 $mV_{p\text{-}p}$ to the amplifier circuit. While monitoring the V_{out} carefully, remove one end of the emitter bypass capacitor and record the V_{out} below. Also observe and record gain differences.

   With emitter bypass capacitor:

   V_{out} = ________________ A_V = ________________

   Without emitter bypass capacitor:

   V_{out} = ________________ A_V = ________________

Fault 2 - Transistor shorted

1. The transistor in the CE amplifier has two main failure modes: The transistor can open or short from emitter to collector. You will explore the shorted failure mode here.

2. Turn off the circuit power. Place a shorting wire from the emitter to the collector of the transistor. Apply circuit power. Monitor the output signal and record the voltages below.

   V_{out} ________________ V_E ________________

   V_C ________________ V_{RC} ________________

DISCUSSION

Section I

1. In Procedure step 6, you measured the phase shift between the signal input on the base to the signal output on the collector. Describe, using the functional operation of the transistor, the reason for the phase shift you measured.

2. In Procedure step 4, using the oscilloscope with AC coupling, the average value of the amplifier signal was 0 V. That is because the positive peak value was equal in magnitude and opposite in polarity to the negative peak value.

   A. In Procedure step 5, the average value of the output signal was not zero. Calculate the average value from the data of Figure 17.3.

      V_{avg} = ___________

   B. What is the circuit parameter value represented by the average value V_{avg} of part A? Explain why you obtained the collector waveform values recorded in Figure 17.3.

3. In Procedure step 5, you measured the AC signal on the emitter of the transistor, which should have been essentially 0 VAC. Explain why this value should have been obtained.

Section II

1. In Fault 1 you discovered that the emitter bypass capacitor can greatly affect the gain of the amplifier. What operational measurement would you make to determine if the capacitor was open? Why?

2. In Fault 2 you found that sometimes the AC signal may disappear completely. You must then rely on DC voltage measurements for trouleshooting. Which of the DC voltages checked do you think was the most obvious for showing a shorted transistor? Why?

Quick Check

1. What would be the gain of the amplifier in Figure 17.1 if the value of R_C were changed to 5 kΩ? (Assume no change in quiescent operating point.)

2. What should be the normal AC voltage read at the emitter of the transistor amplifier of Figure 17.1?

3. If the amplifier of Figure 17.1 were exactly midpoint biased (V_{CZ} = 6 V), what would be the maximum undistorted peak output voltage swing?

   (a) 4 V (b) 5 V
   (c) 6 V (d) 10 V

4. If capacitor C_3 in Figure 17.1 shorted, the transistor would likely be _____.

18

COLLECTOR FEEDBACK BIASED AMPLIFIER

INTRODUCTION

In Experiment 14 you discovered how to use collector feedback to bias a transistor circuit. In this experiment you will use this circuit to amplify an AC signal. In Section I you will calculate the gain for the circuit and compare your measured gain with your calculated gain.

In the troubleshooting section, you will see the effects on the AC output voltage if the transistor opens. You will also observe the effects of open and shorted coupling capacitors on the output voltages.

REFERENCE

Principles of Electronic Devices and Circuits - Chapter 5, Section 5.5

OBJECTIVES

When you complete this experiment, you will:

✓ Better understand amplifier gain

✓ Be able to do basic amplifier troubleshooting

EQUIPMENT AND MATERIALS

DC power supply
Function generator
Circuit protoboard
NPN transistor, 2N3904 or equivalent
Resistors: 5.1 kΩ, 8.2 kΩ, 470 kΩ
1 μF capacitor [2]

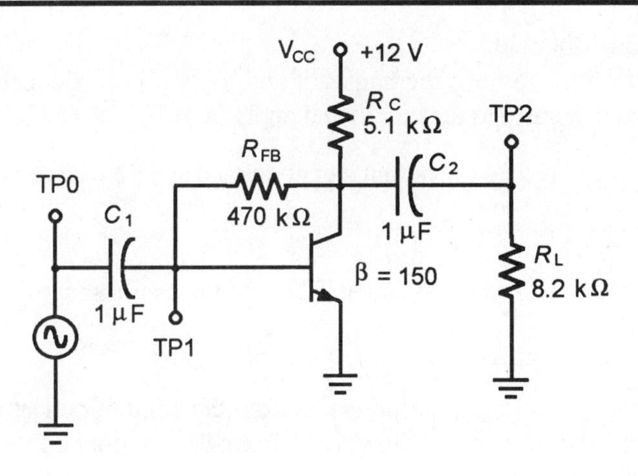

Figure 18.1

BEFORE STARTING

The circuit you will be working with has a relatively high gain. This will require an input signal of 20 $mV_{p\text{-}p}$. If your function generator won't adjust to that level, use the Input Signal Divider of Figure 18.2 between your function generator and your amplifier

1. Build the circuit of Figure 18.1. Apply +12 VDC to your circuit. Do not connect or apply the AC signal at this time. Measure and record the following DC circuit values:

   I_C = ______________ V_{CE} = ______________

2. Connect your AC function generator to your circuit (Use the signal divider if required). Set the function generator to provide a signal of 1 kHz at 20 $mV_{p\text{-}p}$ at TP1.

   Using your oscilloscope, measure and record the input signal at TP1 (V_{in}) and the output signal at TP2 (V_{out}).

   V_{in} = ______________ V_{out} = ______________

   From these values, calculate the circuit loaded voltage gain (A_{VL}).

   A_{VL} = ______________

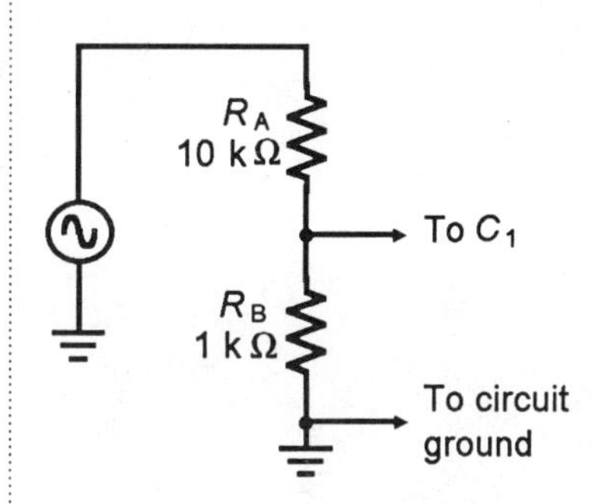

Figure 18.2

3. From the data of step 2, calculate the theoretical loaded voltage gain of your circuit (A_{VL}).

   Calculated A_{VL} = ______________

4. Set your oscilloscope to AC coupling and monitor the output of your circuit (TP2). Increase the input signal to the circuit slowly, stopping as soon as you detect any output clipping or limiting. For this value of output, record the value of the positive peak and the value of the negative peak.

   $+V_{pk}$ = ______________ $-V_{pk}$ = ______________

   Did clipping start simultaneously on both positive and negative peaks? If not, which portion of the output signal showed clipping first?

Step 4 completes the first section of your experiment. If you are proceeding to the troubleshooting, you may leave the circuit connected. Turn off all circuit power, AC and DC.

SECTION II TROUBLESHOOTING

Fault 1 - Transistor open

1. In Figure 18.1, remove the transistor and apply power. Connect the function generator to the circuit and apply 20 mV_{p-p} at 1 kHz.

2. Measure and record the output signal voltage at TP2 with the oscilloscope.

   V_{TP2} = ___________

3. Measure and record the signal at TP1 with the oscilloscope.

   V_{TP1} = ___________

4. Make the DC voltage measurements from the point of collector connection to ground and the voltage from TP1 to ground (V_{BE}) with your oscilloscope.

   V_C = _______________ V_{BE} = _____________

Fault 2 - C_1 open

1. Turn off circuit power. Install the transistor and disconnect one end of C_1.

2. Apply DC and signal power.

3. Measure and record the signal voltage at the following test points:

   V_{TP2} __________ V_{TP1} __________ V_{TP0} __________

Fault 3 - C_2 shorted

1. Turn off circuit power. Reconnect C_1 and place a shorting wire across C_2.

2. Measure and record the signal voltage at TP2.

   V_{TP2} = ___________

Fault 4 - C_2 open

1. Turn off circuit power, remove the shorting wire and lift one end of C_2. Apply circuit power, measure and record the signal voltage at TP2.

   V_{TP2} = ___________

2. Measure the signal voltage at the collector.

   V_C = ___________

DISCUSSION

Section I

1. In procedure step 2, you measured circuit gain. In step 3 you calculated the expected gain. If the two values were the same, how would you explain the equivalence?

2. In procedure step 4, you drove the amplifier to the point of just detecting clipping or limiting of the output signal. What does equal (symmetrical) clipping say about the quiescent operating point? Was the clipping you observed in step 4 cutoff clipping or saturation clipping? Explain why you chose your answer to the last question.

Section II

Fault 1 - Transistor open

1. For this fault you found that the output voltage at TP2 was 0 V while the input voltage at TP1 was 20 mV. What do these two measurements signify to you?

2. Why do you think it helpful to measure the DC collector voltage?

Fault 2 - C_1 open

1. You should have found that the output voltage at TP2 was missing. This time your measurements showed you that the input voltage at TP1 was also missing. However, the signal was present at TP0. You found that C_1 being open will prevent the signal from ever reaching the transistor input. Does this suggest a pattern for troubleshooting to you? Explain.

Fault 3 - C_2 shorted

The measured output voltage at TP2 was seen to be superimposed on a DC level. If C_2 were operating normally, the DC voltage would be blocked. What possible problem with oscilloscope switching could prevent you from seeing this?

Fault 4 - C_2 open

In Fault 4 you should have found that the output voltage was missing. This time, when you measured the collector voltage, you found the output voltage present. This signal was on a DC voltage level and is hard to see with the oscilloscope. Would it be appropriate to measure this signal voltage with the oscilloscope coupling switch on AC?

Quick Check

1. What is the formula for voltage gain in a collector feedback biased amplifier?

2. We can expect our calculated A_V and our measured A_V to always be extremely close.

   True False

3. When troubleshooting collector feedback amplifiers, you often find that coupling capacitors are often found at fault.

   True False

4. What is the value of I_{CSAT} for the circuit of Figure 18.1?

   (a) 1.57 mA (b) 2.40 mA
   (c) 2.90 mA (d) 3.50 mA

19

COMMON COLLECTOR AMPLIFIER

INTRODUCTION

The common collector or emitter follower amplifier provides current gain and a nominal voltage gain of slightly less than unity. This amplifier is also characterized by a moderately high input impedance and low output impedance.

In this experiment you will verify voltage gain for several loads to illustrate the output impedance character of the, amplifier and, through measurement, you will determine the input impedance of the CC amplifier. The troubleshooting section will enable you to relate measured circuit values to voltage divider faults in the CC amplifier.

REFERENCE

Principles of Electronic Devices and Circuits - Chapter 5, Section 6

OBJECTIVES

In this experiment you will:

✓ Examine the effect of small load resistors on the voltage gain of the emitter follower

✓ Verify the phase relationship between the input and output signal voltages of the common collector and amplifier

✓ Determine the input impedance of a common collector amplifier

✓ Understand and learn how to recognize failures of the bias resistors

EQUIPMENT AND MATERIALS

DC power supply	NPN transistor, 2N3904 or equivalent
Function generator	Resistors: 470 Ω, 1 kΩ [2], 10 kΩ [2], 100 kΩ, 2.2 MΩ
Dual trace oscilloscope	Potentiometer, 10 kΩ
Circuit protoboard	Capacitors: 1 μF, 470 μF

Your first procedure steps will be measurement of the common collector amplifier phase shift and then voltage gain with three different load resistor values to let you observe the output impedance characteristics of the common collector amplifier. Following these measurements, you will make measurements to experimentally determine the input impedance of your

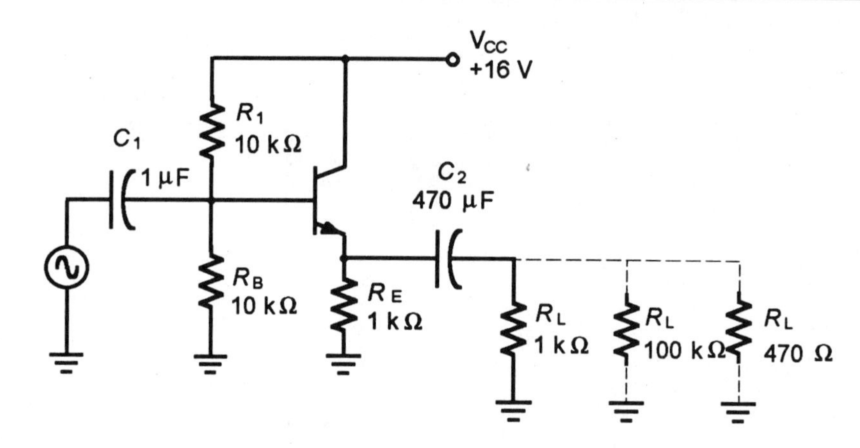

Figure 19.1

1. Build the circuit of Figure 19.1, using first the 100-kΩ load resistance. Apply DC power to the circuit. Use the function generator to apply a sinusoidal input signal of 1.5 $V_{p\text{-}p}$ at 1 kHz.

2. Use your oscilloscope to measure, and then record the values of the input signal (V_{in}) and the output signal at R_L (V_{out}).

   V_{in} = ______________ V_{out} = ______________

   Calculate and record the amplifier voltage gain.

   A_{VL} = ______________

3. Connect your oscilloscope to measure the phase of V_{in} versus V_{out}. Record the phase shift you measure.

   Phase shift = __________

4. Turn off the circuit power. Replace the 100-kΩ load with a 1-kΩ resistor. Reapply circuit power. Measure and make sure that the input signal to the amplifier is 1.5 $V_{p\text{-}p}$.

   V_{in} = ______________

   Measure the signal output at R_L, record V_{out}, and calculate the voltage gain.

   V_{out} = ______________ A_{VL} = ______________

5. Turn off circuit power. Replace the 1-kΩ load resistor with a 470-Ω resistor. Reapply circuit power and repeat the measurements of Step 4.

   V_{in} = __________ V_{out} = __________ A_{VL} = __________

   Remove circuit power and disconnect the load resistance.

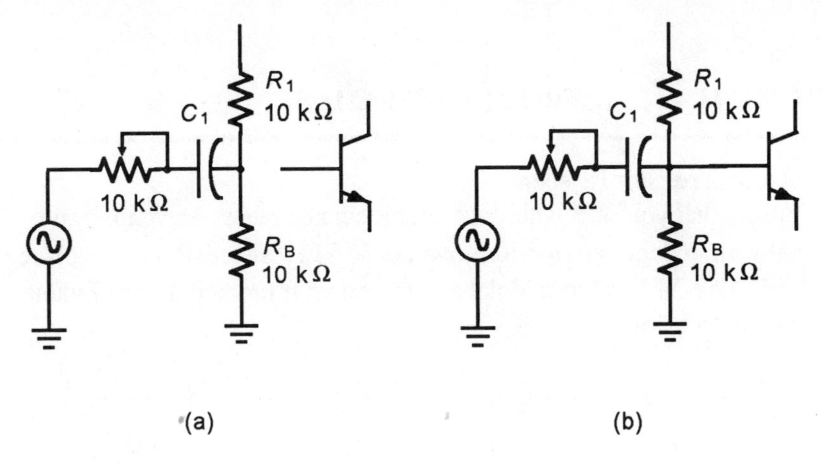

Figure 19.2

6. Ensure that the circuit power is off. Disconnect the base lead of the transistor from the junction of C_1 and the voltage divider. Connect a 10-kΩ potentiometer in series with the function generator to C_1 as shown in Figure 19.2a.

7. Apply circuit power. Set the function generator to have an output of 2 $V_{p\text{-}p}$ at 1 kHz, measured with the oscilloscope. Set the scope vertical range selector to obtain as close to a full screen display as possible. After setting the function generator, move the scope connection to read the AC voltage across R_B of the divider. Do not change the setting of the scope vertical range selector. Adjust the 10-kΩ potentiometer for a scope display of exactly one-half the value set for the input signal.

8. Turn off circuit power. Remove the potentiometer from the circuit and measure its resistance. Record this value as Z_D. It is the value of the divider impedance.

   Z_D = ________________

9. Replace the potentiometer in the circuit. Reconnect the transistor base lead to the junction of C_1 and the voltage divider as shown in Figure 19.2b. Repeat exactly the measurement procedure of step 7. This time the resistance of the potentiometer is the total impedance of the divider and amplifier. Record your measured value of Z_T.

   Z_T = ________________

10. Repeat the measurement steps of 7 and 8, and average the values of the readings made.

    Calculate the input impedance of the amplifier from the measured and averaged data of steps 7 and 8, using the formula below.

$$Z_A = \frac{Z_T \cdot Z_D}{Z_D - Z_T}$$

Step 10 completes Section I measurements. If you are continuing to the troubleshooting section, leave your circuit connected.

Beginning with step 6, you will be measuring the input impedance of the CC amplifier. This is not a difficult measurement; however, it cannot be made directly as you would measure a resistor. If you look at the circuit of Figure 19.1, you will see that at capacitor C_1, there are two impedances in parallel: the bias voltage divider and the amplifier input impedance. You want to obtain the AC or active impedance of the amplifier so that the measurement must be made in an operating circuit.

The way that you will do this is to first, with the transistor disconnected, determine the divider impedance Z_D. Then with a second measurement, obtain the total impedance with the transistor connected. The difference of the two impedance measurements will allow calculation of the amplifier input impedance.

SECTION II TROUBLESHOOTING

Fault 1 - Bias resistor R_1 open

1. Ensure that you start with your circuit connected as shown in Figure 19.1, but with no load resistor. Remove the 10-kΩ resistor (R_1) and replace it with a 2.2-MΩ resistor. Make and record your predicted circuit values as requested below.

   V_E = ______________ v_e = ______________

2. Apply circuit power and an input signal of 1.5 $V_{p\text{-}p}$ at 1 kHz. The bias resistor R_1 of Figure 19.1 fails open. What do you think will happen to the output AC and DC voltages? Record your answer below; then measure and record the results.

   V_E = ______________ v_e = ______________

Fault 2 - Bias resistor R_2 shorted

1. Turn off circuit power. Replace R_1 with the current 10-kΩ resistor. Remove R_B, 10 kΩ, and replace it with a 2.2-MΩ resistor.

2. Estimate the emitter AC and DC values and record below.

   V_E = ______________ v_e = ______________

   Apply circuit power and AC signal input. Measure and record the voltages, AC and DC at the emitter paying attention to the AC signal waveshape.

   V_E = ______________ v_e = ______________

DISCUSSION

Section I

1. In your procedure you should have determined that the common collector amplifier had a moderately high input impedance and was capable of driving a fairly small load resistance. What purpose does this serve in electronics?

2. You also found that the voltage gain (A_v) of the emitter follower was at or a little less than unity (1). You found that this A_v remained close to unity for several different load resistances. In your opinion why is this an advantage?

3. Your measurements showed that the output signal voltage is the same phase as the input signal voltage. Can this be an advantage? If yes, how?

4. Why is the output signal of the common collector amplifier just slightly less than the input signal?

Section II

1. In Fault 1 you found that if the bias resister R_1 opens, you lose both V_E and v_e. Explain why this happens.

2. With Fault 2, R_6 open, you should have recorded a large DC voltage on the emitter and observed a heavily clipped AC signal. What conclusions about the state of the transistor can you draw from your measurements?

Quick Check

1. The common collector amplifier is also called _____.

   (a) an audio amplifier (b) a collector follower
   (c) an emitter follower

2. The common collector Z_{in} is _____.

   (a) medium (b) high
   (c) low

3. The common collector amplifier has a signal phase inversion between the V_{in} and V_{out} of _____.

   (a) 180 degrees (b) 270 degrees
   (c) 0 degrees

4. A common collector amplifier has a nominal voltage gain of ___________.

20

COMMON BASE AMPLIFIERS

INTRODUCTION

The common base amplifier is characterized by a low input impedance and good voltage gain but no current gain at all. The input current is virtually equal to the output current. In Section I you will make measurements and some calculations to verify the DC and AC characteristics of the common base amplifier. In Section II you will explore the effects on the output voltage when the coupling capacitor C_2 shorts, C_2 opens, or R_E opens.

REFERENCE

Principles of Electronic Devices and Circuits - Chapter 5, Section 5.7

OBJECTIVES

In this experiment you will:

✓ Verify through circuit measurements, the voltage gain and phase shift of the common base amplifier

✓ Experimentally determine the input impedance of a common base amplifier

✓ Explore three common failures of a common base amplifier

EQUIPMENT AND MATERIALS

Dual DC power supply
Dual-trace oscilloscope
Digital multimeter
Function signal generator
Circuit protoboard
NPN transistor, 2N3904 or equivalent
Resistors: 100 Ω, 1 kΩ, 2.7 kΩ, 10 kΩ, 20 kΩ, 2.2 MΩ
Potentiometer, 100 Ω
1 μF capacitor [2]

SECTION I **FUNCTIONAL EXPERIMENT**

1. Build the circuit of Figure 20.1. Resistors R_A and R_B are a voltage divider to lower the input signal to your amplifier without excessively loading the signal generator.

2. Apply DC power to your amplifier. Connect the signal generator to the circuit and set the generator for an input of 20 $mV_{p\text{-}p}$ across R_B, using the oscilloscope to measure this voltage. Record the input signal level.

   V_{in} = ____________

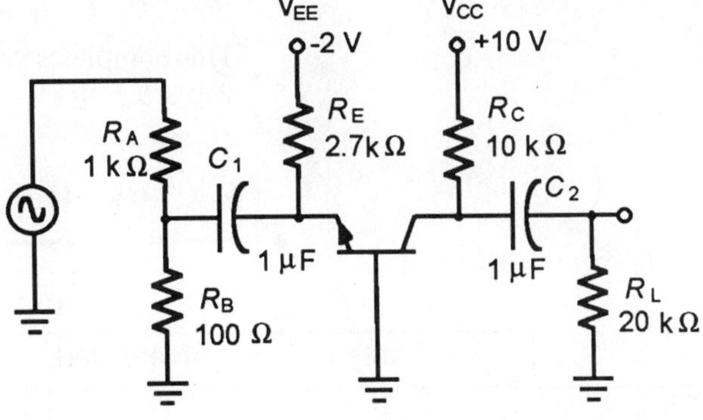

Figure 20.1

3. Using the oscilloscope, measure the amplifier output across the load, and record the measured value: V_{out} = ____________

   Calculate and record the amplifier gain: A_{VL} = ____________

4. Connect the oscilloscope channel 1 to the amplifier input, and channel 2 to the amplifier output. Measure and record the amplifier phase shift.

   Phase shift = ____________

5. In the following steps you will be measuring the amplifier input impedance. Make the settings and measurements as accurately as possible.

6. Turn off the amplifier power. Calculate and record your value of the amplifier input impedance.

   Z_{in} = ____________

   Set a 100-Ω potentiometer to the value you calculated for the input impedance. Connect the 100-Ω potentiometer between the junction of R_A and R_B and C_1 as shown in Figure 20.2. Temporarily connect a jumper across the potentiometer connection to short the potentiometer.

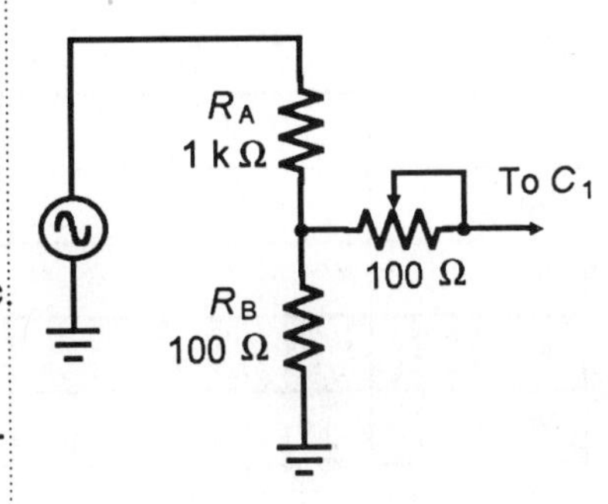

Figure 20.2

7. Apply circuit power. Use the oscilloscope connected across divider resistor R_B to measure the input signal, and adjust the generator for an input of 20 $mV_{p\text{-}p}$. Record your input signal value. Measure and record the amplifier output signal.

   V_{in} = ____________ V_{out} = ____________

8. Remove the jumper shorting the potentiometer. Recheck the signal voltage across R_B and readjust the generator for exactly 20 $mV_{p\text{-}p}$. Set the scope to measure the amplifier output and adjust the potentiometer for an output of exactly half the value recorded in step 7.

Important: Each time the potentiometer is adjusted, in step 8, the signal at divider resistor R_B must be checked and readjusted for the value of 20 $mV_{p\text{-}p}$.

Repeat the adjustments until the correct output is obtained and no adjustments are required.

9. Turn off circuit power. Remove the potentiometer and measure the set resistance. Record the measured resistance below. This is the input impedance of the amplifier. The value measured should be relatively close to the value you calculated in step 6.

Z_{in} = ___________

This completes your circuit measurements. Your circuit may be left connected if you are going to continue to the troubleshooting section.

SECTION II TROUBLESHOOTING

Fault 1 - R_E open

Build the amplifier circuit of Figure 20.1. Replace the 2.7-kΩ emitter resistor (R_C) with a 2.2-MΩ resistor. Estimate the effect this failure would have on the circuit. Record your estimated voltages in the *Estimated* column of the table provided here. Apply circuit power. Connect and set the signal generator to supply 20 mV_p across R_B. Measure and record the listed voltage values in the *Measured* column.

	Estimated	Measured
Vout		
Ve		
Vc		
V_E		

Fault 2 - C_2 shorted

Turn off circuit power. Replace the emitter resistor with the correct 2.7-kΩ resistor. Place a jumper wire to short the output capacitor C_2. Estimate the effect of this failure on the circuit parameters shown in the table at the right. Record your estimates. Apply circuit power and input AC signal. Measure and record the values in the *Measured* column.

	Estimated	Measured
Vout		
V_E		
Ve		
VC		

Fault 3 - C_2 open

Turn off circuit power. Remove the jumper shorting C_2. Disconnect one end of C_2 from the transistor collector. Estimate the effect of an open C_2 on the circuit parameters listed in the table at the right. Record your estimated values. Apply circuit power and input AC signal of 20 mV_{p-p}. Measure and record the circuit values in the *Measured* column of the table.

	Estimated	Measured
Vout		
V_E		
Ve		
VC		

DISCUSSION

Section I

1. You should have found that the gain of the common base amplifier is high and the input impedance is very low. What effect do you think the low input impedance would have on most input signals to the amplifier?

2. Even though the common base amplifier has a high gain, why do you think it is not very popular for use as an audio amplifier?

3. Considering that the impedance of most coaxial cables used to carry RF signals is in the range of 50 to 75 Ω, do you feel that there would be an application in RF use for the common base amplifier? Give an example to explain your answer.

Section II

Fault 1 - R_E open

In Fault 1 we found that if R_E opens, there is no output voltage. Why do you think this happens?

Fault 2 - C_2 shorted

In Fault 2 the output coupling capacitor (C_2) was shorted. You should have found that the output signal was superimposed on the DC voltage level of V_C. In troubleshooting there are two ways of finding this problem. Can you think of a problem you might have in troubleshooting this failure?

Fault 3 - C_2 open

You found that when the output coupling capacitor, C_2, becomes open, the signal path is interrupted and no output voltage will be seen. What do you think the output voltage would look like if C_2 were leaky instead of open or shorted?

Quick Check

1. The common base amplifier is used mostly as an audio amplifier.

   True False

2. There is a 180-degree phase inversion between the input and output signals of the common base amplifier.

   True False

3. The low input impedance of the common base amplifier makes it ideal as a voltage amplifier.

   True False

4. The gain of the common base amplifier is relatively high.

   True False

21

MULTISTAGE AMPLIFIERS

INTRODUCTION

Many times the gain required from an amplifier cannot be obtained by only one amplifier stage. Therefore, several amplifier stages are cascaded together so that the individual gain of each amplifier is multiplied to get an overall gain.

In this experiment you will build a two-stage cascaded amplifier, make measurements, and draw conclusions based on those measurements.

The troubleshooting section will examine the effects of capacitor failures on the multistage amplifier. You will install faults and make circuit measurements to be able to relate measured values to component failures.

REFERENCE

Principles of Electronic Devices and Circuits - Chapter 5, Section 5.9

OBJECTIVES

In this experiment, you will:

✓ Add to your knowledge of multistage amplifiers

✓ Be able to relate measured values to circuit failures

EQUIPMENT AND MATERIALS

DC power supply
Digital multimeter
Dual-trace oscilloscope
Function generator
Circuit protoboard
NPN transistor [2], 2N3904 or equivalent
Resistors: 82 Ω, 180 Ω, 330 Ω, 1 kΩ, 3.3 kΩ, 3.9 kΩ, 4.7 kΩ, 5.6 kΩ, 33 kΩ, 56 kΩ
Capacitors: 1μF [3], 47 μF [2]

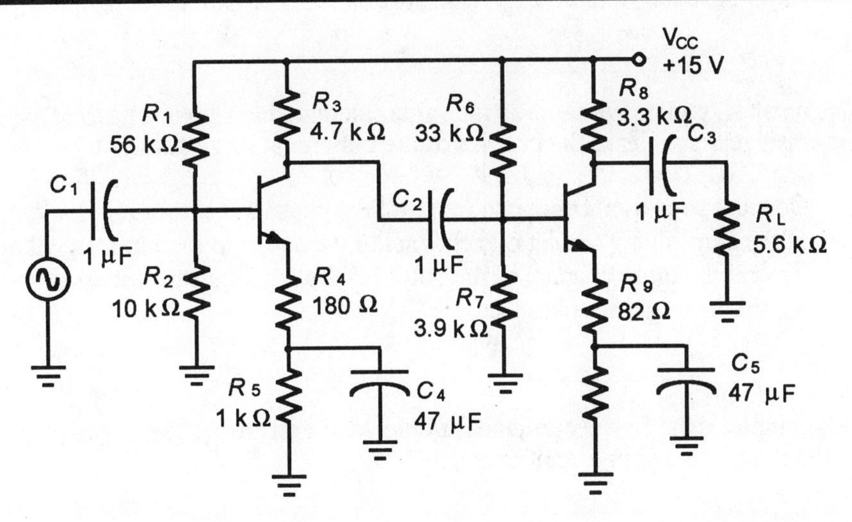

Figure 21.1

1. Construct the two-stage amplifier of Figure 21.1.
2. Apply DC power. Do not connect the function generator at this time.

   Measure the DC voltages at the base, collector, and emitter of the first and second stages of the amplifier. Record your values in Table 21.1.

	DC Values			AC Values			
	V_B	V_E	V_C	V_b	V_c	V_e	V_{out}
Stage 1							
Stage 2							

Table 21.1

3. Connect the function generator to provide an input to your amplifier at C_1. Set the generator to 20 $mV_{p\text{-}p}$ at 1 kHz.

   Note: If your function generator will not adjust to a sufficiently low signal, connect the input voltage divider network of Figure 21.2 to your circuit.

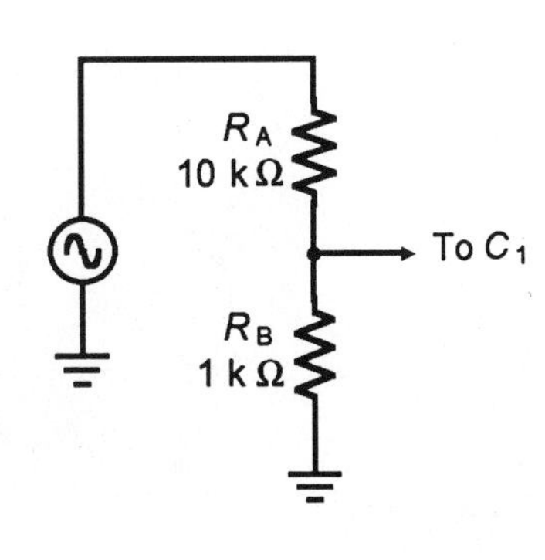

Figure 21.2

4. Using the oscilloscope, measure the peak-to-peak AC signal at the base, emitter and collector of each amplifier stage, and measure the amplifier output across the load resistor. Record your measured values in Table 21.1.
5. From your measured data, calculate the loaded voltage gain of each stage and record below.

   A_{VL} (Stage 1) = __________ A_{VL} (Stage 2) = __________

6. Calculate the overall voltage gain of the amplifier by taking the product of the individual stage gains from step 5.

7. From the data of Table 21.1, calculate the overall gain of the amplifier.

   A_{VT} = ___________

This result should be the same as the value calculated in step 6. The total gain of a multistage amplifier is the product of the individual stage gains.

8. Connect your oscilloscope to measure the phase shift from input to output of the amplifier. Connect one channel of the scope to the input signal at C_1, and the other channel to the output at the load resistor. Measure and record the amplifier phase shift.

   Phase shift = ___________

For stable 'scope triggering in step 8, set the 'scope to trigger on the channel connected to the amplifier output.

This completes the first section measurements. If you are proceeding to the troubleshooting, leave your circuit connected.

SECTION II TROUBLESHOOTING

Troubleshooting amplifiers is very similar to troubleshooting the single stage amplifier. There are some differences, however. One major difference is the coupling capacitors. Like a leaky or shorted bypass capacitor, coupling capacitors alter the signal characteristics. Also, since they isolate the V_c of the first stage from the V_b of the second stage, network bias is affected. Another difference of the amplifier from the single-stage is that the subsequent stages act as the load for the previous stage. Therefore, any problem in later stages affects the operation of the previous stages. This experiment, then, focuses on these two major differences by examining failures of the coupling capacitors.

Fault 1 - C_2 open

1. Turn off circuit power. Open C_2 by disconnecting one end from the capacitor at Q_1. Reapply circuit power and the AC signal of 20 $mV_{p\text{-}p}$ at 1 kHz. Measure and record in Table 21.2 the AC and DC voltages at the collector of Q_1 and the base of Q_2.

2. Reconnect C_2 and measure V_c of Q_1 and V_b of Q_2. Do you observe any differences in the signal?

Fault 2 - C_2 shorted

Turn off circuit power. Remove capacitor C_2 and replace it with a jumper wire (connecting the collector of Q_1 to the base of Q_2). Reapply circuit power and the input signal. This simulates a leaky or shorted capacitor. Measure and record the appropriate data in Table 21.2. Also record the DC levels.

	DC Values			AC Values		
Fault	V_B	V_E	V_C	V_b	V_c	V_e
C_2 open						
C_2 shorted						

Table 21.2

DISCUSSION

Section I

1. Calculate the unloaded voltage gain of the amplifier first stage. How does it compare to your measured value?

2. The overall gain you measured for your multistage amplifier could be approximated by a single-stage unswamped amplifier. What reasons would you give for using the multistage amplifier? What reasons would you give for using the multistage amplifier as a better choice?

Section II

1. What effect did an open coupling capacitor have on the circuit? Also, explain how and why the signal changed when C_2 was reinserted into the network.

2. Explain what effect a leaky coupling capacitor would have on the operation of the multistage amplifier.

3. Explain the measurements you observed when C_2 was shorted. What, for example, happened to the gain of the amplifier, or to the output signal waveshape?

Quick Check

1. The term h_{fe} refers to _____.

   (a) A_V (b) beta
   (c) impedance (d) slew rate

2. C_2 is used to couple the first stage to the second stage.

   True False

3. An open emitter bypass capacitor _____ the gain of an amplifier?

   (a) increases (b) decreases
   (c) has no effect on

4. A multistage amplifier increases signal strength in small steps.

   True False

5. In a multistage amplifier, faults in one stage can affect the operation of other stages.

   True False

22

CLASS A POWER AMPLIFIERS

INTRODUCTION

As its name notes, the emphasis of the power amplifier is power gain. It is most often found in the final stages of multistage amplifiers. Some important features of the class A amplifier are the current drain, maximum power dissipation by the transistor, the stage efficiency and full power output (maximum unclipped signal the amplifier can deliver).

In this experiment you will calculate and measure power output and efficiency of a class A Power amplifier.

The troubleshooting section of this experiment will simulate two amplifier faults and you will, through measurements made, be able to relate failures to circuit measured values.

REFERENCE

Principles of Electronic Devices and Circuits - Chapter 6, Section 6.3

OBJECTIVES

In this experiment you will:

✓ Determine by measurement the efficiency of a class A power amplifier

✓ Understand the effect of a swamping resistor on the signal linearity of a large signal amplifier

✓ Simulate faults and be able to determine their effect on amplifier parameters

EQUIPMENT AND MATERIALS

DC power supply
Digital multimeter
Dual-trace oscilloscope
Function generator
NPN transistor, 2N3904 or equivalent
Resistors: 220 Ω, 820 Ω, 1 kΩ, 3.3 kΩ [2], 6.8 kΩ, 33 kΩ
Capacitors: 1 μF [2], 470 μF

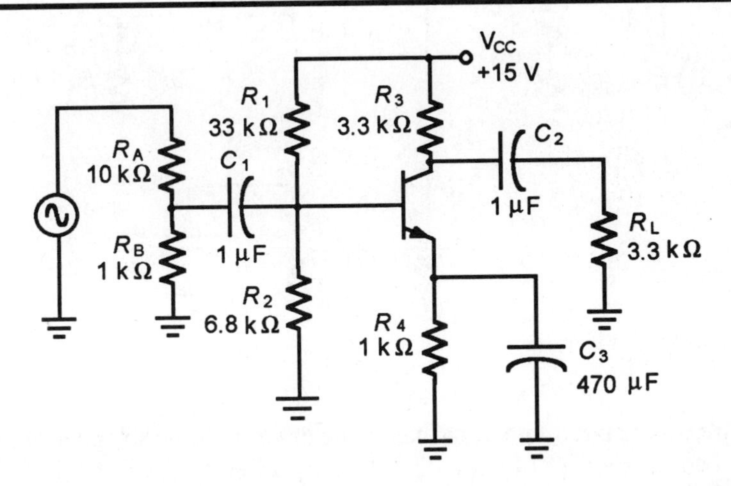

Figure 22.1

1. Construct the circuit in Figure 22.1.
2. Apply DC power and with no signal applied, measure and record in Table 22.1 the values of I_{CQ}, V_{CEQ}, and the total current drawn by the amplifier (I_{CC}).
3. Connect the function generator and adjust for a signal of 20 $mV_{p\text{-}p}$ at 1 kHz at the base of the transistor.

   Using the oscilloscope, observe the load voltage while increasing the AC input signal. Continue to increase the input signal until clipping is observed. Take note of the nonlinear distortion of the signal, in fact the signal begins to squash and elongate before clipping in reached. This is due to the changes in r'_e.

NOTE: Although, at the power levels of your circuit you won't do any damage, it is usually not desirable to operate an amplifier in saturation clipping for long intervals.

4. Reduce the input signal until the output signal is at its maximum value without clipping.
5. Measure and record in Table 22.1 the peak-to-peak output voltage ($V_{p\text{-}p}$).
6. Calculate and record in Table 22.1 the DC power supplied to the amplifier (P_{DC}).
7. Calculate and record in Table 22.1 the total power delivered to the load (P_L).
8. Calculate and record in Table 22.1 the efficiency of the amplifier.

Equation for step 6:

$$P_{DC} = V_{CC} \cdot I_{CC}$$

Equations for step 7:

$$V_{Lp} = \frac{V_{Lp\text{-}p}}{2}$$

$$V_{LRMS} = 0.707 V_{Lp}$$

$$P_L = \frac{V_{LRMS}^2}{R_L}$$

Equation for step 8:

$$\text{Efficiency} = \frac{P_L}{P_{DC}} \times 100\%$$

I_{CQ} =
V_{CEQ} =
$V_{p\text{-}p}$ =
I_{sat} =
P_{DQ} =
P_L =
P_{DC} =
Eff =

Table 22.1

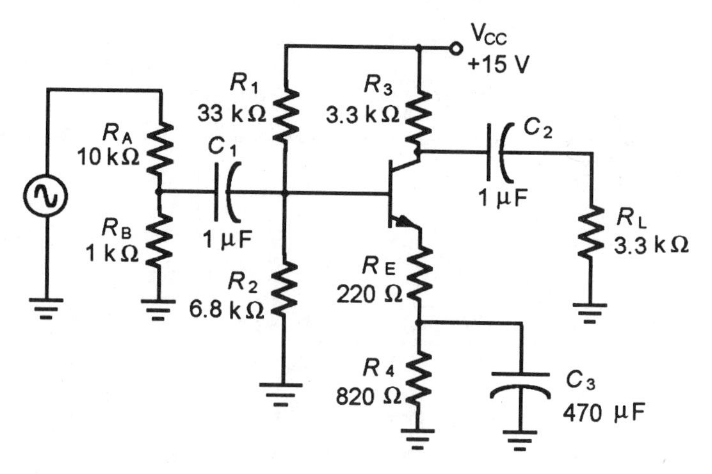

Figure 22.2

9. Turn off the circuit power. Change the emitter circuit to add a swamping resistor as shown in Figure 22.2. Reapply circuit power and the AC signal input.

10. Connect the oscilloscope to monitor the amplifier output. Adjust the input signal amplitude until the output just starts to clip. Compare the signal waveshape to that observed in step 3. Does it appear to be less distorted?

This completes the measurements of this section.

SECTION II TROUBLESHOOTING

Fault 1 - Output capacitor C_2 is shorted

1. Starting with the circuit of Figure 22.1, place a jumper wire to short the output capacitor, C_2.

   Apply circuit power, measure and record the following DC voltages:

   V_C = __________ V_{CE} = __________ V_{RL} = __________

   Apply an AC input signal of 20 mV_{p-p} at 1 kHz. Measure and record the peak-to-peak voltages.

   V_C = ______________ V_{RL} = ______________

2. Turn off circuit power. Remove the jumper shorting capacitor C_2.

Fault 2 - Emitter bypass capacitor C_3 is shorted

1. With circuit power off, connect a jumper wire to short C_3 Apply DC power, measure and record the following DC voltages:

   V_B = __________ V_C = __________ V_E=__________

2. Apply an input signal of 20 mv_{p-p} at 1 kHz. Measure the record the following peak-to-peak AC voltages:

   V_C = ______________ V_{RL} = ______________

Observe the waveshape of the output signal at the collector and load. Does this waveshape imply anything about the state of the transistor?

3. Turn off the power and disconnect your circuit.

DISCUSSION

Section I

1. What effect did including a swamping resistor have on your measurements? Explain the effect.

2. The amplifier efficiency you calculated in step 8 should have been much less than the ideal class A efficiency of 25 percent. With reference to your text, can you identify measured circuit parameters that contributed to the low efficiency?

Section II

1. What happens to the signal at the load when the bypass capacitor shorts. Why does this occur?

2. What does nonlinear distortion look like and explain why it occurs in a class A amplifier

3. What one measurement could you make to be certain to identify a shorted output capacitor? Explain why this measurement would be certain.

Quick Check

1. The class A amplifier is highly efficient.

   True　　　　False

2. The maximum, theoretical, efficiency of a class A amplifier is _____.

   (a) 50%　　　　(b) 25%
   (c) 75%　　　　(d) 33.3%

3. Overdriving an amplifier is a good idea.

   True　　　　False

4. Distortion is caused by changes in _____.

   (a) r'_e　　　　(b) R_E
   (c) R_L　　　　(d) r_c

23

CLASS B PUSH-PULL AMPLIFIERS

INTRODUCTION

In this experiment you will construct a voltage divider biased and a diode (current mirror) biased class B push-pull amplifier. This will allow you to observe crossover distortion and its elimination by the diode biased amplifier. You will also demonstrate that changing class B bias to class AB eliminates crossover distortion. You will also determine the efficiency of your class B amplifier.

Simulated fault measurements will be made in Section II of the experiment. You will fault a portion of the circuit and make measurements to see the effects of the fault.

REFERENCE

Principles of Electronic Devices and Circuits - Chapter 6, Sections 6.4 and 6.5

OBJECTIVES

In this experiment you will:

✓ Observe crossover distortion

✓ Demonstrate the AC and DC operating characteristics of the class B complementary symmetry amplifier

✓ Demonstrate class B amplifier faults

EQUIPMENT AND MATERIALS

DC power supply
Oscilloscope
Digital multimeter
Function generator
Circuit protoboard

Diode [2], 1N914 or equivalent
NPN transistor, 2N3904
PNP transistor, 2N3906
Resistors: 220 Ω, 1 kΩ [2], 1.8 kΩ [2]
Potentiometer: 1-kΩ ten-turn trimpot
Capacitors: 1 μF [2], 100 μF

SECTION I FUNCTIONAL EXPERIMENT

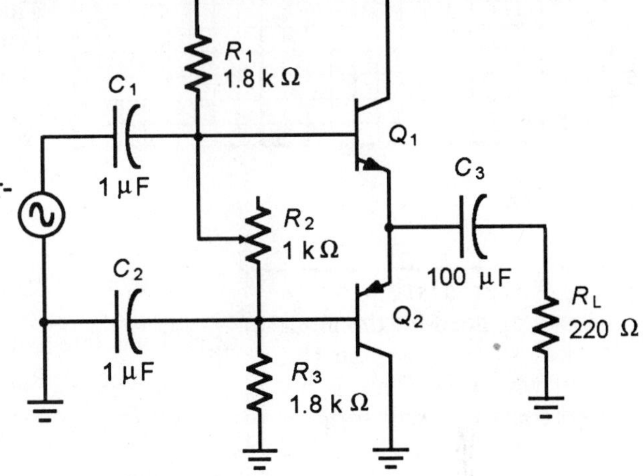

Figure 23.1

Crossover Distortion

1. Construct the circuit in Figure 23.1.

2. Adjust the potentiometer (R_2) for a total resistance of 50 Ω or less between the bases of Q_1 and Q_2.

3. Connect a milliamp meter to measure the collector current of Q_1.

4. Connect the DC power supply to your circuit and set it for 12 V.

CAUTION

Step 5 below must be performed carefully to avoid exceeding the current limit of Q_1 and Q_2.

5. Slowly adjust the potentiometer (R_2) while observing the milliammeter. Adjust R_2 to obtain a current value of 0.25 mA. The exact value is not critical and can be in the range of 0.1 to 0.25 mA. Following the adjustment of R_2, record your current meter reading.

   I_{CEQ} = ____________ (0.1 to 0.25 mA)

6. Turn the DC power supply off and disconnect the current meter.

7. Reapply 12 VDC. Use your digital voltmeter; measure and record the V_{CEQ} of each transistor.

   Q_1 V_{CEQ} = ____________

   Q_2 V_{CEQ} = ____________

8. Connect the function generator to your circuit and set the generator to provide a 1-kHz sine wave at 2 $V_{p\text{-}p}$.

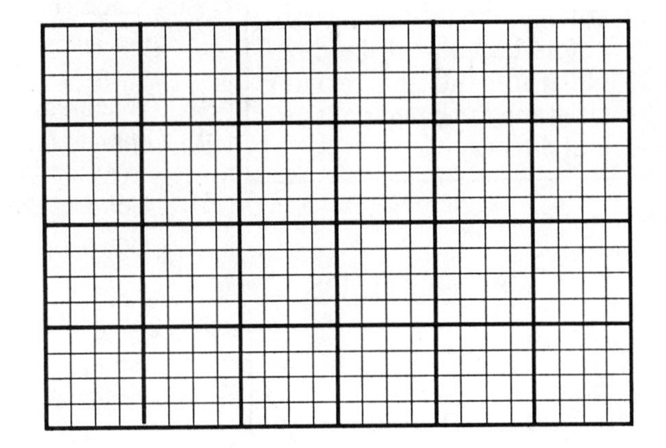

Graph 23.1

9. Using the oscilloscope, observe the signal across the load resistor. The signal you observe should have crossover distortion.

10. Draw the signal in Graph 23.1.

The following procedure step allows you to provide Class AB biasing of your voltage divider biased circuit and observe the elimination of crossover distortion.

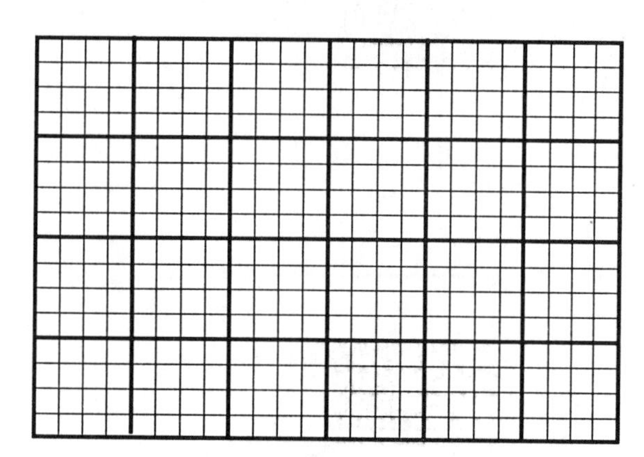

Graph 23.1

11. Repeat Procedure steps 3 through 10, except in step 5 adjust the collector current to 5 mA, and plot your signal in Graph 23.2. Do the adjustment carefully; follow the procedure exactly. You will be increasing the transistor forward bias, and too big an increase can put the transistor into full conduction with no limit on the collector current.

This completes the first part of the experiment. Disconnect all power and disassemble the circuit.

Crossover distortion is caused by need for the input signal to increase to the point where the Class B biased transistor will turn on and start conduction. Increasing the bias to a point where the transistor is barely in conduction (Class AB) allows the transistor to follow the input signal without the delay that appears as crossover distortion. This is not easily accomplished with voltage divider biased transistors because of the difficulty in selecting resistor sizes and the danger of thermal runaway.

The Diode Biased Amplifier

1. Connect the circuit in Figure 23.2.

2. Calculate and measure the DC level at the emitter junction of Q_1 and Q_2 (point A). Record your data in Table 23.1.

3. Calculate the remaining parameters shown in Table 23.1 and enter the results in the *Calculated* column.

Reminder: I_{CC} is equal to the sum of the current mirror bias network (I_D) plus the amplifier transistor collector current (I_{CEQ}):

$$I_{CC} = I_D + I_{CEQ}$$

Since your circuit may not be operating at maximum power to the load, use an estimated load voltage value of 9 $V_{p\text{-}p}$ to calculate load power.

4. When your calculations are complete, make the I_{CC} measurement before connecting the function generator to your circuit.

5. Connect the function generator to your circuit and set the generator to provide a 1-kHz signal at 2 $V_{p\text{-}p}$.

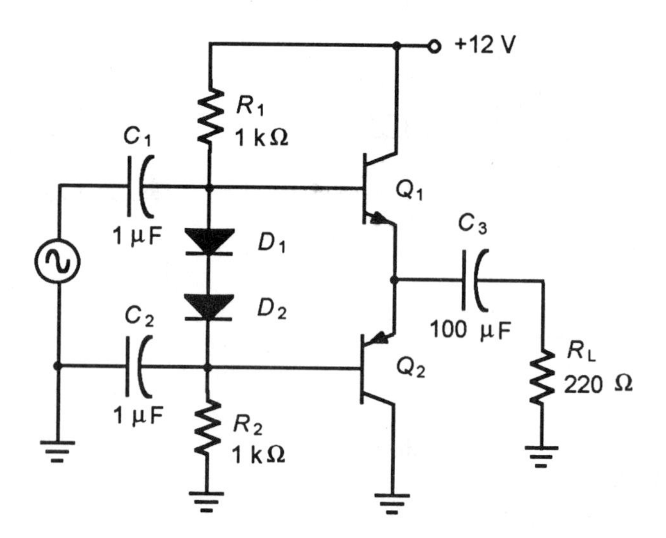

Figure 23.2

	Calculated	Measured
Q_1 V_{CEQ}		
Q_2 V_{CEQ}		
I_{CEQ}		
P_L		
I_{CC}		
P_{DC}		
Eff		

Table 23.1

6. Connect the oscilloscope to measure the output voltage across the load. While observing the scope display, adjust the function generator output amplitude to obtain the maximum undistorted (no clipping) load voltage.

7. Measure the output voltage across the load and sketch the load signal waveform in Graph 23.3.

8. Turn off the generator and power supply. Complete the power calculations for the output load power, the circuit DC power, and the amplifier efficiency from your measured values. Enter the results in Table 23.1.

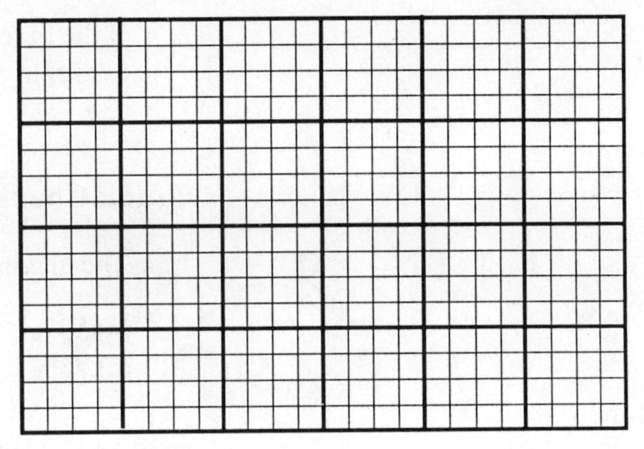

Graph 23.2

You may leave the circuit connected if you are proceeding to the troubleshooting section.

SECTION II TROUBLESHOOTING

Fault 1 - D_1 shorted

Place a jumper wire across D_1. Measure the DC levels and signals at the base of Q_1 and Q_2, and the output junction before the coupling capacitor. Are the DC levels normal?

Fault 2 - Injected fault

Have your instructor or lab partner inject a problem into the circuit. Try covering the fault with a piece of tape to hide the fault, and make your conclusions based on your measurements only.

DISCUSSION

Section I

1. Explain the term *push-pull*. How does this term describe the class B amplifier?

2. Describe crossover distortion. What does it look like and how can it be eliminated?

3. Considering the measured efficiency of your amplifier versus the theoretical maximum value, what things would you suggest to increase your amplifier efficiency?

Section II

1. Describe the procedure you used to troubleshoot the fault that your instructor or partner injected into the circuit.

2. If the base to emitter junction of Q_1 opened, what signal would you expect to observe at the output?

Quick Check

1. The maximum efficiency of a class B amplifier is approximately _____.

   (a) 50%
   (b) 63.3%
   (c) 25%
   (d) 78.5%

2. A class B amplifier is normally biased above cutoff to eliminate crossover distortion.

   True
   False

3. The class B amplifier consists of two transistors each conducting for 270 degrees of the input signal.

   True
   False

4. Provided the transistors are biased equally, the DC level at the collector emitter junction of a class B amplifier is _____.

   (a) half the value of V_{CC}
   (b) 10% of V_{EE}
   (c) 0.7 V above ground
   (d) 10% below V_{CC}

5. An abnormal voltage reading at the emitter junctions (point A) indicates _____.

   (a) an open biasing resistor
   (b) an open coupling capacitor
   (c) a shorted or saturated transistor
   (d) an excessive load

24

DARLINGTON AMPLIFIERS

INTRODUCTION

The Darlington amplifier is a common transistor network. In the common collector configuration it has characteristics of a very high current gain, a voltage gain of 1, a high input impedance, and a low output impedance. Therefore, the Darlington amplifier is often used to isolate a high input source from a low input load.

In this experiment you will construct two darlington pair amplifiers: a common emitter and a common collector amplifier. You will also calculate DC and AC levels, inject an AC signal, and measure the current gain.

REFERENCE

Principles of Electronic Devices and Circuits - Chapter 6, Section 6.5

OBJECTIVES

In this experiment you will:

✓ Measure the current gain of a darlington pair

✓ Demonstrate the characteristics of the darlington pair

✓ Learn the difference between the common collector and the common emitter darlington pair

EQUIPMENT AND MATERIALS

DC power supply
Digital multimeter
Oscilloscope
Function generator
Circuit protoboard
NPN transistor [2], 2N3904 or equivalent
Resistors: 680 Ω, 820 Ω, 3.3 kΩ, 150 kΩ, 300 kΩ, 500 kΩ, 1 MΩ [2]
Capacitors: 1 μF [3], 470 μF

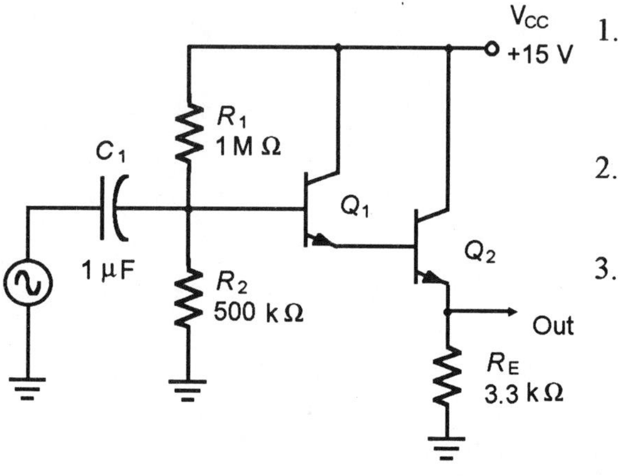

Figure 24.1

Common Collector Amplifier

1. Build the circuit of Figure 24.1. Connect the +15 VDC power. Do not connect the AC function generator at this time.

2. Measure and record in Table 24.1 the base voltage of Q_1 and Q_2, and the emitter voltage and current of Q_2.

3. Connect the AC function generator and set it to supply a 1-$V_{p\text{-}p}$ signal at 1 kHz.

   Measure and record in Table 24.1 the peak-to-peak values of the input and output signal. From these measurements calculate the amplifier voltage gain and record the value in Table 24.1.

In these procedure steps you will be measuring the current gain (A_i) of your amplifier. Because the expected gain is large, typically in the range of 1000 to 20,000, a direct measurement of the input current (I_{in}) is not possible. Therefore, some calculations are required to determine the value of I_{in}. Your accuracy in this measurement is limited, but will serve to give you a fair measure of current gain and illustrate the high input impedance of the Darlington amplifier.

Figure 24.2a shows the input portion of the amplifier and the source that will be used to make the measurement.

Figure 24.2b shows the equivalent AC impedances of the input circuit.

Figure 24.2c shows the currents and two voltage measurements to be made.

4. Because the current values will be derived from voltage measurements, it is necessary to know the actual values of the circuit resistances. Measure and record the values of resistors R_A, R_1, R_2, and R_E.

   R_A = ______________ R_1 = ______________

   R_2 = ______________ R_E = ______________

5. Connect the input circuit to your amplifier as shown in Figure 24.2a. Set the generator to provide an exact 1 $V_{p\text{-}p}$ at 500 Hz using your digital multimeter.

6. Measure and record the AC voltage drops across R_A (V_1) and across divider resistor R_2 (V_2).

   V_1 = ______________ V_2 = ______________

	Common Collector (Figure 24.1)	Common Emitter (Figure 24.3)
V_{B1}		
V_{B2}		
V_{E2}		
I_{E2}		
V_{in}		
V_{out}		
A_V		

Table 24.1

7. Calculate the total input current (I_T) and the divider current (I_D) using the formulas of Figure 24.2c.

I_T = __________________

I_D = __________________

8. Calculate the amplifier current (I_A) from formula c of Figure 24.2c.

I_A = ______________

9. The approximate input impedance of your amplifier can be calculated from formula d of Figure 24.2c.

Z_{in} = ______________

10. Measure the output AC signal across R_E.

V_{out} = ______________

11. Calculate I_{out} using formula e of Figure 24.2c.

I_{out} = ______________

12. Calculate A_i from the data of steps 8 and 11.

A_i = ______________

This completes your measurements of the common collector amplifier. Turn off the power and disconnect your circuit.

Common Emitter Amplifier

1. Build the circuit of Figure 24.3. Apply 12 VDC power. Do not connect the function generator at this time. Measure and record in Table 24.1 the DC voltages at the bases of Q_1 and Q_2, and the emitter and collector voltages of Q_2.

2. Connect the function generator, set to apply an input signal of 20 $mV_{p\text{-}p}$ at 1 kHz.

3. Measure and record in Table 24.1 the AC signal voltages at the input and output of the amplifier.

   Calculate from these measurements the amplifier voltage gain and record it in Table 24.1.

This completes the measurements of your Darlington amplifiers. Turn off the circuit power and disconnect your circuit.

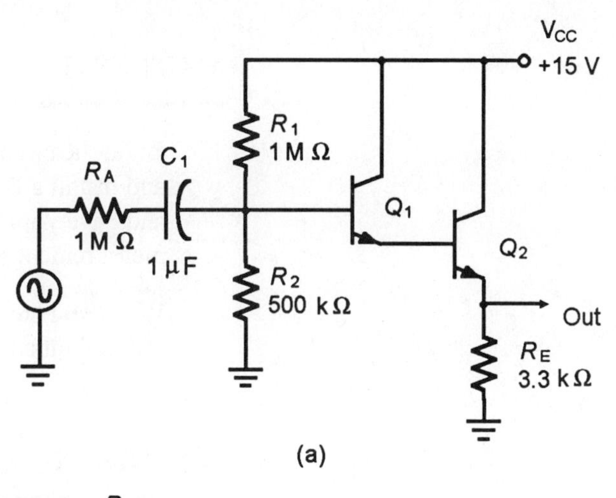

(a)

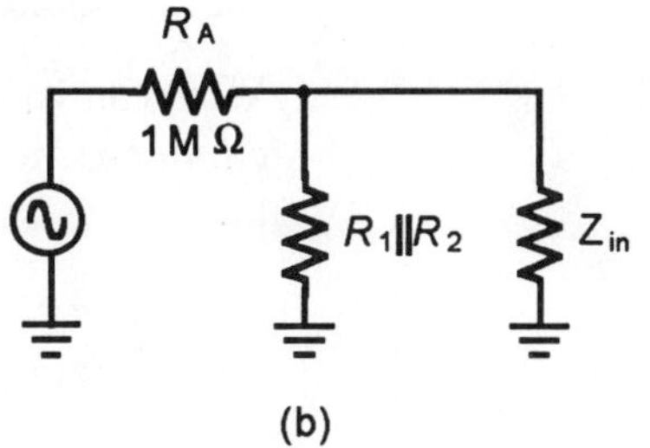

(b)

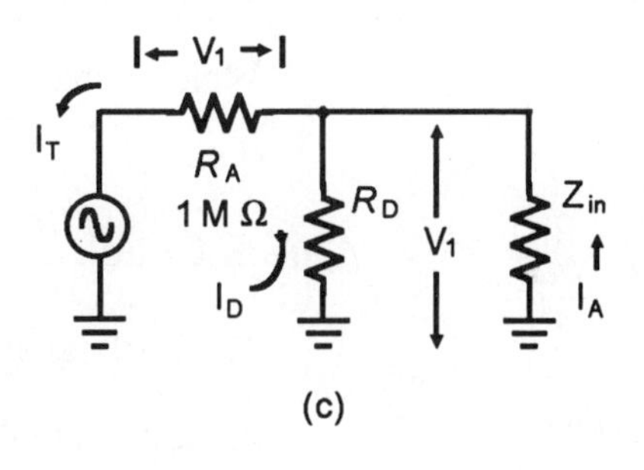

(c)

a. $I_T = \frac{V_1}{R_A}$

b. $I_D = \frac{V_Z}{R_D}$

c. $I_a = I_T - I_D$

d. $Z_{in} = \frac{V_Z}{I_A}$

e. $I_{out} = \frac{V_{out}}{R_E}$

Figure 24.2

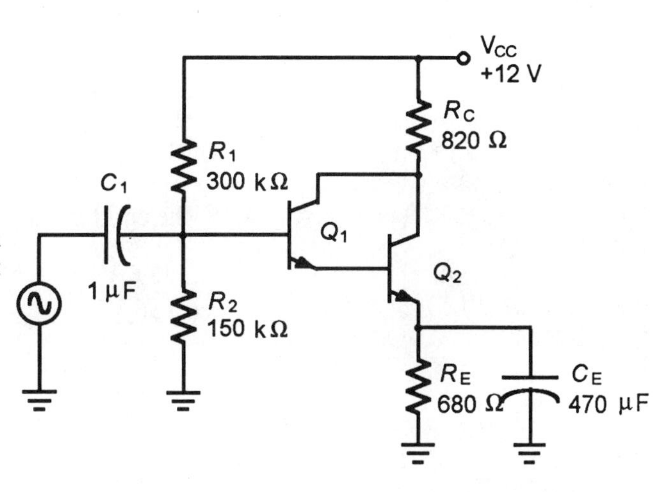

Figure 24.3

SECTION II TROUBLESHOOTING

1. Build the circuit of Figure 24.1. Have your lab partner take your circuit and install a fault. To troubleshoot the circuit, use the schematic diagram and have your lab partner make and give the measurement values for any measurement you request.

   When you have identified the fault, exchange roles; that is, you install a fault and make the measurements for your lab partner.

DISCUSSION

1. What can you say about the current and voltage gains of the common emitter darlington pair amplifier. Explain why the voltage gain is less than 1.
2. How does the voltage gain in the common collector differ from the voltage gain of the common emitter darlington pair?
3. Which of the two circuits has the lesser voltage gain. Why?
4. Which of the darlington pairs has the greater current gain. Why?

Quick Check

1. The Darlington pair is often used to _____.

   (a) match a high impedance source with a low impedance load
   (b) increase the frequency response of an amplifier
   (c) match a low impedance source with a high impedance load
   (d) increase the voltage gain of a signal

2. The darlington pair has high current gain.

   True False

3. The darlington pair is a good example of a class A amplifier.

   True False

4. The emitter voltage of Q_2 if figure 24.1 is approximately _____.

   (a) 3.6 V (b) 36 mV
   (c) 36 V (d) 12 V

25

CLASS C AMPLIFIERS

INTRODUCTION

In a class C amplifier, the transistor conducts for less than 180 degrees of the input cycle. Because the transistor conducts for a only a small portion of the input signal, the efficiency can be very high. To obtain a sinewave output, the class C amplifier uses a parallel LC resonant circuit.

In this experiment you will construct a class C amplifier and will measure and record its characteristics. You will also simulate circuit faults and, through measurement, will relate measured parameters to failures.

REFERENCE

Principles of Electronic Devices and Circuits - Chapter 6, Section 6.6

OBJECTIVES

In this experiment you will:

✓ Measure and understand the DC characteristics of the class C amplifier

✓ Determine the AC characteristics of the class C amplifier

✓ Demonstrate a method to observe the center frequency of the tuned amplifier

✓ Learn to relate measured circuit values to circuit faults

EQUIPMENT AND MATERIALS

DC power supply
Digital multimeter
Oscilloscope
Function generator
Circuit protoboard
NPN transistor, 2N3440 or equivalent
Resistors: 1.5 kΩ, 270 kΩ
Capacitors: 470 pF [2], 0.01 μF [2]
15 mH inductor

SECTION I FUNCTIONAL EXPERIMENT

1. Construct the circuit in Figure 25.1.

2. Apply DC power to the amplifier. Measure and record the following DC levels:

   V_B = __________ V_E = __________ V_C = __________

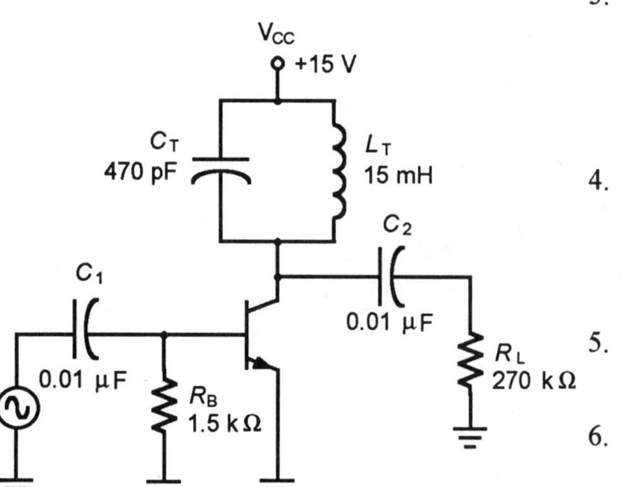

Figure 25.1

3. Calculate and record the resonant frequency of the LC circuit (this is the center or mid-frequency of the amplifier).

   Calculated f_r = __________

4. Adjust the function generator to a 1.5 $V_{p\text{-}p}$ amplitude and an operating frequency equal to the center frequency calculated in step 3. Use the sinewave function.

5. Connect the function generator to the input of the amplifier.

6. Connect the oscilloscope to the output, and change the function generator frequency until a maximum output is obtained. Measure and record this frequency.

   Measured f_r = __________

7. Without changing the function generator frequency, adjust the output amplitude of the generator to the largest value that won't produce an output signal at the collector. Measure and record this peak-to-peak minimum signal level.

   V_{in} = __________ (With no output signal)

8. Increase the function generator level while observing the amplifier output. Adjust the function generator signal level to the minimum value that just produces a complete sinusoidal output signal. Measure and record the peak-to-peak input and output signal levels and record these values.

   V_{in} = __________ (With output signal)

   V_{out} = __________

9. Using your scope to measure the signal, set the input signal to 2 $V_{p\text{-}p}$. Set the scope to monitor the amplifier output, and decrease the function generator frequency to approximatey one-half the frequency measured in step 6. While monitoring the amplifier output, tune the function generator frequency to obtain the maximum output of your output frequency.

   f_{in} = __________ f_{out} = __________

NOTE
Your amplifier should be capable of operation as a frequency doubler if the tank Q is a reasonable value. You may find that the output amplitude is a bit less in frequency doubling since the tank circuit is receiving a pulse input every other cycle.

10. Reset the function generator frequency and amplitude to the values measured in steps 6 and 8. Tune the generator frequency to ensure that you are at the frequency that produces the maximum output sinewave signal. With the scope connected to the amplifier output, tune the generator to the tank corner frequencies f_1 and f_2.

    Measure and record the two corner frequencies.

    f_1 = ______________ f_2 = ______________

11. From the corner frequencies measured in step 10, calculate the tank bandwidth (BW) and Q. Record these values.

    BW = ______________ Q = ______________

Remember

Frequencies f_1 and f_2 are the two frequencies: f_1 below and f_2 above resonance where the output signal amplitude is 0.707 of the maximum signal at resonance.

Step 11 completes the measurements for this section. If you are proceeding to the troubleshooting section, leave your circuit connected.

SECTION II TROUBLESHOOTING

Fault 1 - Change in tank circuit capacitance

1. If the value of C_T in the amplifier circuit of Figure 25.1 decreased to one-half the value, what effect would this have on the circuit operation and parameters? Write your prediction here:

   ____

   ____

2. With the circuit power off, install a second 470 pF capacitor in parallel with C_T. Apply circuit power and input signal. Measure the circuit parameters to determine the effect of this fault.

   ____

   ____

Fault 2 - Tank circuit capacitor open

1. With the circuit power off, remove both 470 pF capacitors of the tank circuit. What effect do you think an open capacitor will have on the circuit operation? Record your prediction here:

   ____

   ____

2. Apply circuit power and input signal. Measure the amplifier circuit para meters to determine the effect of this fault. Explore the effect of a range of input frequencies, particularly above 100 kHz. Record your measurements here:

   ____

   ____

Caution!

As you measure the effect of an open tank capacitor on the amplifier circuit, do not apply low input frequencies. The value of the inductive reactance is the only limit on the transistor saturation current. Keep input frequencies above 2 kHz.

DISCUSSION

Section I

1. Discuss the operation of the class C amplifier. Where might this type of amplifier be used, and what are some of its characteristics?

2. In steps 7 and 8, you should have found that your class C amplifier response to input signal level was not linear. That is, the signal input had to be above a specific value before any output could be obtained. Considering the bias of the amplifier, explain why the class C amplifier behaved in this fashion.

3. In step 9, you should have found that your class C amplifier would operate as a frequency doubler. Explain why the tuned class C amplifier will function as a frequency doubler.

Section II

1. For Fault 1, discuss the reason for the change in resonant frequency in your class C amplifier for a change in the tank circuit capacitance. What single measurement would isolate the fault to the tank circuit?

2. In Fault 2, it is likely that you found a new resonant frequency for your amplifier tank circuit. Discuss the reasons that the tank circuit could (and would) have a new resonant frequency even though the capacitor was open.

Quick Check

1. The ideal class C amplifier has an efficiency close to _____.

   (a) 50% (b) 25.5%
   (c) 78.6% (d) 100%

2. A class C amplifier is used _____.

   (a) where specific frequencies need amplification
   (b) whenever a large bandwidth is necessary
   (c) to match a high impedance source with a low impedance load
   (d) to increase the reactance of a circuit

3. The class C amplifier is conductive for _____.

   (a) 360 degrees (b) 270 degrees
   (c) 180 degrees (d) < 180 degrees

4. The LC tank is used to_____.

(a) increase the gain
(b) decrease the gain
(c) produce a sinewave output
(d) increase the bandwidth

5. The tuned class C amplifier is a linear amplifier.

True
False

6. The bandwidth of a tuned Class C amplifier is dependent upon _____.

(a) transistor beta
(b) tank circuit Q
(c) type of circuit bias
(d) It is always a fixed value

26

DIRECT-COUPLED AMPLIFIERS

INTRODUCTION

The direct coupled amplifier is used where low frequencies need to be amplified. In this experiment you will construct a direct coupled amplifier and will measure AC and DC levels and amplifier gain.

The troubleshooting section will let you explore the effect of transistor failures on a direct coupled amplifier. You will simulate device failures and make measurements to see the effect of the fault.

REFERENCE

Principles of Electronic Devices and Circuits - Chapter 6, Section 6.7

OBJECTIVES

In this experiment you will:

✓ Gain knowledge of the direct coupled amplifier

✓ Understand the difference between direct and capacitively coupled amplifiers

✓ Learn to relate measured circuit values to a transistor failure

EQUIPMENT AND MATERIALS

DC power supply [2]
Digital multimeter
Oscilloscope
Function generator
Circuit protoboard
NPN transistors [2], 2N3904 or equivalent
Resistors: 2.2 kΩ [2], 4.7 kΩ, 10 kΩ, 18 kΩ

SECTION I FUNCTIONAL EXPERIMENT

1. Refer to the circuit of Figure 26.1. To operate properly with the function generator, your direct-coupled amplifier must have the AC signal referenced to ground. Basically, this means that your function generator must have an output that swings positive and negative around a 0-V (ground) potential. The alternative would be to bias the function generator so that the positive and negative alternations would change above and below a reference potential, for example +1 V rather than a 0-V ground potential.

   This circuit criterion could be met using a PNP Darlington amplifier like your text diagram of Figure 7.50. Your amplifier instead uses an emitter bias so that the base of Q is at ground potential. Notice that although the first stage looks somewhat like a common base circuit, it is really a common emitter amplifier.

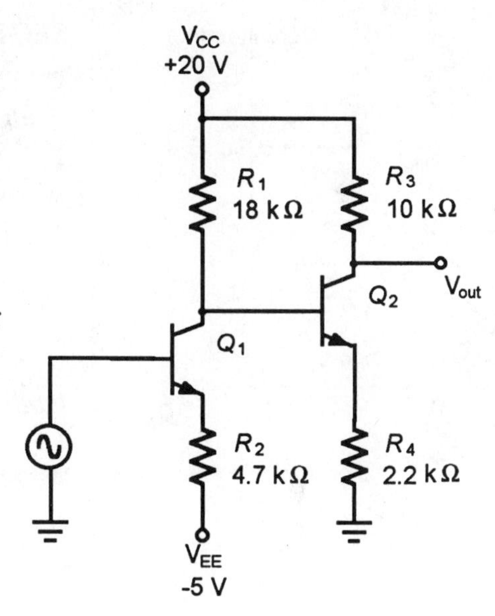

Figure 26.1

2. Construct the circuit of Figure 26.1.

   Note: If your generator does not provide a ground return path, connect a 2.2-kΩ resistor in parallel with the generator (from the base of Q_1 to ground).

3. Connect DC power to your circuit. Use a 2.2-kΩ resistor in place of your function generator. Measure and record the following DC voltages of your amplifier:

   V_{B1} = ________ V_{E1} = ____________ V_{C1} = __________

4. Set your function generator to provide an output of 1 kHz at 200 $mV_{p\text{-}p}$, and connect it to your amplifier.

In step 4, if your function generator has an offset control, this must be adjusted so that there is no DC offset of your signal. Your amplifier will amplify any DC offset just as well as an AC signal.

5. Measure the input and output signals for each stage of your amplifier.

   Stage 1:

   V_{in} = ______________ V_{C1} = ____________

   Stage 2:

   V_{B2} = ______________ V_{C2} = ____________

6. From your data of step 5, calculate the individual stage gains and the total amplifier gain.

   A_{V1} = ________ A_{V2} = ____________ A_{VT} = __________

7. Your amplifier is capable of operation down to DC. Although a specific measurement is not required in this step, adjust your function generator to the lowest frequency you can comfortably work with on your scope, and verify that the amplifier gain is the same as that of step 6.

8. To measure your amplifier operation with a DC input, disconnect the signal generator and 2.2-kΩ resistor, if used.

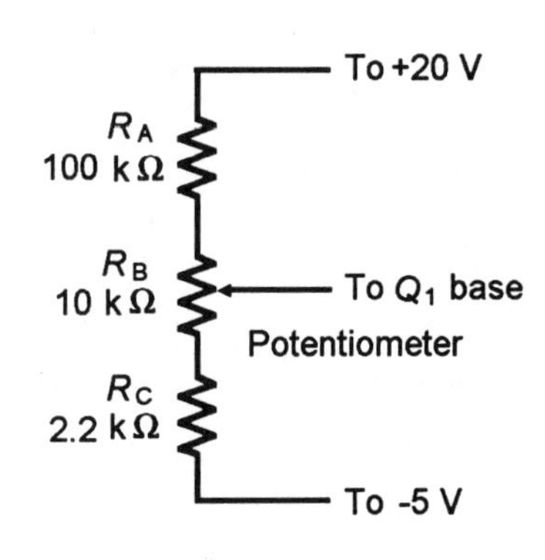

Figure 26.2

9. Connect the voltage divider network of Figure 26.2. Monitor the DC input (to 100 mV positive or negative) and measure the DC level change at the amplifier.

SECTION II TROUBLESHOOTING

Fault 1 - Transistor Q_2 has a base-emitter short.

1. With the circuit of Figure 26.1, ensure that power is off. Jumper the B-to-E junction of Q_2. Replace power and inject a 200-mV$_{p\text{-}p}$, 1-kHz signal at the base of Q_1. Measure the AC and DC levels.

   Note these measurements in a table or notes of your own.

2. Remove power.

Fault 2 - Transistor Q_1 has a base-emitter short.

1. Turn off circuit power. Remove the jumper at Q_2 and jumper Q_1 the base to the emitter.

2. Measure the circuit DC and AC voltage levels and record in a table or notes of your design.

DISCUSSION

Section I

1. The gain of your direct coupled amplifier was significantly less than that of a capacitively coupled amplifier like that of Experiment 21. Identify one major factor responsible for the low gain, and explain why this factor caused the low gain.

2. It is a given that an advantage of a direct coupled amplifier is the ability to operate from DC to some high frequency.

   Considering your amplifier of Figure 26.1, identify and explain what you believe are two disadvantages of the direct coupled amplifier.

Section II

1. What effect did the B-to-E short of Q_2 have on the overall circuit? Did the Q_2 failure affect the signal at the collector of Q_1?

2. Using your measured data for Fault 2 (base-to-emitter short of Q_1), identify and give your rationale for each of the following:

   a. If you could use only one piece of test equipment for troubleshooting, what would you choose?

b. What is the minimum number of different measurements you would need to confirm the specific failure of Q_1 (base-to-emitter short)?

3. Describe the method you employed to isolate the problem your instructor or lab partner inserted into your amplifier.

Quick Check

1. The direct coupled amplifier is used when low frequency signals need to be amplified.

   True False

2. In a direct coupled amplifier, the dc collector voltage of the first stage does not affect the base voltage of the second stage.

   True False

3. The dc level at the base of Q_2 was approximately _____.

   (a) 10.2 V (b) 4.0 V
   (c) 3.6 V (d) 2.3V

4. The load that the second stage of Figure 26.1 presents to the first stage is _____.

   (a) $B \cdot r'_{e_2}$ (b) R_4/B
   (c) $R_4 + B$ (d) $(R_4 + r'_{e_2}) \cdot B$

5. Typically you would expect the gain of a direct coupled amplifier to be greater than that of an equivalent capacitively coupled amplifier.

   True False

27

JFET CHARACTERISTICS

INTRODUCTION

Remember that the JFET is a voltage-controlled device because there is no current through the gate. Because the JFET has essentially no input gate current, it is a very high input impedance device and has application where an exceptionally high input impedance is required. An example is the input stage in an electronic voltmeter.

In this experiment you will construct two circuits, one to measure pinch-off voltage and the other to measure the transconductance curve of the JFET.

REFERENCE

Principles of Electronic Devices and Circuits - Chapter 7, Section 7.1

OBJECTIVES

In this experiment you will:

✓ Experimentally determine the relationship between V_{DS}, I_D, and V_P

✓ Experimentally determine the relationship of I_D versus V_{GS}

EQUIPMENT AND MATERIALS

DC power supply [2]
Digital multimeter [2]
Circuit protoboard
N-channel JFET, 2N3819 or MPF102 or equivalent
Resistor, 100 Ω
Potentiometer, 20 kΩ

SECTION I FUNCTIONAL EXPERIMENT

Measuring Pinch-Off Voltage, V_P

1. Construct the circuit in Figure 27.1.
2. Set the DC power supply for zero volts, and connect the power supply to your circuit. Measure and record in Table 27.1 the value of I_D.
3. Carefully increase the power supply output until V_{DS} is 500 mV ±20 mV. Record I_D and V_{DS} in Table 27.1.
4. Repeat step 3 for all the values of V_{DS} in Table 27.1.
5. Once you have completed the table, graph the data onto Graph 27.1. Your graph should resemble that of your text Figure 27.5.
6. From your graph, determine the value of V_P and record it below.

V_P = ___________

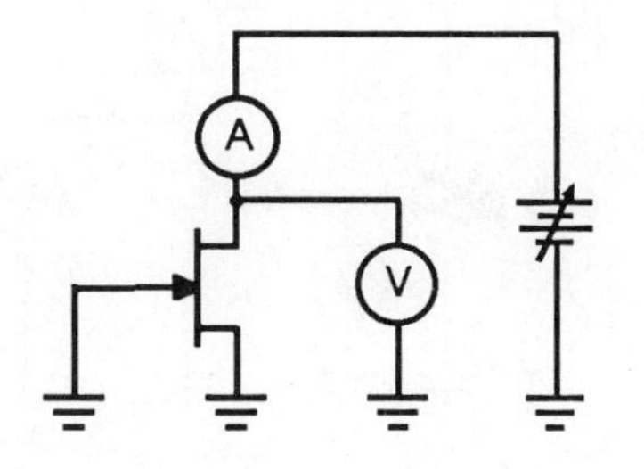

Figure 27.1

V_{DS}	I_D
0.0	
0.5	
1.0	
1.5	
2.0	
2.5	
3.0	
3.5	
4.0	
4.5	
5.0	
5.5	
6.0	
6.5	
7.0	
7.5	
8.0	
8.5	
9.0	
9.5	
10.0	
12.5	

Table 27.1

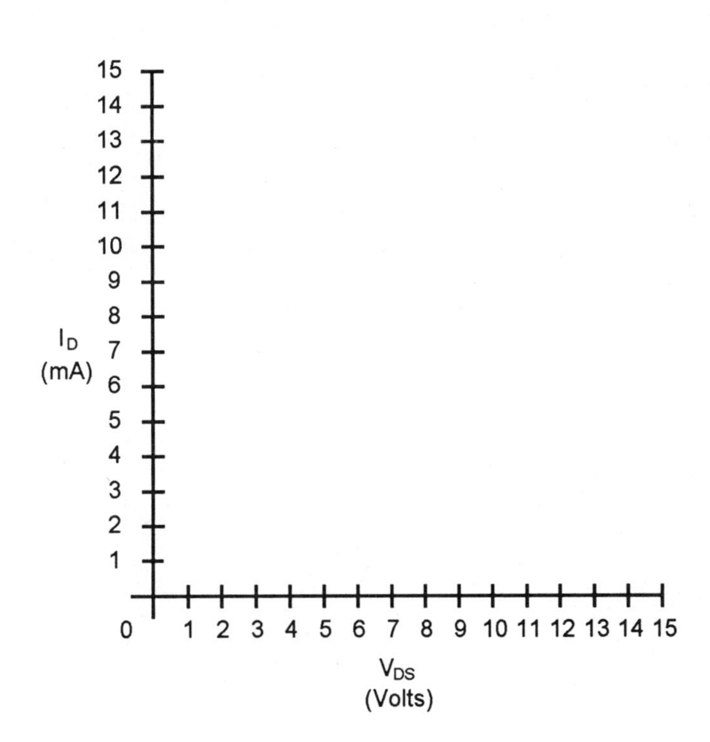

Graph 27.1

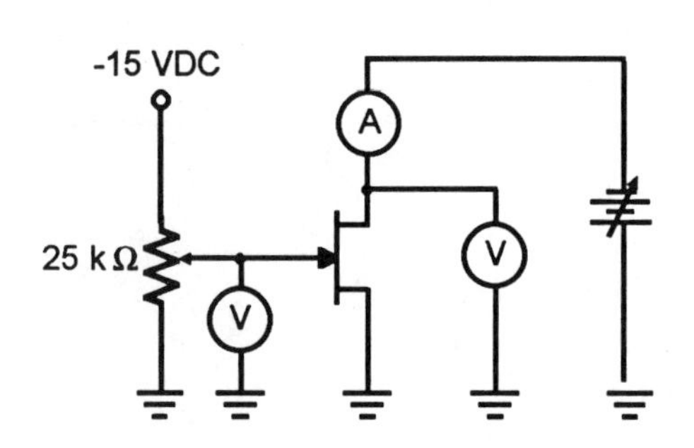

Figure 27.2

Measuring Cutoff Voltage, $V_{GS(off)}$

1. Construct the circuit in Figure 27.2.
2. Adjust V_{DD} to zero volts and adjust the gate resistor until V_{GS} is also zero.
3. Increase V_{DD} until pinch-off is reached.
4. Increase V_{GS} in the negative direction until I_D goes to zero. This is $V_{GS(off)}$.

   $V_{GS(off)}$ = ___________

5. Increase V_{GS} to each of the values in Table 27.2. Measure and record in Table 27.2 the value of I_D for each V_{GS} value.

   Note: Omit those values that are greater (more negative) than the $V_{GS(off)}$.

6. Plot the data in Table 27.2 onto Graph 27.2. Your graph should resemble the graph in Figure 7.9 of your textbook.

V_{GS}	I_D
-1.0	
-1.5	
-2.0	
-2.5	
-3.0	
-3.5	
-4.0	
-4.5	
-5.0	
-5.5	
-6.0	
-6.5	
-7.0	
-7.5	
-8.0	
-8.5	

Table 27.2

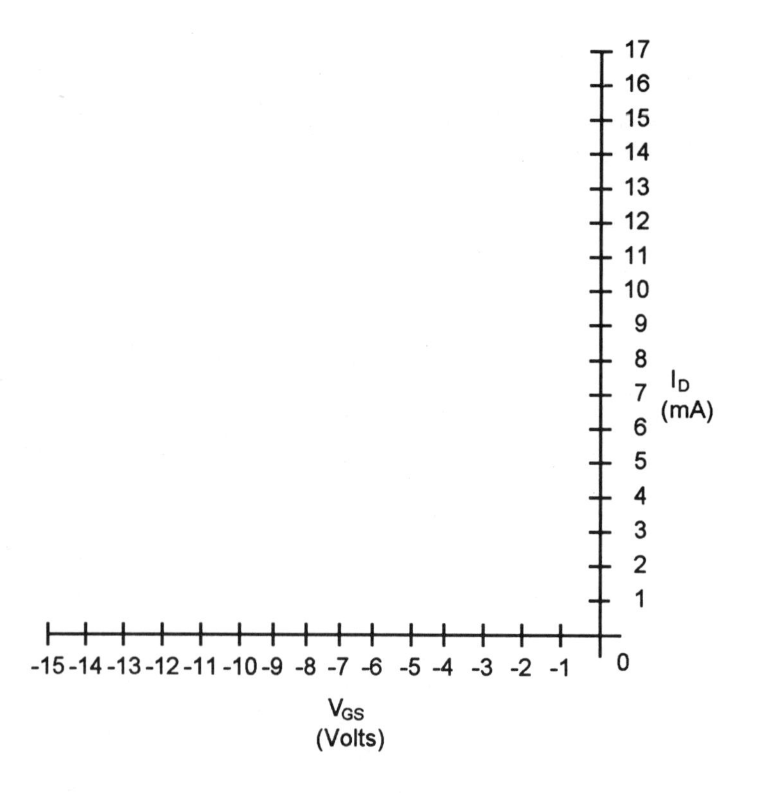

Graph 27.2

DISCUSSION

1. Referring to the data in Table 27.1 and Graph 27.1, was there a point where the I_D no longer increased as V_{DS} increased? What is this characteristic called? At what point did this occur?

2. Describe the term *pinch-off*. What affects this characteristic?

3. Referring to the data in Table 27.2 and Graph 27.2, at which point does $V_{DS} = V_P$?

4. Explain the terms V_{DS}, V_{GS}, V_{DD}, and V_{SS}. How do they relate to a BJT?

5. Was any current measured when $V_{GS} = V_{GS(off)}$?

6. Why is it necessary for V_G to be negative in relationship to V_S? Can any damage occur the JFET if the V_{GS} goes positive? Explain.

Quick Check

1. The drain, source, and gate of the JFET correspond to the collector, emitter, and base of the BJT.

   True False

2. The arrow in a schematic drawing of a JFET points toward the P-type material.

   True False

3. _____ indicates the area where I_D stabilizes and any increase in V_{DS} will not result in an increase in I_D.

   (a) Rolloff (b) Cutoff
   (c) Pinch-off (d) Transconductance

4. The term that defines the relationship between I_D and V_{GS} is called _____.

   (a) rolloff (b) cutoff
   (c) pinch-off (d) transconductance

28

SELF-BIASED JFET

INTRODUCTION

Self-bias of a JFET is a simple, yet reasonably effective, bias method. Compared to gate bias, self-bias provides a degree of current stabilization not possible with gate bias. In this experiment, you will determine the value of source resistance needed to bias a common source stage. Then you will construct and test the self-biased JFET circuit.

In the troubleshooting section you will simulate component failures and, through measurement of circuit parameters, observe the fault effect on the circuit.

REFERENCE

Principles of Electronic Devices and Circuits - Chapter 7, Section 7.2

OBJECTIVES

In this experiment you will:

✓ Experimentally determine the DC characteristics of a self-biased JFET

✓ Verify your ability to produce a predicted Q point

✓ Determine the effect of a failed component on circuit parameters

EQUIPMENT AND MATERIALS

DC power supply
Digital multimeter
Circuit protoboard
N-channel JFET, 2N5459 or equivalent
Resistors: 2.2 kΩ, 100 kΩ, 1 MΩ

SECTION I FUNCTIONAL EXPERIMENT

In the procedure steps that follow, you will determine the value of source resistance to bias a common source stage, construct the circuit, and measure the DC circuit parameters to verify your selection value.

1. From the data sheet for a 2N5459 JFET, obtain and record the following minimum and maximum parameters.

   Minimum:

   $V_{GS(off)}$ = ____________ I_{DSS} = ____________

   Maximum:

   $V_{GS(off)}$ = ____________ I_{DSS} = ____________

   From these data you will calculate a set of minimum and a set of maximum tables of I_D versus V_{GS}. Then using your tabled data, you will draw two transconductance curves (minimum and maximum) for the JFET.

2. Calculate for the minimum and maximum levels in Table 28.1 the values of I_D for three values of V_{GS} that fall in the range between $V_{GS} = 0$ and $V_{GS(off)}$. Use the formula $I_D = I_{DSS}\left(1 - \frac{V_{GS}}{V_{GS(off)}}\right)^2$.

3. Plot your tabled data on Graph 28.1. Your plotted data should resemble your text Figure 7.19.

4. Select a load line that will pass through approximately the center of the minimum value transconductance curve and extend the straight line through the maximum value transconductance curve. Following example 7.5 in your text, determine the value of R_s represented by that line and record the resistance value.

   Calculated R_s = ____________

   Standard R_s = ____________

Minimum		Maximum	
V_{GS}	I_D	V_{GS}	I_D
0	(I_{DSS})	0	(I_{DSS})
($V_{GS(off)}$)	0	($V_{GS(off)}$)	0

Table 28.1

Select the closest standard resistor value to your calculated resistance and record above. For example, if your calculated value was 509 Ω, use 470 Ω. Plot the standard resistance value on your transconductance curves and read from each curve the Q point I_D value. Record these below.

Minimum I_D = ___________

Maximum I_D = ___________

You now know the two limit values of I_D for your circuit with the value of R_S you have determined.

5. Construct the circuit of Figure 28.1 using your standard resistor R_S for the source resistance.

6. Apply circuit power. Measure and record the following circuit values:

I_D = _______________ V_S = _______________

V_D = _______________ V_{GS} = _______________

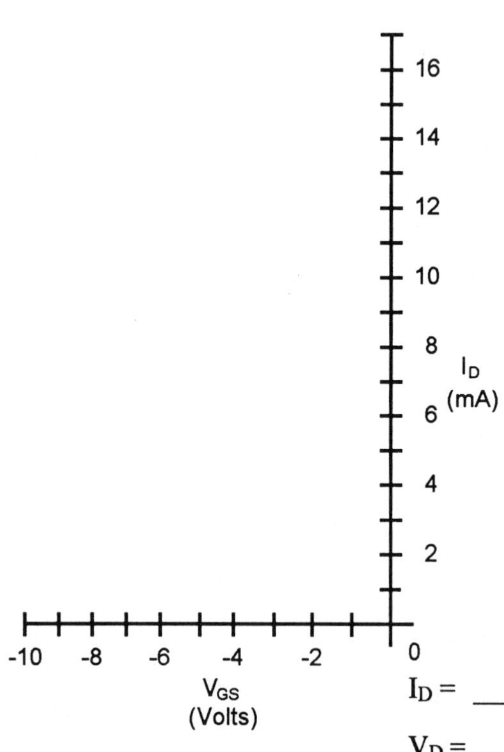

Graph 28.1

7. The value of I_D that you measured should fall between the minimum and maximum values you recorded in step 4. If there is a minor discrepancy as a result of curve accuracy plot or meter tolerance, you may disregard it. In the event of a major difference, recheck your transconductance curve and Q point plot of R_S.

With this work, you have established the ability to verify that self-bias of a JFET can predict and establish a Q point within device limits. This completes the first section of your experiment.

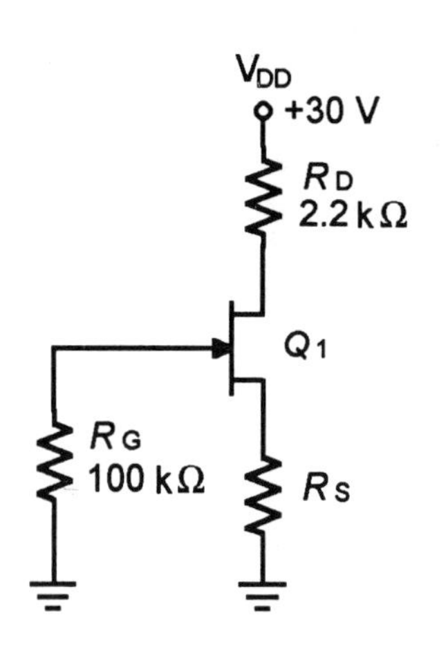

Figure 28.1

SECTION II TROUBLESHOOTING

Fault 1 - Source resistor R_S open

1. With the circuit power off, remove R_S and replace it with a 1-MΩ resistor.

2. Reapply circuit power. Measure and record the following circuit values:

I_D = _______________ V_S = _______________

V_D = _______________ V_{GS} = _______________

Fault 2 - Drain resistor R_D open

1. With circuit power off, remove the 1-MΩ source resistance and replace it with the correct resistor. Remove the 2.2-kΩ drain resistor and replace it with a 1-MΩ resistor.

2. Apply circuit power. Measure and record the following circuit values:

I_D = _______________ V_S = _______________

V_D = _______________ V_{GS} = _______________

DISCUSSION

Section I

1. You went through a fairly extensive process to establish a self-biased JFET. Describe what you gained from taking this approach to circuit exploration.

2. Could you put any 2N5459 JFET in your circuit of Figure 28.1 with the value of R_S that you determined and be certain of the knowledge of Q point? Discuss the reasons for your answer.

Section II

Fault 1 - Source resistor R_S open

Is there another component failure that would produce the same readings you obtained for this fault, for example, an internal open in the JFET source lead? Discuss the effect that this device failure would have on the measured circuit values.

Fault 2 - Drain resistor R_D open

Do your measured data for this fault prove the failure of R_S? Discuss this and describe, if appropriate, any additional measurements necessary to isolate the failure to R_D.

Quick Check

1. Self-biasing places the gate at a fixed voltage.

   True　　　　False

2. Self-biasing is an improvement over _____.

   (a) gate bias　　　　(b) emitter bias
   (c) current-source bias　　　　(d) voltage-divider bias

3. A JFET is usually operated in forward bias mode.

   True　　　　False

4. In a self-biased mode, the voltage at the gate is _____.

   (a) equal to V_S　　　　(b) greater than V_S
   (c) zero　　　　(d) equal to V_D

29

VOLTAGE DIVIDER BIAS

INTRODUCTION

Like the self-biased JFET, the voltage divider is also a good method of stabilizing the Q point. Also, like in the BJT, the V_G will be referenced to ground. Therefore,

$$V_{TH} = \left(\frac{R_2}{R_1 + R_2} \right) \times V_{DD}$$

In this experiment you will construct a voltage divider JFET circuit, and you will calculate and measure V_S, V_G, I_D, and V_D.

REFERENCE

Principles of Electronic Devices and Circuits - Chapter 7, Section 7.3

OBJECTIVES

In this experiment you will:

✓ Become familiar with the characteristics of the voltage divider JFET

✓ Observe the effect of circuit component failures on circuit parameters

EQUIPMENT AND MATERIALS

DC power supply
Digital multimeter
Circuit protoboard
Resistors: 1 kΩ, 2.2 kΩ [2], 4.7 kΩ, 1 MΩ
N-channel JFET, 2N5459 or equivalent

SECTION I FUNCTIONAL EXPERIMENT

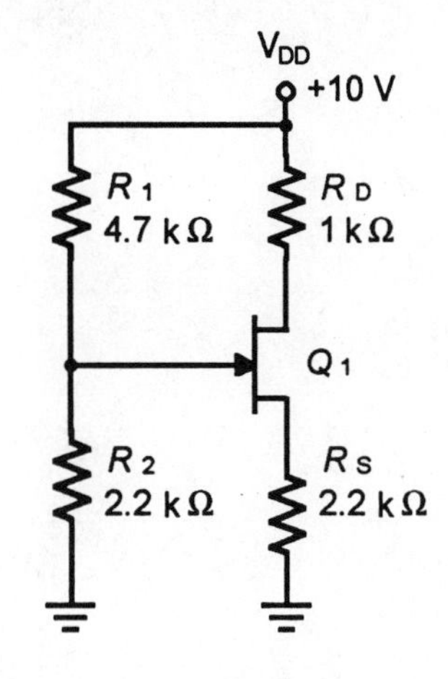

Figure 29.1

1. Construct the circuit in Figure 29.1.

2. Apply DC power. Measure and record the following circuit values.

   V_D = ______________ V_G = ______________

   V_S = ______________ V_{GS} = ______________

   I_D = ______________

3. Your measured Q point should lie on the circuit load line between the minimum and maximum Q points. Using the data of Experiment 28, plot the minimum and maximum transconductance curves on Graph 29.1.

If you did not do Experiment 28, you can follow steps 1 and 2 in Experiment 28 to obtain the data points for the transconductance curves.

4. Plot the load line for your circuit on Graph 29.1. From the load line, read the minimum and maximum I_D values and record below.

   Min I_D = ______________ Max I_D = ______________

   Your measured I_D should fall in the range of values between the I_D values recorded above.

A Q point value that is out of range by a small amount accounted for by meter tolerance or curve accuracy is acceptable. In the event of greater differences, you should recheck your measurements and your plotted data of Graph 29.1.

5. If you want to explore voltage divider biasing further, select a new load line on your transconductance plot of Graph 29.1. Try an ideal I_D of approximately half the value of the circuit of Figure 29.1. Calculate the new value of R_S and take the closest standard resistor size. Record the selected value of R_S below.

   R_S = ______________

6. From the plot of Graph 29.1, determine the new minimum and maximum values of I_D and record below.

   Min I_D = ______________ Max I_D = ______________

7. Install the new R_S in your circuit and measure and record the quiescent circuit values below.

   I_D = ______________ V_S = ______________

   V_D = ______________ V_{GS} = ______________

   You should find that the new quiescent value of I_D falls between the predicted minimum and maximum values.

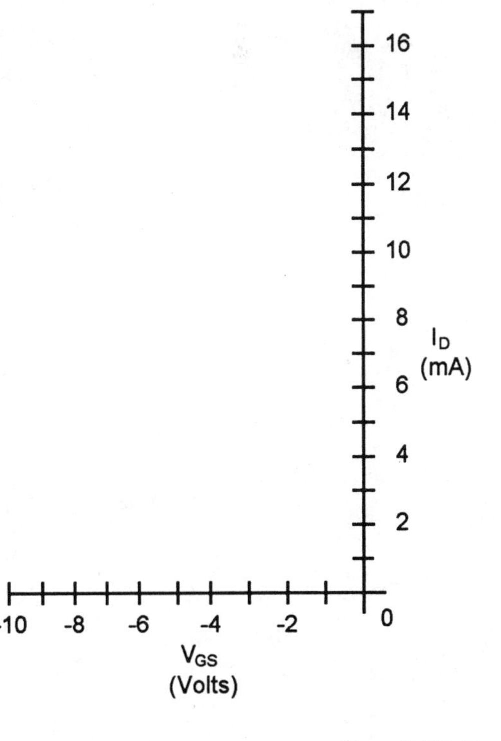

Graph 29.1

SECTION II TROUBLESHOOTING

Fault 1 - Divider resistor R_1 open

1. With circuit power off, remove the 4.7-kΩ (R_1) resistor and replace it with a 1-MΩ resistor.

2. Apply circuit power. Measure and record the below listed circuit DC values.

   I_D = ________________ V_G = ________________

   V_D = ________________ V_{GS} = ________________

   V_S = ________________

Fault 2 - JFET gate-source shorted

1. With circuit power off, remove the 1-MΩ resistor and replace it with the correct R_1 4.7-kΩ resistor. Place a jumper wire from gate to source of the JFET.

2. Apply circuit power. Measure and record the following DC circuit values.

   I_D = ________________ V_G = ________________

   V_D = ________________ V_{GS} = ________________

   V_S = ________________

DISCUSSION

Section I

1. Looking at the predicted minimum and maximum values of I_D (step 4) and the load line you plotted in step 4, can you agree that voltage divider bias produces a smaller range for I_D than self-bias? Discuss this, explaining why voltage divider bias accomplishes this result.

2. In your voltage divider biased circuit of Figure 29.1, the voltage divider resistors are 4.7 kΩ and 2.2 kΩ. Would the circuit function equally well if the divider resistors were 4.7 MΩ and 2.2 MΩ? Discuss this and explain why you think the smaller and larger resistance values might be preferred.

Section II

Fault 1 - Divider resistor R_1 open

You should have found that with R_1 open, the circuit continued to function, although at a different Q point. Discuss why the circuit operates at a different Q point and how you would identify this circuit failure.

Fault 2 - JFET gate-source shorted

You should have found with this fault that the circuit still was functional although the Q point had made a significant shift. Discuss how you could be certain that the circuit had a failure and what additional (if any) measurements you would make to isolate the failure to the JFET.

Quick Check

1. A voltage divider biased JFET has a stable Q point.

   True　　　　　　　　　　False

2. The voltage divider biased JFET has a stable Q point because the slope of the DC lead line is nearly horizontal.

   True　　　　　　　　　　False

3. An open source will result in _____.

   (a) high I_D current　　　　(b) no I_D current
   (c) an increase in V_{DD}　　　　(d) a change in V_G

4. In a voltage divider biased network the gate voltage is referenced to ground.

   True　　　　　　　　　　False

5. In comparing a BJT and JFET, I_{DSS} is the JFET equivalent to the BJT I_{CSAT}.

   True　　　　　　　　　　False

30

CURRENT SOURCE BIASED JFET

INTRODUCTION

Of the various biasing methods for a JFET, current source biasing provides the most predictable and JFET independent Q point. Current source biasing has some limitations in amplifier circuits, and its applications are generally limited to low-drift DC amplifiers and source-coupled differential pairs.

This experiment will allow you to explore a JFET circuit using a BJT for the current source biasing. You will set up a basic circuit, test circuit value, and then predict and test a modified circuit.

In the troubleshooting section you will simulate circuit component faults and make measurements to see the effect of the faults.

REFERENCE

Principles of Electronic Devices and Circuits - Chapter 7, Section 7.3

OBJECTIVES

Through this experiment you will:

✓ Understand the characteristics of current source bias of a JFET

✓ Be able to determine circuit values and predict the JFET Q point

✓ Be able to relate measured circuit parameters to the circuit fault

EQUIPMENT AND MATERIALS

DC power supply
Digital multimeter
Circuit protoboard
N-channel JFET, 2N5459 or equivalent
NPN transistor, 2N3904 or equivalent
Resistors: 4.7 kΩ [2], 10 kΩ, 100 kΩ, 1 MΩ

SECTION I FUNCTIONAL EXPERIMENT

1. Construct the circuit in Figure 30.1.

2. Apply DC power to your circuit. Measure and record the DC circuit values included in Table 3.1.

3. For the circuit of Figure 30.1, calculate the value of I_E and record it in Table 30.1.

4. You should find, within component and equipment measurement accuracy, close agreement with the measured I_D of step 2 and the calculated value of step 3.

5. Calculate and select a new value of R_E (a standard resistor value) to produce an I_D in the range of 0.75 to 0.95 mA. Record your selected value in Table 30.1. With circuit DC power turned off, install the new emitter resistor.

Note: Current source biasing has the limitation that the current source cannot exceed the JFET minimum value of I_{DSS}.

6. Apply DC power. Measure and record in Table 30.1 the required circuit values.

This completes the measurements of Section I.

Figure 30.1

	R_E = 4.7 kΩ	R_E =
I_D		
V_D		
V_{GS}		
V_S		
V_E		
I_E (Calculated)		

Table 30.

SECTION II TROUBLESHOOTING

Fault 1 - BJT inoperative

1. Ensure that DC power is off.

2. Turn the BJT off by shorting the base to emitter. What effect do you predict this will have on the circuit?

____

____

____

In this circuit, the operating characteristics of the JFET are dependent upon the BJT: any problem in the BJT will result in a problem in the JFET circuit. Therefore, a knowledge of the operating characteristics of both types of devices is necessary, so you may want to review the basic operating characteristics of the BJT. Also, careful measurement must be taken to ensure accurate evaluation.

3. Measure the appropriate voltages. What significant changes do you observe? What happened to I_D, for example?

____

____

____

Fault 2 - Emitter resistor open

1. Turn th power off, restore the BJT to operate properly, and replace the emitter resistor with a 1-MΩ resistor. Note below the effect you expect with this fault.

____

____

____

2. Turn on the power and measure the appropriate voltages. Were there any changes to V_B, V_D, or I_D? How do your measurements match your predictions?

____

____

____

DISCUSSION

Section I

1. In the Introduction it was stated that current source biasing provides a Q point that is JFET independent. Discuss this statement and explain your reasons for believing that it is true or not true.

2. In Procedure step 5 it was noted that the current source should not exceed the JFET minimum value of I_{DSS}. Discuss this statement and indicate why you think this limitation is placed on current source bias.

Section II

Fault 1 - BJT inoperative

Refer to the measurements you made for this fault. Do your measured data clearly indicate a BJT failure? In your answer, cite the measured data to illustrate and support your response.

Fault 2 - Emitter resistor open

1. Do your measured data conclusively prove R_E failure? If your answer is *yes*, cite the measurement data. If your answer is *no*, explain what measurements you would make.

Quick Check

1. In a current source biased JFET, is there any noticeable change in I_D as V_{GS} changes?

   Yes No

2. The BJT must be forward biased at all times to ensure a constant current source for the JFET.

   True False

3. What was the gate voltage, V_G, of the JFET?

   (a) -2 V (b) 2 V
   (c) 0 V (d) $V_G = V_C$

4. Current source biasing provides the most stable Q for the JFET.

   True False

5. With a 4.7 kΩ R_E the V_E at the emitter was approximately _____.

   (a) 3.1 V (b) -9.3 V
   (c) 0.7 V (d) -0.7 V

31

COMMON-SOURCE JFET AMPLIFIERS

INTRODUCTION

The common-source JFET amplifier is a good choice for single-ended DC and general-purpose AC amplifiers. The common-source JFET amplifier has the same general characteristics of the common emitter BJT amplifier. However, the voltage gain of the JFET amplifier is less than that of the BJT.

In this experiment you will construct and explore the characteristics of a common-source JFET amplifier. You will measure both the DC and AC circuit values and voltage gain, comparing them to your estimated values.

In the troubleshooting section, you will simulate a component failure and will measure circuit values to observe the effect of the fault.

REFERENCE

Principles of Electronic Devices and Circuits - Chapter 7, Section 7.5

OBJECTIVES

Through this experiment you will:

✓ Understand the operating characteristics of a common-source amplifier

✓ Observe the similarities of the common-source amplifier with those of the BJT common emitter amplifier

✓ Be able to relate measured circuit values to circuit fault

EQUIPMENT AND MATERIALS

DC power supply
Function generator
Dual-channel oscilloscope
Digital multimeter
Circuit protoboard
N-channel JFET, 2N5459 or equivalent
Resistors: 2.7 kΩ, 4.7 kΩ, 10 kΩ, 1 MΩ
Capacitors: 0.1 μF, 10 μF

SECTION I FUNCTIONAL EXPERIMENT

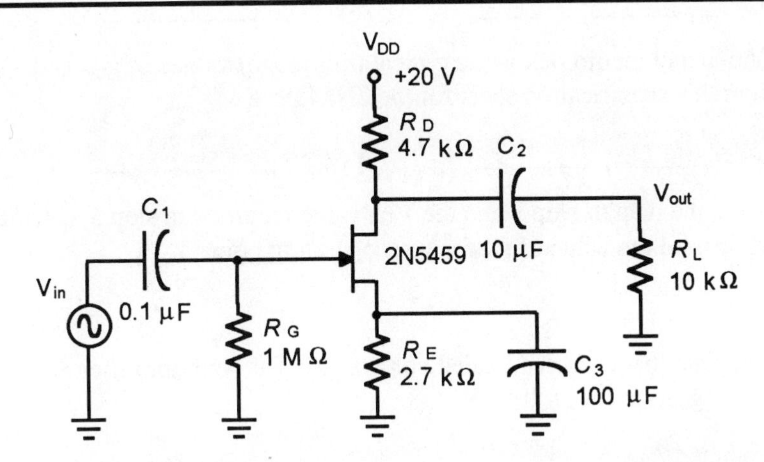

Figure 31.1

1. Build the circuit shown in Figure 31.1. Temporarily omit R_L. Set the DC power supply for a V_{DD} of 12 V.

2. Measure and record the following DC parameters:

   V_G = ______________ V_S = ______________

   V_D = ______________ V_{GS} = ______________

3. Using the values of V_S and R_S, calculate I_D.

   I_D = ______________

4. Set the function generator for an input voltage of 500 $mV_{P\text{-}P}$ at a frequency of 1 kHz. Measure the amplitude and frequency of the input with the oscilloscope.

   V_{in} = ______________ f = ______________

6. Connect your oscilloscope channel 1 to the signal input (junction of C_1, R_G ,and JFET gate) and channel 2 to the amplifier output.

7. Measure and record V_{out} and the phase difference between the signal input and the amplifier output.

   V_{out} = ______________ (unloaded)

   V_{out} = ______________ (loaded)

   Phase shift = ______________ V_S = ______________

   Connect the load resistor (R_L), and measure and record the loaded amplifier output. Also measure and record the AC signal at the JFET source.

Note: This signal should be essentially zero volts.

8. From the data of steps 5 and 7, calculate both the loaded and unloaded voltage gain of your amplifier.

   A_V = ______________ A_{VL} = ______________

9. Obtain and record below the typical (midrange) values of g_{mo} and $V_{GS(off)}$ from this specification sheet for the 2N5459.

   g_{mo} = ______________ $V_{GS(off)}$ ______________

10. Using the data of step 9 and the V_{GS} value recorded in step 3, calculate and record the value of g_m for your operating point.

    g_m = ______________

    Calculate the predicted loaded voltage gain for your amplifier, where $A_{VL} = g_m r_D$ and $r_D = R_D \parallel R_L$

    Predicted A_{VL} = ______________

    How well does your predicted gain compare to your measured loaded gain? ______________________________

    ____

SECTION II TROUBLESHOOTING

Fault 1 - Source bypass capacitor open

1. Ensure that the DC power is off. Simulate the loss of the bypass capacitor by disconnecting the ground lead of C_3 from the circuit.

2. Measure and record the output voltage across R_L and then the AC signal at the JFET source with CH 2 of the oscilloscope. Record the result below:

   V_{out} = ____________ V_S = ____________ A_{VL} = ____________

3. Compare the output values just obtained with those from Section I, steps 7 and 8. Note the decrease in V_{out}.

DISCUSSION

Section I

1. The common source amplifier is the FET equivalent to the BJT common emitter amplifier. If both types produce similar outputs, discuss why it may be advantageous to use an FET-based amplifier in place of a BJT-based type.

2. Considering that the JFET is a nonlinear device, the I_D versus V_{GS} relationship is parabolic. Discuss what is meant by *nonlinear* and what effect the JFET nonlinearity can have on the amplifier output.

Section II

1. The A_V of the amplifier was decreased when the C_3 was removed from the circuit. From your observations during the experiment, describe how you could determine the cause of a reduced output, and the measurement(s) to prove the failure of the source bypass capacitor.

Quick Check

1. If $g_m = 3500$ µmhos and $R_D = 2.2$ kΩ, A_v = __________.

2. The output of the common-source amplifier is in phase with input.

   True   False

3. If C_3 of Figure 31.1 were to open, what changes would you expect to occur to the DC and AC parameters?

   For DC: ______________________________

   For AC: ______________________________

4. What is the FET equivalent to the BJT DC load line?

   ____

32

COMMON-DRAIN JFET AMPLIFIERS

INTRODUCTION

Just as BJT amplifier circuits are classified according to their configuration, and each configuration has its own unique characteristics, the JFET amplifier is also classified according to its configuration.

In this experiment you will construct a common-drain amplifier, the JFET equivalent of the BJT common-collector amplifier, and you will make measurements to evaluate this circuit configuration.

In the troubleshooting section, you will simulate a circuit failure and measure the circuit parameters to see the effect of the fault.

REFERENCE

Principles of Electronic Devices and Circuits - Chapter 7, Section 7.5

OBJECTIVES

Through this experiment you will:

✓ Understand the operating characteristics of a common-drain amplifier

✓ Experimentally verify the output impedance of the common-drain amplifier

✓ Be able to relate a circuit fault to circuit measured values

EQUIPMENT AND MATERIALS

DC power supply
Function generator
Dual-channel oscilloscope
Digital multimeter
Circuit protoboard
N-channel JFET, 2N5459 or equivalent
Resistors: 2.7 kΩ, 10 kΩ, 1 MΩ
Capacitors: 0.1 μF, 10 μF

SECTION I FUNCTIONAL EXPERIMENT

1. Build the circuit shown in Figure 31.1. Set the DC power supply for a V_{DD} of 12 V.

2. Measure and record the following DC parameters:

   V_G = ______________ V_S = ______________

   V_D = ______________ V_{GS} = ______________

   Using the values of V_S and R_S, calculate I_D.

   I_D = ______________

3. Connect your function generator and set it to apply a 500-m$V_{p\text{-}p}$, 1-kHz input signal. Connect your oscilloscope channel 1 to the signal input and channel 2 to the circuit output. Measure and record the peak-to-peak input and output circuit values.

   V_{in} = ______________ V_{out} = ______________

4. Measure and record below the phase difference of the output to the input signal.

   Output phase = ______________

In the following steps, you will make measurements and a calculation to determine the output impedance of your common-drain amplifier. Your amplifier can be represented as a Thevenin source (V_S) with a source impedance (Z_S) driving a load resistance of R_L. See Figure 32.2.

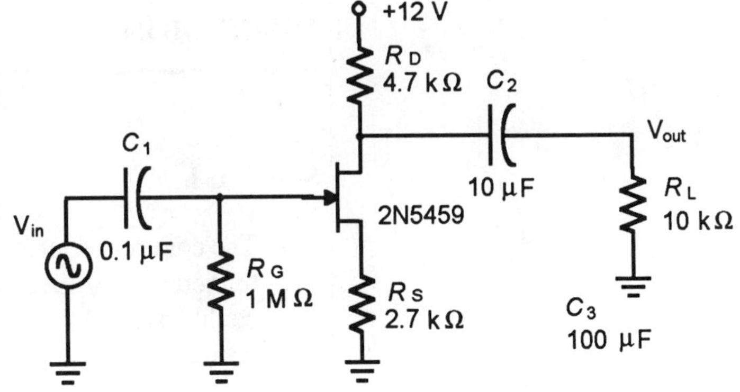

Figure 32.1

5. With circuit power off, disconnect R_L. Apply circuit power. Set the signal input to 500 m$V_{p\text{-}p}$ at 1 kHz. Carefully measure V_{in} and then the no-load V_{out} and record below.

   V_{in} = ______________ V_{out} = ______________ (unloaded)

   The unloaded value of V_{out} is the value of V_S in the diagram of Figure 32.2.

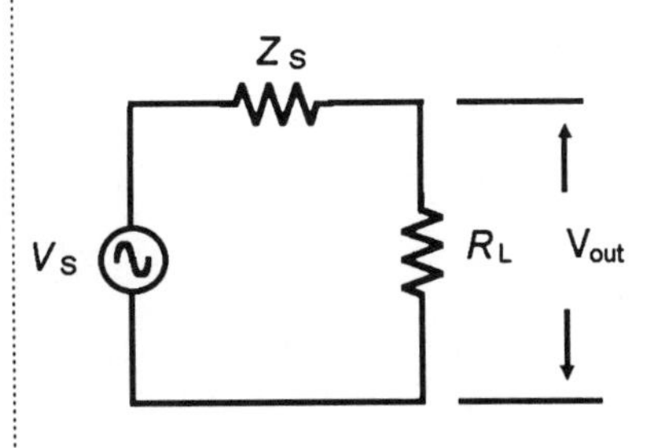

Figure 32.2

6. Turn circuit power off. Connect a 1-kΩ load resistor. Reapply circuit power and input signal. Ensure that the input signal is exactly the same as recorded in step 6. Measure and record the peak-to-peak loaded value of V_{out}.

   V_{out} = ______________ (loaded)

7. Using the formula below, calculate the output impedance (Z_S) of your common-drain amplifier.

$$Z_S = \frac{R_L(V_S - V_{out})}{V_{out}}$$

Z_S = ______________

SECTION II TROUBLESHOOTING

Fault 1 - Gate resistor R_G is open

1. Ensure that you have the circuit of Figure 32.1 connected. R_G is used to keep the gate of the JFET at zero volts to prevent input signal loading. Simulate an open gate resistor by disconnecting the ground lead of R_G from the circuit.

2. Apply circuit power and the 500-m$V_{p\text{-}p}$, 1-kHz input signal. Use your oscilloscope to measure the peak-to-peak signal input and output. Record the measured values below.

   V_{in} = ______________ V_{out} = ______________

3. Compare the output values just obtained with those from Section I, step 3. Note the decrease in V_{out}.

DISCUSSION

Section I

1. The common-source amplifier is usually operated with a bypass capacitor to increase A_V. Because of the nonlinear nature of FET amplifiers, this could lead to distortion in the output. Discuss why there should be no distortion in the output of a common-drain amplifier.

2. Even though the gain of the common-drain amplifier is less than unity, the device can be useful in conjunction with other amplifier types and configurations. Discuss what major characteristics the common-drain amplifier has in common with its BJT equivalent, and how it may be useful.

Section II

The A_V of the amplifier was further decreased when the R_G was opened. From your observations during the experiment, discuss how you could determine the cause of the circuit failure.

Quick Check

1. For the circuit of Figure 32.1, if $g_m = 3500\ \mu S$, $R_S = 2.2\ k\Omega$, determine A_V.

   A_V = ____________

2. The output of the common-drain amplifier is in phase with the input.

   True False

3. What is the FET equivalent to the BJT DC load line? ____________

   ____

4. The gain of the common-drain amplifier could be increased by adding a source bypass capacitor.

   True False

33

BIASING D MOSFETS

INTRODUCTION

The D MOSFET is a depletion configuration device. It has a source-drain conducting channel. The D MOSFET is unique because it can operate in both the depletion and enhancement mode.

You will, in this experiment, explore two different bias forms. The first is self-bias, where the MOSFET is configured to operate in the depletion mode. The second bias form is the simplest and more typical no-bias form made possible by the enhancement capability of the D MOSFET.

The second section of the experiment is troubleshooting. In this section, you will simulate faults and will make measurements of the circuit under fault conditions in order to observe the effects of component failure.

REFERENCE

Principles of Electronic Devices and Circuits - Chapter 8, Section 8.4.

OBJECTIVES

In this experiment you will:

✓ Understand the characteristics of the D MOSFET

✓ Be able to relate measured circuit parameters to a circuit fault

EQUIPMENT AND MATERIALS

DC power supply
Digital multimeter
Circuit protoboard
N-channel D MOSFET, 2N3796 or equivalent
Resistors: 820 Ω, 3.9 kΩ, 9.1 kΩ, 47 kΩ, 910 kΩ , 1 MΩ

SECTION I FUNCTIONAL EXPERIMENT

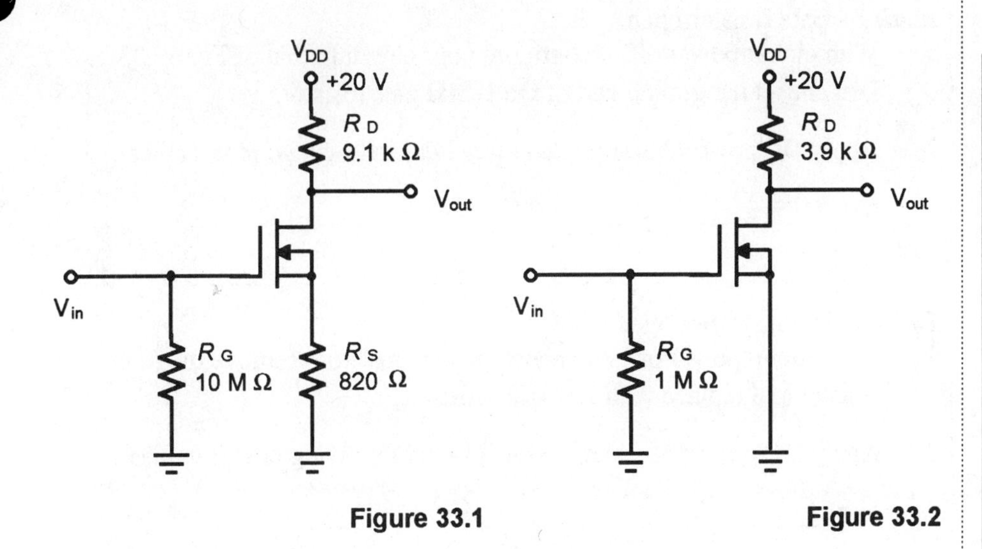

Figure 33.1

Figure 33.2

Because the D MOSFET is a depletion-type device like the JFET, it may be biased using gate bias, self-bias, or voltage divider bias. Because of its capability of operating in the enhancement mode, it can also be operated with no bias. Your first circuit will use self-bias, and the second the simple no bias.

CAUTION!

MOSFETS are subject to damage by electrostatic discharge (ESD). Use care when handling these devices and follow ESD handling rules.

1. Construct the circuit of Figure 33.1. Apply DC power. This circuit should operate with a small V_{GS} value of approximately 0.5 V.

2. Measure and record the following circuit values.

   I_D = ________________ V_S = ________________

   V_D = ________________ V_{GS} = ________________

3. Turn off the DC power. Modify your circuit to that of Figure 33.2. This is the no-bias case, and I_{DQ} will be I_{DSS}.

4. Apply DC power to your circuit. Measure and record the following circuit values.

   I_D = ________________ V_S = ________________

   V_D = ________________ V_{GS} = ________________

5. The following Procedure steps will permit you to observe the enhancement mode of the D MOSFET.

6. Turn circuit power off. Add the voltage divider of Figure 33.3 to the gate of your circuit.

7. Reapply DC power. Measure the record the following circuit values.

   I_D = ________________ V_S = ________________

   V_C = ________________ V_{GS} = ________________

   Can you confirm enhancement-mode operation by the increased value of I_D?

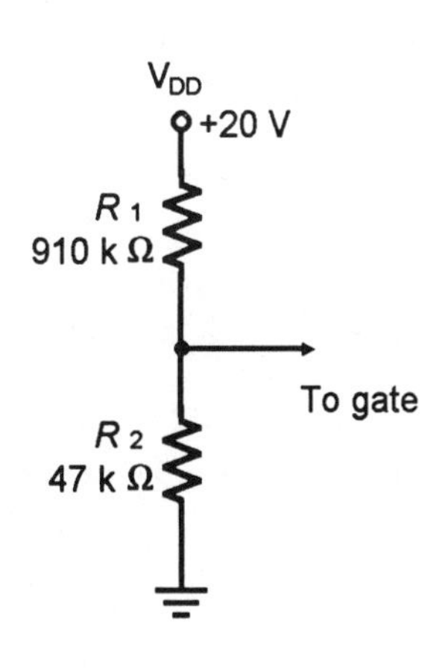

Figure 33.3

SECTION II TROUBLESHOOTING

Fault 1 - Gate resistor open

1. With circuit power off, reconfigure your circuit to that of Figure 33.1. Disconnect the ground end of the 1-MΩ gate resistor.

2. Apply DC power. Measure and record the following circuit values.

   I_D = ______________ V_S = ______________

   V_D = ______________ V_{GS}= ______________

Fault 2 - Drain resistor open

1. With circuit power off, reconnect the gate resistor. Remove the drain resistor and replace with a 1-MΩ resistor.

2. Apply DC power. Measure and record the following circuit values.

   I_D = ______________ V_S = ______________

   V_D = ______________ V_{GS} = ______________

DISCUSSION

Section I

1. Discuss the advantages and the disadvantages of the biasing forms covered in this experiment.

2. Discuss the advantages and disadvantages of the D MOSFET compared to the JFET.

Section II

Fault 1 - Gate resistor open

Did the open gate resistor cause the D MOSFET to stop conducting? Discuss the circuit response to this failure mode by explaining how the open gate resistor affected the D MOSFET operation.

Fault 2 - Drain resistor open

Did your measured data for this failure prove the failure to be a specific component? Describe, as appropriate, measurement to prove the failure of the drain resistor.

Quick Check

1. In the zero-bias circuit, R_G must be at least 1 MΩ.

   True　　　　　　　　　　False

2. In a voltage divider, $V_{DD} = 12$ V, $R_1 = 100$ MΩ, $R_2 = 22$ MΩ. Determine V_{R2}.

   V_{R2} = ____________

3. The advantage of zero bias over voltage divider bias is _____.

   (a) greater gain　　　　　　(b) thermal stability
   (c) simplicity of construction　　(d) lower gain

4. When a D MOSFET is operated in a zero-bias circuit, the value of I_{DQ} is ______.

   (a) $I_{D(off)}$　　　　　　(b) I_{DSS}
   (c) $I_{D(on)}$

5. Both the D MOSFET and E MOSFET will function properly in a zero-bias circuit.

   True　　　　　　　　　　False

34

MOSFET AMPLIFIERS

INTRODUCTION

The capability of operating with no bias and a gate signal of either polarity lets the D MOSFET function as an amplifier with very simple circuitry. Its main disadvantage is that it has a higher noise performance than does the JFET.

In this experiment you will construct and test a depletion-mode MOSFET amplifier. You will observe and verify the MOSFET advantage over a JFET as well as the general FET square law effect when operating a large signal.

The troubleshooting section will let you examine the amplifier circuit parameters with a fault inserted. You will be able to see the effect on the AC and DC circuit values for your simulated failure.

REFERENCE

Principles of Electronic Devices and Circuits - Chapter 8, Sections 8.4 and 8.5.

OBJECTIVE

Through this experiment, you will:

✓ Understand the operating characteristics of MOSFET common source amplifiers

✓ Observe the similarities of the MOSFET common-source amplifier compared with those of JFET amplifiers

EQUIPMENT AND MATERIALS

DC power supply
Dual-trace oscilloscope
Digital multimeter
Function generator
Circuit Protoboard
N-channel D MOSFET, 2N3796 or equivalent
Resistor, 4.6 kΩ
Capacitors: 0.1 μF, 10 μF

SECTION I FUNCTIONAL EXPERIMENT

1. Build the zero-biased D MOSFET common-source amplifier circuit shown in Figure 34.1. Set the DC power supply for a V_{DD} of 20 V.

2. Measure and record the following DC parameters:

   V_G = __________ V_D = ______________ I_D = ____________

   Since $V_{GS} = 0$, it follows that $I_D = I_{DSS}$

3. Connect your function generator to provide a 500 $mV_{p\text{-}p}$ sinewave input signal at 1 kHz to your circuit input. Connect the oscilloscope channel 1 to the input signal at C, and channel 2 to the circuit output at C.

   Measure and record the peak-to-peak values of the input signal and output signal and the phase difference between the input and output.

   V_{in} = ______________ V_{out} = ____________

   Phase difference = ___________

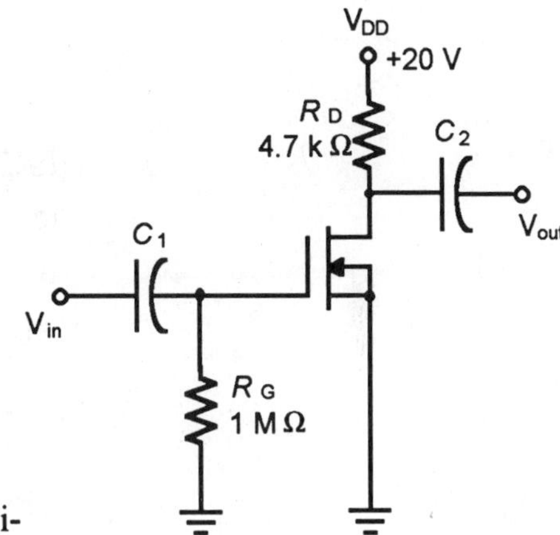

Figure 34.1

4. From your data of step 3, calculate the gain of your MOSFET amplifier.

   A_V = 

5. Observe carefully the amplifier input and output waveforms. The output should be a close replica of the input signal—a uniform sinusoidal waveform.

6. With your oscilloscope connected as in step 3, increase the input signal amplitude to 2.5 $V_{p\text{-}p}$. If this causes output signal clipping, reduce the input to obtain as large a signal as possible without clipping. Examine the output waveform. Is it still a uniform, undistorted sine wave? Which part of the output is exaggerated—the positive or negative portion?

7. Step 6 illustrates the square law distortion of an FET device. The distortion is not readily apparent when operating a small signal, but it is more clearly observed in large signal operation.

SECTION II TROUBLESHOOTING

Fault 1 - Gate resistor open

1. Turn off circuit power. Disconnect the ground end of the 1-MΩ gate resistor. Disconnect the function generator.

2. Apply DC power. Measure and record the DC values of I_D and V_D.

   I_D = ______________ V_D = ____________

Use a second DMM; connect it to the gate of the MOSFET. Leave this meter connected for all of the following measurements. Measure and record V_G, V_D, and I_D.

V_G = __________ V_D = ______________ I_D = ______________

3. With R_G still disconnected, disconnect the DMMs. Connect your function generator to supply a V_{in} signal to 500 mV_{P-P} at 1 kHz. Connect your oscilloscope as you did in Section I, step 3, and repeat the measurements of step 3.

   V_{in} = ______________ V_{out} = ______________

   Phase difference = ___________

DISCUSSION

Section I

1. The MOSFET common source amplifier is another alternative to the BJT devices. Discuss the possible advantages of the MOSFET over the JFET and BJT amplifiers.

2. Refer to your data observations of Procedure steps 5 and 6 and the 2N3796 drain characteristic curves. Discuss the effect of large signal distortion. Is this explained by the drain characteristic curves, and if so, how?

Section II

Fault 1 - Gate resistor open

1. You should have found that the data measured in Procedure step 2 were the same as in step 2 in Section I. Discuss this and indicate why you should have expected the same results.

2. Compare the data of step 4 and Section I step 3. Are these data sufficient to prove a failed gate resistor? Would the DC measurements of step 2 be sufficient?

   In your response, indicate what additional tests you would make, if necessary, to isolate this fault.

Quick Check

1. If g_m = 3500 μmhos and R_D = 4.7 kΩ, determine A_V.

   A_V = ___________

2. The output of the common-source amplifier is in phase with the input.

   True False

3. If R_G of the circuit of Figure 34.1 were open, only the MOSFET circuit DC parameters would be affected.

   True False

4. For the circuit of Figure 34.1, an enhancement MOSFET could be substituted for the 2N3796 with no change in AC parameters.

   True False

35

BJT SWITCHES

INTRODUCTION

Transistor switches can be useful as replacements for mechanical switches. Even though the transistor switch is not perfect, it still has advantages over the mechanical switch. A solid state switch eliminates the problems of moving elements and mechanical contacts, thereby providing reliability and greater operating speed.

This experiment will examine the transistor switch and some of the parameters that affect circuit performance. Switching speeds and delay times will be examined. You will also verify that the addition of a speed-up capacitor can improve switching speed. In the troubleshooting part of this experiment, you will simulate two faults in the transistor switch circuit and will measure the resulting circuit parameters.

REFERENCE

Principles of Electronic Devices and Circuits - Chapter 9, Sections 9.1 and 9.2

OBJECTIVES

After completing this experiment you will:

✓ Be able to determine whether a BJT is operating correctly as a switch

✓ Have experimentally verified the switching speed limitations of a transistor switch

✓ Know how to improve the switching speeds of a BJT switch using speed-up capacitors

✓ Be able to relate measured circuit parameters to failed components

EQUIPMENT AND MATERIALS

DC power supply
Dual-trace oscilloscope
Circuit protoboard
Digital multimeter

NPN transistor, 2N3904 or equivalent
Small-signal diode, 1N914 or similar
Resistors: 1.5 kΩ, 4.7 kΩ, 47 kΩ, 1 MΩ
47 pF capacitor

SECTION I FUNCTIONAL EXPERIMENT

1. Connect the circuit in Figure 35.1. Do not apply an input signal at this time.

2. Measure and record the following:

   V_B = __________ V_E = ______________ V_C = ____________

3. Voltage measurements should indicate that the transistor is in cutoff.

4. Apply +5 V to the base of Q_1. Measure and record the following:

   V_B = __________ V_E = ______________ V_C = ____________

5. Voltage measurements should indicate that the transistor is saturated. Based on these observations, would you say that Q_1 is behaving like a switch?

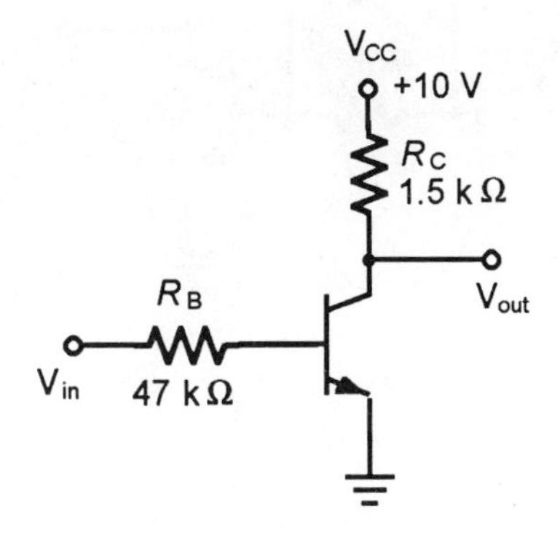

Figure 35.1

In the following Procedure steps you will measure the transistor switch response times to a square wave input. The input signal needs to be a square wave that transitions from 0 V to approximately +5 V and back. Your signal generator operating with a square wave output will not deliver the required waveform since its square wave output swings from a negative level to a positive level operating around a 0-V reference.

If your generator does not have a TTL output, you will need to feed the generator output through a negative shunt clipper (Experiment 10). See Figure 35.2.

6. Connect the signal input to your switch circuit. Set the generator for an output frequency of 100 kHz. The input signal levels at R_B should be close to +5 V and 0 V.

7. Connect a dual-trace oscilloscope channel 1 to the signal input at R_B and channel 2 to the switch output at the collector of the transistor. Apply DC power to the circuit. Adjust the oscilloscope to trigger on channel 1, and adjust the sweep timing to display 1 cycle of input signal.

   Note: You may need to select + or - trigger levels and use the sweep multiplier to obtain the measurements of the following Procedure steps.

   Your oscilloscope display should resemble the diagram of Figure 35.3.

   Your measurement of the transistor switch timing will be the measure of the times t_{on} and t_{off} as shown in Figure 35.3. This will simplify your measurements a bit, yet still serve to verify and illustrate the time delays in the switch circuit. Time t_{on} is the sum of t_d and t_r, and t_{off} is the sum of t_s and t_f. Refer to your text for descriptions of these BJT switch times.

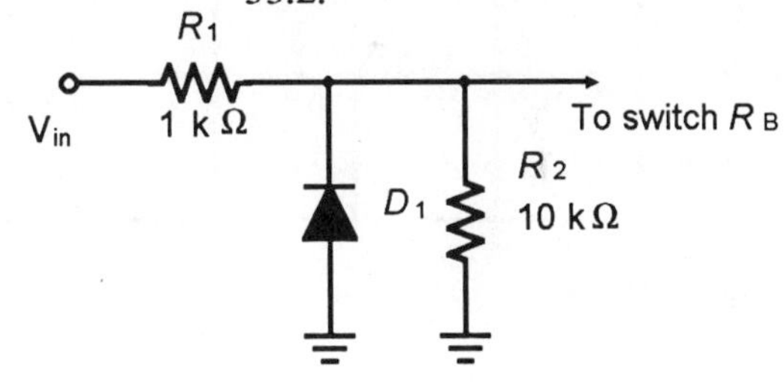

Figure 35.2

8. Measure and record below the times t_{on} and t_{off}.

   t_{on} = ________________ t_{off} = ________________

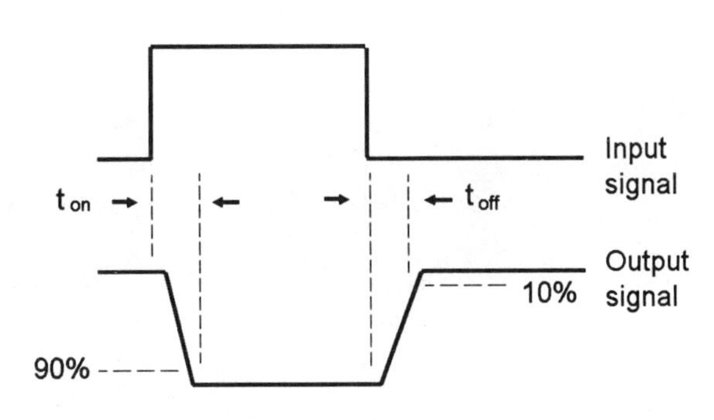

Figure 35-3

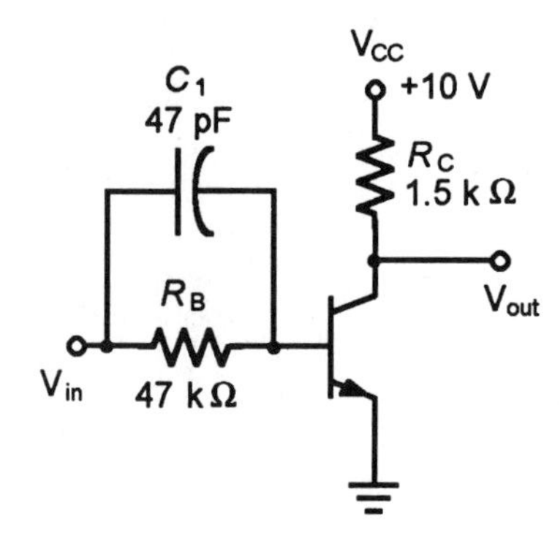

Figure 35.4

9. Turn off circuit power. Install a 47 pF speed-up capacitor in parallel with R_b. See Figure 35.4.

10. Reapply circuit power and repeat the t_{on} and t_{off} measurements of step 8. Record your measured values below.

    t_{on} = ______________________ t_{off} = ______________

11. Compare the output waveforms. What conclusion can you draw about adding a speed-up capacitor to the circuit?

    ____

    ____

SECTION II TROUBLESHOOTING

Fault 1 - Collector resistor open

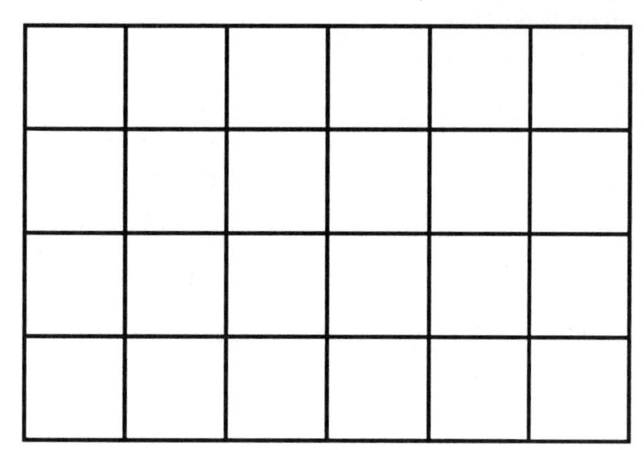

Graph 35.1

1. With circuit power off, and using the circuit of Figure 35.2, remove the 1.5-kΩ collector resistor and replace it with a 1-MΩ resistor.

2. Apply circuit power. Apply the same input signal you used in Section I, step 7. Connect your oscilloscope channel 1 to the input signal at R_6, and channel 2 to the output at the transistor collector. Set channel 2 to DC coupling. Sketch the output waveform in Graph 35.1, noting the collector DC level(s).

Fault 2 - Base-emitter junction leaky

1. With circuit power off, replace R_C with the correct 1.5-kΩ resistance. Connect a 4.7-kΩ resistor from the transistor base to emitter.

2. Apply circuit power and input signal. Using your oscilloscope set to DC coupling, measure the signal at the transistor base and the collector. Sketch the base waveform in Graph 35.2 and the collector waveform in Graph 35.3. Note the minimum and maximum DC signal levels.

   Remove the 4.7-kΩ B-E resistor and observe the signal at the base again. Compare the scope waveform with your prior sketch of the base waveform.

Base Waveform

Graph 35.2

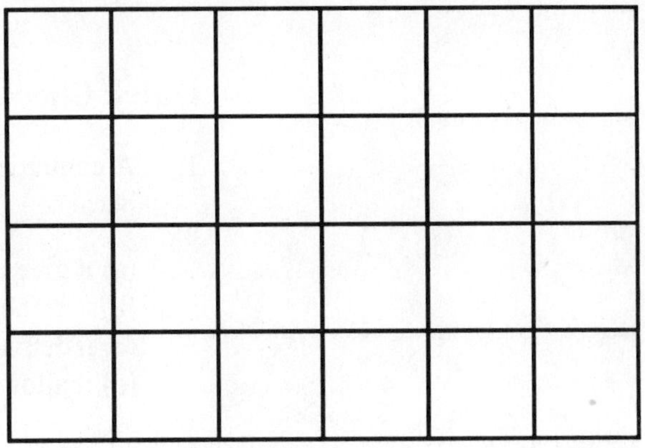

Collector Waveform

Graph 35.3

DISCUSSION

Section I

1. Describe the benefits obtained by using a speed-up capacitor in the transistor switch circuit. Consider and discuss switch timing and output waveform.
2. The 2N3904 transistor has a maximum I_C rating of 200 mA. With this limitation, what are some applications you can suggest for a BJT switch?

Section II

Fault 1 - Collector resistor open

Looking at your measured data for Fault 1, do you think the data are sufficient to prove the failure of the collector resistor?

If your answer is *yes*, explain how the data prove this conclusion. If your answer is *no*, cite the additional measurements you would make to prove the failed component.

Fault 2 - Base-emitter junction leaky

The measured data for this fault would point to a transistor failure. What additional measurements would you make to prove the transistor failure and the failure mode?

Quick Check

1. A common-emitter configuration is most often used in transistor switches because _____.

   (a) it gives current and voltage gain needed for switching
   (b) it is the most economical to build
   (c) it delivers slow transition time
   (d) it allows the use of a speed-up capacitor

2. A transistor cannot be an ideal switch because _____.

   (a) it can never be truly fully closed
   (b) it can never be truly fully open
   (c) it dissipates energy
   (d) all the above

3. One way of speeding up the switching rate of a transistor is _____.

   (a) keep it in deep saturation
   (b) do not let it fully reach saturation
   (c) do not let it reach full cutoff
   (d) keep it fully cut off at all times

4. A transistor switch without a speed-up capacitor _____.

   (a) operates faster than one with a capacitor
   (b) has base-emitter delay time
   (c) offers a short circuit path across R_B
   (d) reduces the heavy saturation of turn-on

5. Recovery time can best be described as _____.

   (a) propagation delay
   (b) a healing process for the transistor
   (c) the time required for the speed-up capacitor to discharge
   (d) the time required for a transistor to oscillate freely without damping

36

BJT SCHMITT TRIGGER

INTRODUCTION

In today's world of digital and industrial electronics, there is a requirement for circuitry that can serve to detect or provide a squaring function. Schmitt triggers can satisfy that requirement. They allow a slow-transition waveform to be converted to a quick transition pulse waveform.

In this experiment you will construct a BJT Schmitt trigger circuit and observe its operation. You will be observing non-square wave input signals and the resulting square wave output signals. Switching characteristics such as upper and lower threshold points will be calculated and verified.

In the troubleshooting portion of this experiment you will examine two of the functional limitations of the BJT Schmitt trigger.

REFERENCE

Principles of Electronic Devices and Circuits - Chapter 9, Section 9.3

OBJECTIVES

After completing this experiment, you will be able to:

✓ Build a BJT Schmitt trigger circuit

✓ Calculate and analyze upper and lower threshold voltage values

✓ Determine the hysteresis voltage of a BJT Schmitt trigger

✓ Experimentally determine and understand two of the functional limits of a discrete Schmitt trigger circuit

EQUIPMENT AND MATERIALS

DC power supply
Oscilloscope
Function generator
Circuit protoboard
NPN transistor [2], 2N3904 or equivalent
Resistors: 1.2 kΩ, 5.1 kΩ, 8.2 kΩ, 15 kΩ, 47 kΩ

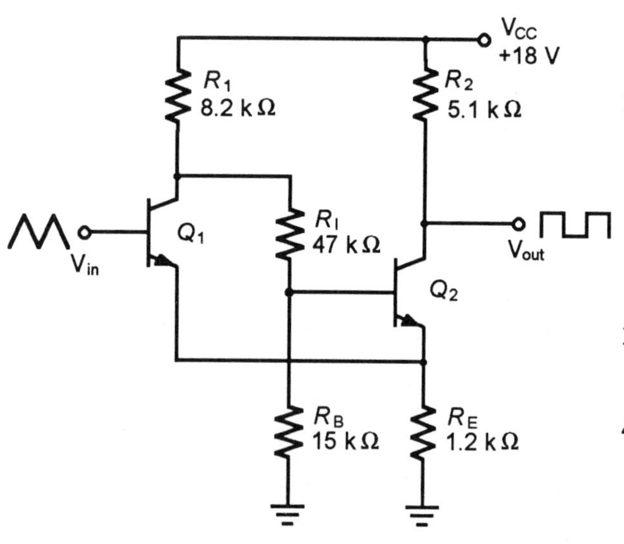

Figure 36.1

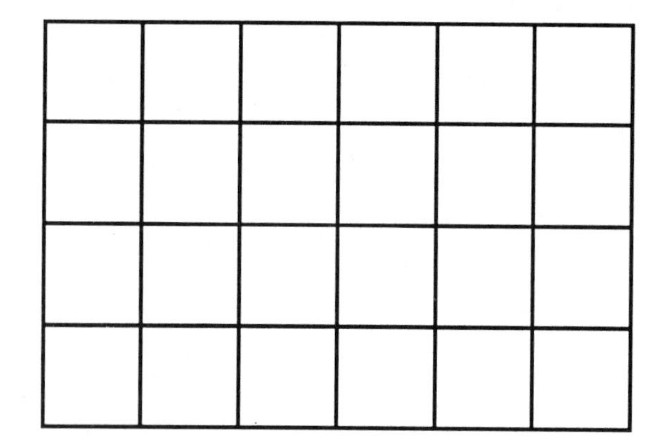

Graph 36.1

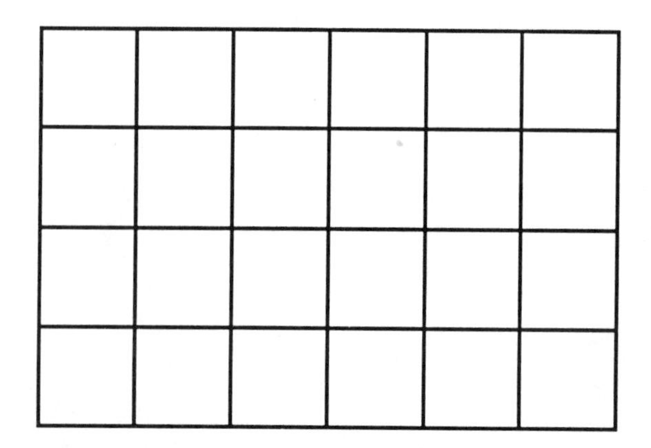

Graph 36.2

1. Construct the circuit of Figure 36.1. Do not apply any input signal.

2. Measure and record the following:

   V_{B1} __________ V_E __________

   V_{C1} __________ V_{B2} __________

   V_{E2} __________ V_{C2} __________

3. The measured values should indicate that Q_1 is in cutoff and that Q_2 is in saturation.

4. Calculate the upper and lower threshold voltages, using the followign equations:

$$UTP = \frac{V_{CC} \times R_b}{R_1 + R_i + R_b} = \_\_\_\_\_\_\_\_$$

$$LTP = \frac{V_{CC} \times R_E}{R_1 + R_E} = \_\_\_\_\_\_\_\_$$

5. The voltage level between the UTP and the LTP can be identified as the hysteresis. What is the hysteresis for this Schmitt trigger?

$$V_H = V_{utp} - V_{ltp} = \_\_\_\_\_\_\_\_$$

6. Apply a 10-$V_{p\text{-}p}$, 1-kHz triangular waveform to the base of Q_1. Monitor this waveform on channel 1 of the oscilloscope.

7. Monitor the output at the collector of Q_2 on channel 2 of the scope.

8. Sketch the waveforms in Graph 36.1.

9. Describe the effect the Schmitt trigger has on a triangular waveform input.

10. Change the input waveform to a 10-$V_{P\text{-}P}$, 1-kHz sine wave. Observe the output signal at the collector of Q_2.

11. Sketch the waveforms in Graph 36.2.

SECTION II TROUBLESHOOTING

1. With the function generator set to provide a 1-kHz triangular wave output at 10 $V_{p\text{-}p}$, decrease the input voltage level until the output signal disappears. Measure and record the maximum peak-to-peak input signal that won't produce an output signal.

   V_{in} = ___________ (no output)

2. Increase the input to 10 $V_{p\text{-}p}$. Begin increasing frequency until the output waveform distorts. Measure and record the lowest-frequency input signal that produces a distorted output signal.

   f = ___________

This may require increasing the frequency significantly.

DISCUSSION

Section I

1. Briefly give a functional description of the operation of the circuit of Figure 36.1. Be sure that your description includes the signal values at which the BJTs change state, and explain why.
2. Describe a household operation that either uses or alternatively would be improved by the use of a BJT Schmitt trigger circuit.

Section II

1. Discuss the operation of your schmitt trigger with low level signals. Consider the output waveform with a triangular wave input in the range of 3 to 5 V_p.
2. Describe the effect of increasing the input frequency observed in step 2.

Quick Check

1. Initially, a Schmitt trigger operates under the condition where _____.

   (a) both transistors are in cutoff
   (b) both transistors are in saturation
   (c) Q_1 is saturated and Q_2 is in cutoff
   (d) Q_1 is in cutoff and Q_2 is saturated

2. Schmitt triggers can be used as _____.

(a) linear amplifiers
(b) oscillators
(c) waveshaping circuits
(d) variable power sources

3. The upper threshold point is _____.

(a) usually higher than the LTP
(b) always higher than the LTP
(c) usually lower than the LTP
(d) always lower than the LTP

4. Hysteresis can be defined as _____.

(a) a lagging effect
(b) a difference between the two trip points
(c) neither (a) nor (b)
(d) either (a) or (b)

37

JFET SWITCHES

INTRODUCTION

The advantages of solid state devices over mechanical switches was pointed out in Experiment 35. The JFET, like the BJT, can switch load current on or off and provide output waveform transitions, although it is not necessarily the device of choice for these applications.

The JFET characteristic of having a variable resistance from source to drain that can conduct current in either direction, and change from a low resistance (on) to a large resistance (off), gives the JFET the capability of performing some switch functions not possible with the BJT.

In this experiment you will set up and examine a basic JFET switch and then apply the JFET in a chopper circuit to illustrate the special switch capability of an FET device.

Due to the circuit simplicity and your prior troubleshooting of the basic JFET, no troubleshooting is provided in this experiment.

REFERENCE

Principles of Electronic Devices and Circuits - Chapter 9, Section 9.4

OBJECTIVES

After completing this experiment, you will:

✓ Be able to construct a JFET switching circuit

✓ Be able to compare the frequency response of a JFET to that of a BJT switch

✓ Understand the operation of the JFET in the FET unique chopper application

EQUIPMENT AND MATERIALS

DC power supply
Function generator [2]
Oscilloscope
Circuit protoboard
N-channel JFET, 2N5459 or equivalent
Small-signal diode, 1N914 or similar
Resistors: 10 kΩ, 15 kΩ, 33 kΩ

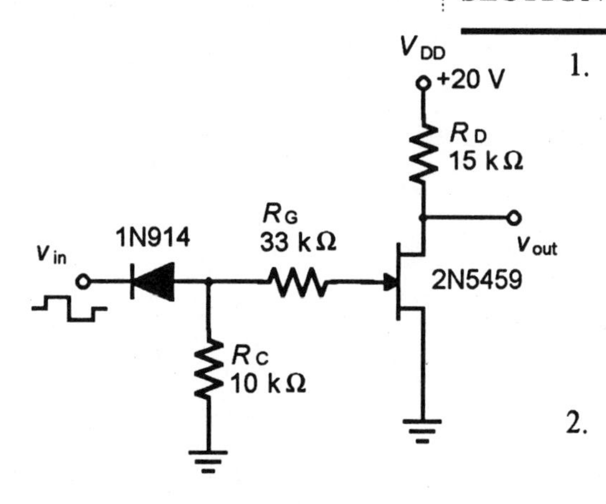

Figure 37.1

1. Construct the circuit of Figure 37.1.

Note: In controlling the JFET as a switch, the gate cannot be positive with respect to the source. The gate signal must change from 0 V to $-V_{GS(off)}$. You should recognize the positive series clipper between the signal source and the JFET gate. This will allow you to use a squarewave signal source that transitions from a positive to a negative value around a 0-V reference.

2. Apply DC power to the circuit. Set the function generator to supply a squarewave output at 10 kHz of 12 V_{p-p}. Connect your oscilloscope channel 1, DC coupled, to the JFET gate and channel 2 to the circuit output at the drain.

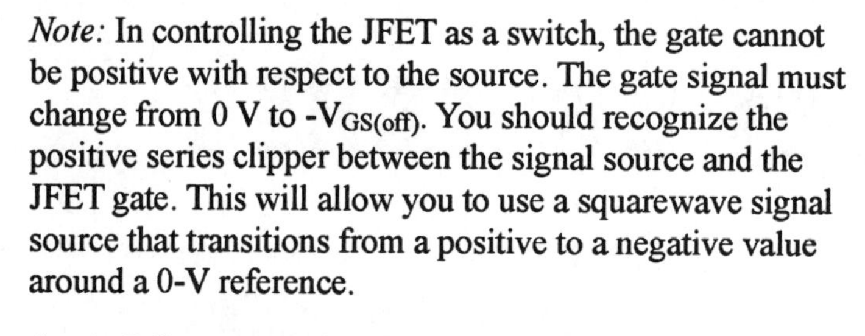

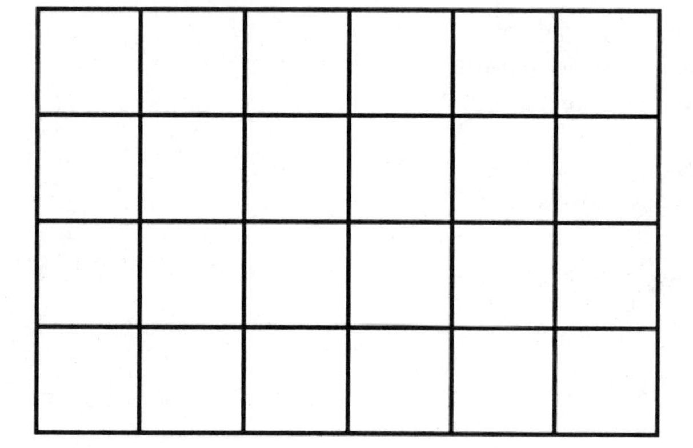

Graph 37.1

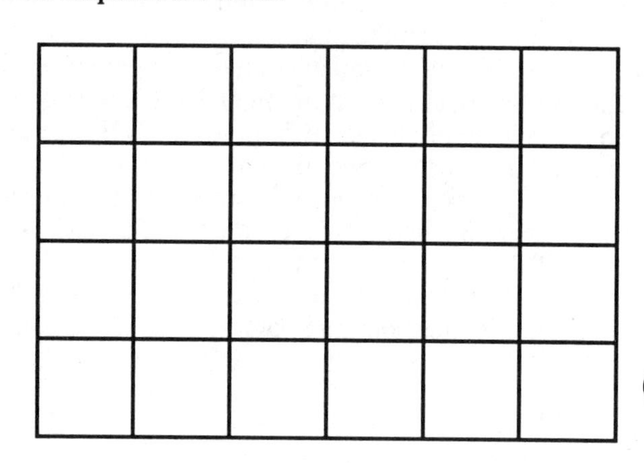

Graph 37.2

Sketch the input and output waveforms in Graph 37.1, noting the positive and negative peak values.

3. Increase input frequency to the point where the output has clearly degenerated from a square wave (approximately 100-150 kHz). Record the input frequency below; sketch the input and output waveform in Graph 37.2.

Freq. in = ___________

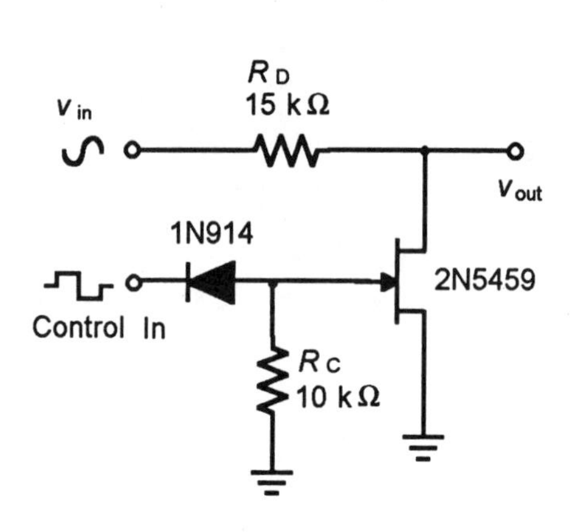

Figure 37.2

4. Remove circuit power and the signal generator. Modify your circuit to that of Figure 37.2.

5. Connect the control signal squarewave source set to provide a 12 V_{p-p} square wave at 10 kHz to the series clipper. Set the second generator to provide a sinewave of 1 V_{p-p} at 1 kHz. Connect this source to the 15-kΩ resistor leading to the drain of the JFET. Use your oscilloscope to monitor the chopper circuit output at the drain of the JFET. Adjust the scope to obtain approximately 1 cycle of the sinewave input.

Step 5 requires the use of two function generators: one to provide a sinewave signal and the second a squarewave control signal.

6. To see the effect of the JFET passing and stopping the input sine wave, change the control frequency to 1 kHz and the sinewave signal input to 10 kHz. Adjust the oscilloscope to sweep and sync at the squarewave frequency and you should observe a burst of 10-kHz signal followed by no signal and then a burst of the 10-kHz signal.

Important

These adjustments will likely require careful adjustment of the scope trigger control. Your oscilloscope display should resemble the signal shown in Figure 37.3.

The 1-kHz sine wave is chopped (or sampled) at a 10-kHz rate. The signal amplitude is zero when the JFET is switched on and the normal amplitude when the JFET is off.

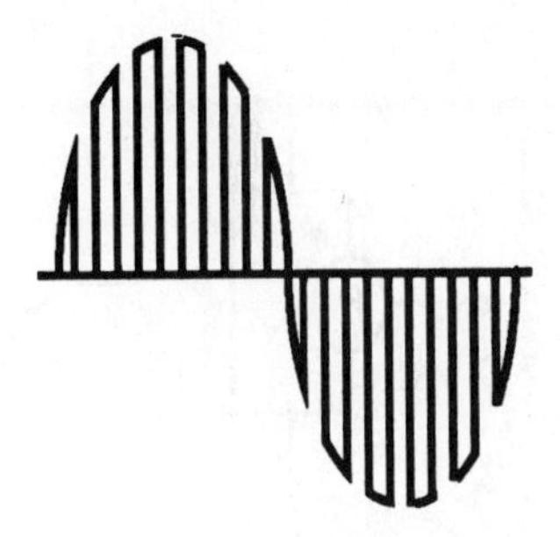

Figure 37.3

DISCUSSION

1. In Procedure steps 4 through 6, you observed the JFET perform a chopper function. Could you perform this function using a BJT? Discuss your answer and explain the JFET characteristics that make this function possible.

2. Compare your basic JFET switch data (steps 2 and 3) to your data of Experiment 35. Explain which device you would select, and why, to perform the basic switch function.

3. Notice that the circuit of Figure 37.2 resembles in some respects the shunt diode clipper. Compare and contrast the JFET chopper and the shunt diode clipper.

Quick Check

1. A JFET switch will perform better at higher frequencies than a BJT switch.

   True   False

2. A basic difference between a JFET and a BJT is that _____.

   (a) JFETs are far more expensive
   (b) JFETs are current-controlled and BJTs are voltage-controlled
   (c) JFETs are voltage-controlled and BJTs are current-controlled
   (d) JFETs are much larger than BJTs

3. The source, gate and drain terminals on a JFET compare to what terminals on a BJT?

   (a) base, emitter, and collector   (b) collector, base, and emitter
   (c) base, collector, and emitter   (d) emitter, base, and collector

4. One advantage JFET switches have over mechanical switches is that _____.

   (a) JFETs are slower operating   (b) JFETs are faster operating
   (c) JFETs have contact bounce   (d) JFETs will cause arcing

5. The JFET bilateral characteristics allow its use in switch applications not possible with a BJT.

   True   False

38

UJT RELAXATION OSCILLATORS

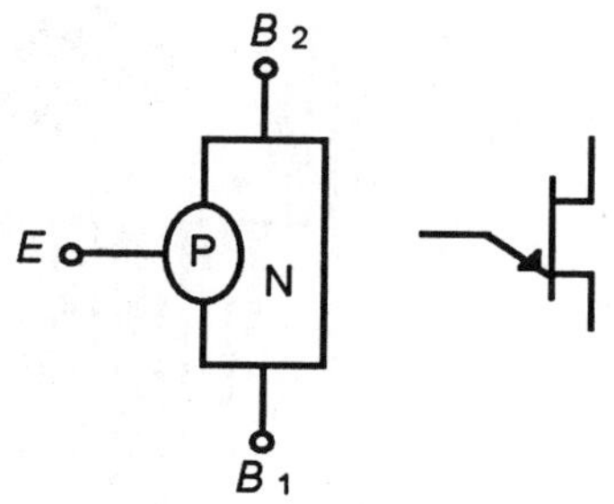

INTRODUCTION

The unijunction transistor (UJT) is a three-terminal device having only one PN junction. It is composed of a silicon bar with ohmic contacts at each end, designated as base 1 (B1) and base 2 (B2). Fused to the silicon bar is a heavily doped emitter (E). The basic construction of the UJT and the schematic symbol are shown in Figure 38.1. Figure 38.2 illustrates the emitter characteristic curve of a UJT. Note the area in which emitter current is increasing while emitter voltage is decreasing. This is called the *negative resistance region* and is a unique characteristic of the UJT.

In this experiment you will explore the UJT by making resistance measurements of the device. Then you will construct a UJT relaxation oscillator and observe and record circuit data.

The troubleshooting section will permit you to see the effect of a simulated fault on the circuit operation and circuit values.

Figure 38.1

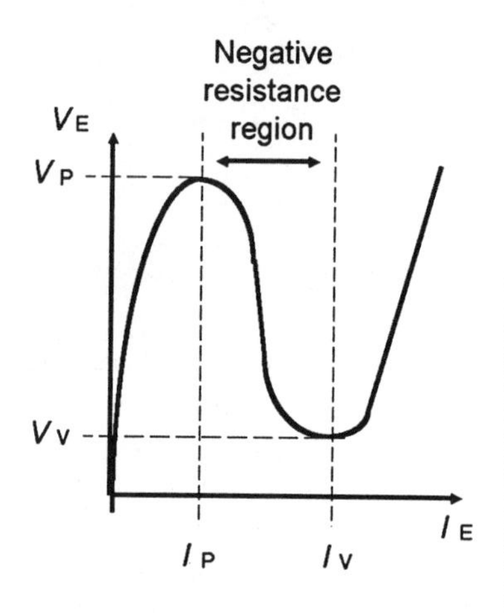

Figure 38.2

REFERENCE

Principles of Electronic Devices and Circuits - Chapter 9, Section 9.5

OBJECTIVES

Upon completing this experiment, you will be able to:

✓ Measure R_{BB}

✓ Determine the intrinsic standoff ratio (η)

✓ Determine the output frequency of a UJT relaxation oscillator

✓ Determine the effect of changing the timing component values on output frequency

EQUIPMENT AND MATERIALS

DC power supply
Oscilloscope
Digital multimeter
Circuit protoboard

UJT, 2N2646 or equivalent
Resistors: 47 Ω, 1 kΩ, 12 kΩ, 47 kΩ
Capacitors: 0.01 µF, 0.047 µF

SECTION I FUNCTIONAL EXPERIMENT

From the measured values of step 3 and Table 38.1, you can determine that there is a certain amount of interbase resistance and that a PN junction exists between the emitter and B1 and the emitter and B2.

1. Identify the B1, B2, and emitter leads on the UJT.
2. Set the meter to the OHMS function, 10-k range.
3. Measure and record the values in Table 38.1.

Positive Ohmmeter Lead to:	Negative Ohmmeter Lead to:	Resistance Reading
Base 2	Base 1	
Base 1	Base 2	
Emitter	Base 1	
Base 1	Emitter	
Emitter	Base 2	
Base 2	Emitter	

Table 38.1

4. Connect the circuit shown in Figure 38.3. The component values should be $R_1 = 12\ k\Omega$, $R_2 = 1\ k\Omega$, $R_3 = 47\ \Omega$, and $C_1 = 0.047\ \mu F$.

5. Using the oscilloscope, measure the peak voltage at the emitter.

   V_p = ____________

   Describe the waveform. ________________________________

   ____

6. Using the formula $\eta = V_p/V_{BB}$, determine the intrinsic standoff ratio. Once the intrinsic standoff ratio is found, this will enable you to calculate V_p for any value of V_{BB}.

   η = ____________

7. Calculate the period (T) of the output waveform using the formula $T = R_1 \times C_1$, and record this in Table 38.2. Do this for all indicated values of R_1 and C_1.

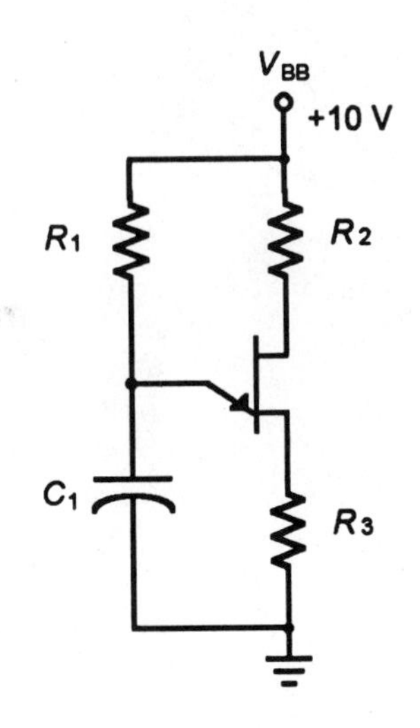

Figure 38.3

R_1 Value	C_1 Value	Calculated T	Calculated f	Measured f	Measured T
12 kΩ	0.047 μF				
12 kΩ	0.01 μF				
47 kΩ	0.01 μF				
47 kΩ	0.047 μF				

Table 38.2

8. Calculate the frequency (f) of the output waveform and record this in Table 38.2. Do this for all values of R_1 and C_1.

9. Using the indicated values and oscilloscope, measure and record the period and frequency in Table 38.2.

SECTION II TROUBLESHOOTING

1. Remove C_1 from the circuit. Use the oscilloscope to measure the period of the emitter waveform.

   T = ___________

2. Compare this value to that recorded in Table 38.2.

3. Replace C_1, and lightly touch the emitter lead with your finger. Record your observation.

DISCUSSION

Section I

1. Why is the region of the characteristic curve between V_p and V_v called *negative resistance*?

2. The intrinsic standoff ratio for a 2N2646 UJT is listed as a range from 0.56 to 0.75. Explain how this range of values might come about.

Section II

Stray capacitance can have adverse effects on the operation of electronic circuits. Explain how it affected the UJT relaxation oscillator in this experiment.

Quick Check

1. Assuming a value of $\eta = 0.60$, calculate the values of V_p for the values of V_{BB} listed below:

   $V_{BB} = 16$ V, V_p = ___________

   $V_{BB} = 5$ V, V_p = ___________

2. In a UJT relaxation oscillator, if the value of capacitance doubles, the output frequency _____.

(a) doubles
(b) halves
(c) remains the same
(d) fluctuates wildly

3. What effect does increasing V_{BB} have on the frequency of a UJT relaxation oscillator?

(a) Increases frequency
(b) Decreases frequency
(c) No effect on frequency
(d) Stops oscillation

4. In reference to the characteristic curve, which voltage causes the UJT to turn off?

5. Why is a UJT not likely to be used in an audio amplifier circuit?

39

BJT RAMP GENERATORS

INTRODUCTION

One of the easiest ways to generate a ramp is with a bipolar junction transistor in parallel with a capacitor. When the transistor is turned off, the capacitor will charge and will discharge during the time the transistor is turned on. This will allow a slow charge time but will give a very rapid discharge time. The output ramp will look like a curved sawtooth waveform.

You will construct a BJT ramp generator in the first part of this experiment. Then you will observe and measure the circuit values to aid your understanding of the circuit operation.

In the troubleshooting section, you will simulate circuit faults and make measurements to see the effect of the fault on the circuit values.

REFERENCE

Principles of Electronic Devices and Circuits - Chapter 9, Section 9.6

OBJECTIVES

In this experiment you will:

✓ Add to your understanding of generating a ramp voltage using a transistor in parallel with a capacitor

✓ Verify the output of a ramp generator utilizing the switching of a transistor to provide the charge and discharge path for a capacitor

✓ Understand how to generate a constant ramp using a sinewave input

✓ Draw the output of a ramp generator

EQUIPMENT AND MATERIALS

DC power supply
Dual-trace oscilloscope
Function generator
Circuit protoboard

NPN transistor, 2N3904 or equivalent
Resistors: 330 Ω [2], 47 kΩ, 1 MΩ
Capacitors: 0.01 μF, 0.033 μF

SECTION I FUNCTIONAL EXPERIMENT

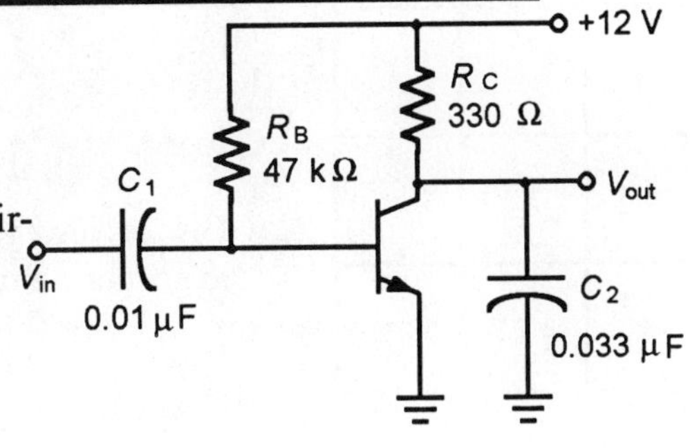

Figure 39.1

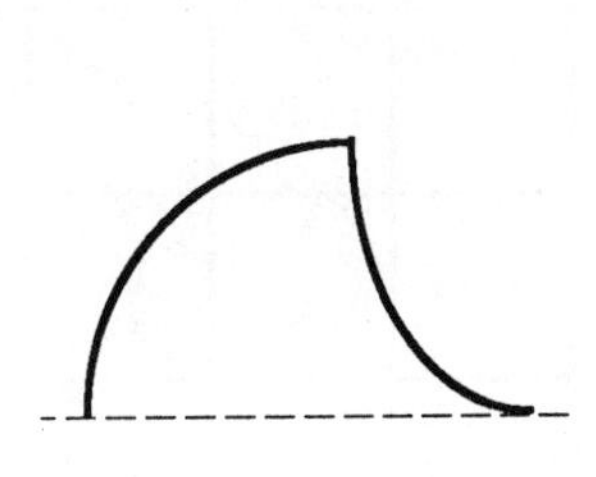

Figure 39.2

1. Build the circuit as shown in Figure 39.1. Adjust the DC power supply to +12 V.
2. Connect the function generator to the circuit input. Adjust the function generator for a sine wave of 8.5 kHz at 3.5 V_P. Be sure to note that this is peak voltage. Use your oscilloscope to set up this voltage.
3. Adjust your oscilloscope channel 2 for 1 volt/division.
4. Connect channel 2 of the oscilloscope to the circuit output. Adjust the sweep/division to display one complete waveform. (See Figure 39.2.)

   Sketch the input and output waveforms in Graph 39.1.
5. Adjust the frequency generator toward 9 kHz while observing the waveform. You should observe that the signal ramp does not peak out as high as it did previously.

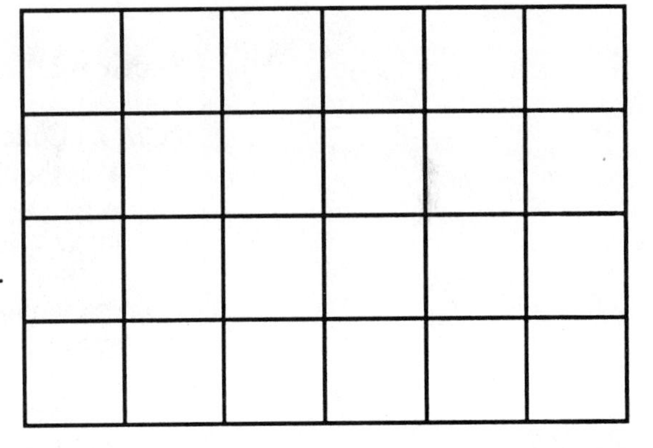

Graph 39.1

SECTION II TROUBLESHOOTING

Fault 1 - Collector resistor open

1. With circuit power off, remove the 330-Ω collector resistor and replace it with a 1-MΩ resistor. Apply circuit DC power and input signal.
2. Set your oscilloscope channel 2 to DC coupling. Use channel 1 to monitor the input signal and channel 2 at the circuit output. Sketch the input and output waveforms in Graph 39.2, and note on the sketch the peak value of each waveform.

Fault 2 - Output capacitor leaky

1. Turn off the circuit power. Replace the collector resistor with the correct 330-Ω resistor. Connect a second 330-Ω resistor in parallel with the output capacitor, C_2. Reapply circuit power and the input signal.
2. With your oscilloscope connected in the same way as in Fault 1, observe and sketch the waveforms in Graph 39.3 (next page), and note the peak signal values.

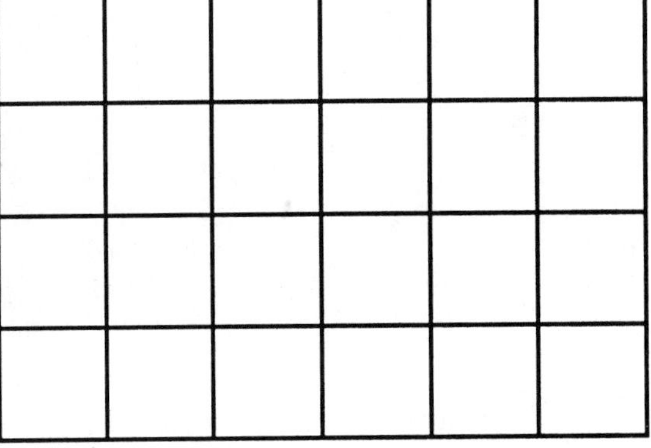

Graph 39.2

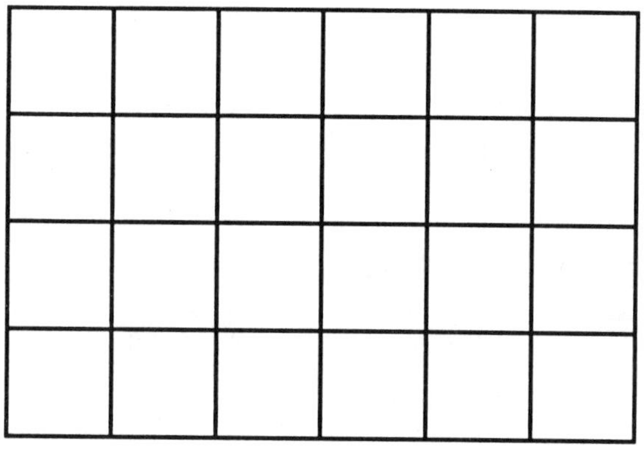

Graph 39.3

DISCUSSION

Section I

1. Discuss how you would use a pulse generator in this circuit to generate a ramp.
2. Discuss what effect the incoming frequency has on the ramp. How does the incoming signal cause the waveform. Discuss the turning off and on of the transistor.

Section II

Fault 1 - Collector resistor open

You should have observed a very low ramp output waveform. Could this result be caused by the BJT? Discuss with your answer the additional measurements you would make to isolate the failed component.

Fault 2 - Output capacitor leaky

This fault assumed a very leaky, almost shorted, capacitor.

a. Do you feel that the output waveform and amplitude pointed to the capacitor being defective?

b. If the capacitor had a very small leakage (equivalent to 10 kΩ), would the fault have been as easy to diagnose? What do you think the output waveform would have looked like in this case?

Quick Check

1. The transistor is parallel with the capacitor to furnish a charge path only.

   True False

2. When the transistor is turned off, the capacitor charges through the resistor.

   True False

3. The negative portion of the sine wave or a negative pulse applied to the base of the transistor will cause a constant ramp output.

   True False

4. The negative voltage at the base of the transistor biases the transistor into cutoff.

   True False

40

FREQUENCY EFFECTS IN BJT AMPLIFIERS

INTRODUCTION

To this point in your transistor amplifiers, it has been assumed that capacitances in the amplifier were always a short circuit to AC and an open circuit to DC. In this experiment you will now see the practical limits that circuit capacitances impose on the amplifier in limiting both low and high frequency responses.

You will measure both the low and high corner (critical) frequencies for a typical common emitter amplifier. You will also, from your measured data, make a Bode plot for your amplifier and, through circuit calculation, identify the capacitors that control the frequency limits for your amplifier.

In the troubleshooting section, you will simulate circuit faults and through your measurements, see the effect on the amplifier AC and DC parameters.

REFERENCE

Principles of Electronic Devices and Circuits - Chapter 10, Section 10.3

OBJECTIVES

In this experiment you will:

✓ Add to your understanding of determining the bandwidth of a small-signal BJT amplifier

✓ Verify the effect of the coupling and bypass capacitors on the amplifier frequency limits

✓ Calculate and draw the Bode plot of a small-signal amplifier

EQUIPMENT AND MATERIALS

DC power supply
Digital multimeter
Dual-trace oscilloscope
Function generator
Circuit protoboard
NPN transistor, 2N3904 or equivalent
Resistors: 1 kΩ, 2.7 kΩ [2], 3.3 kΩ, 12 kΩ
Capacitors: 0.33 µF, 10 µF, 22 µF

SECTION I FUNCTIONAL EXPERIMENT

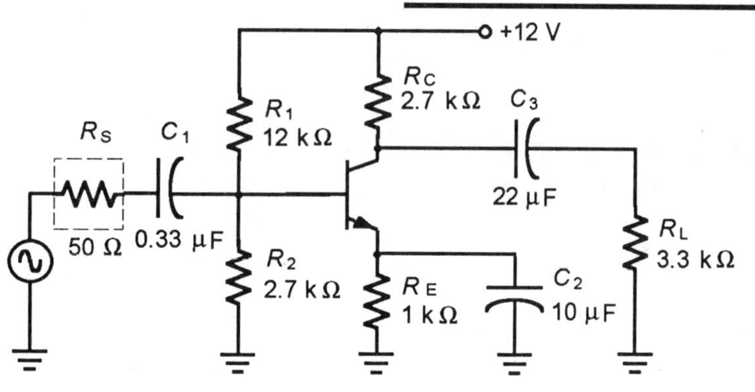

Figure 40.1

1. Build the circuit as shown in Figure 40.1. Connect the DC power supply set to +12 V to your circuit.

2. Measure the DC voltages of your circuit with a DMM and record the voltages below.

   V_B = ________

   V_E = ________

   V_C = ________

3. Using your measured values, calculate the expected amplifier gain.

   A_v = ____________

4. The data sheet for the 2N3904 transistor shows that the collector junction capacitance (C_{OB}) has a value of 4 pF. With this information you should be able to calculate your Miller capacitance values.

   A_v = ____________

   C_{in} (Miller) = __________ C_{out} (Miller) = ________

5. Calculate the critical frequencies for the following networks:

   Critical low frequency for the bypass capacitor = ____________

   Critical frequency for the input terminal = ____________

   Critical frequency for the input shunt network = ____________

6. Adjust the function generator to a midpoint between your two calculated dominant critical frequencies. Set the input signal to obtain an output of approximately 3 $V_{p\text{-}p}$, measured with your oscilloscope.

7. Measure and record the mid-band peak output voltage as V_{mid} in Table 40.1.

   Remember that the critical frequency is 3 dB down from the peak or at 0.707 times the peak voltage.

8. Adjust the frequency generator up in frequency until the voltage level drops to the calculated critical value. This is the upper critical frequency. Record this frequency in the *High Band* section of Table 40.1.

9. Increase the generator frequency to obtain the frequencies for values of $0.5V_{mid}$ and $0.25V_{mid}$. Also record these frequencies in the *High Band* section of Table 40.1.

10. Decrease the generator frequency to find the lower critical frequency point ($V_{out} = 0.707V_{mid}$). Record this frequency in the *Low Band* section of Table 40.1.

11. Decrease the generator frequency to obtain the frequencies for the $0.5V_{mid}$ and $0.25V_{mid}$ values. Record these frequencies in the *Low Band* section of Table 40.1.

12. Calculate the amplifier bandwidth and record in Table 40.1.

13. Plot in Graph 40.1 the lower band and upper band response of your amplifier using the data of Table 40.1.

V_{mid} = ____________

Low Band		High Band	
V_{out}	freq.	V_{out}	freq.
$0.707V_{mid}$		$0.707V_{mid}$	
$0.5V_{mid}$		$0.5V_{mid}$	
$0.25V_{mid}$		$0.25V_{mid}$	

BW = ____________

Table 40.1

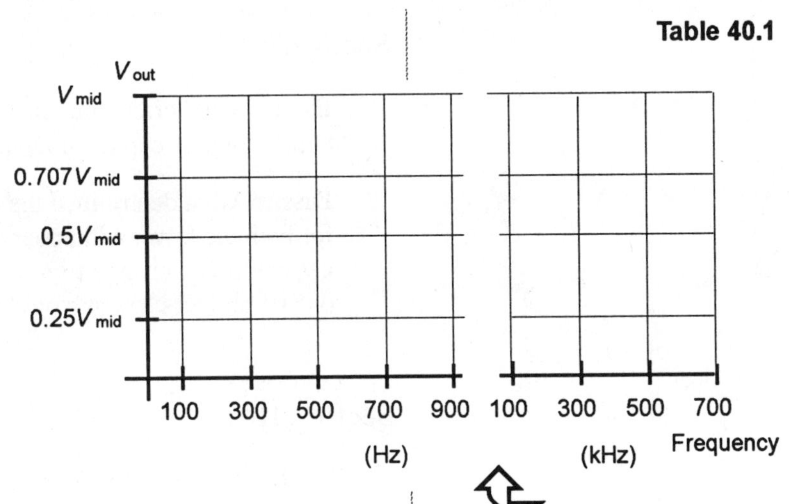

Graph 40.1

Note: The graph omits the midband frequencies to permit you to see the upper and lower frequency response areas more clearly.

SECTION II TROUBLESHOOTING

Fault 1 - Emitter bypass capacitor shorted

1. Ensure that the circuit power is off. Place a jumper wire in parallel with the emitter resistor and capacitor.

2. Apply DC power. Adjust your function generator to a frequency in the amplifier midband. Observe the output signal wave form and measure the peak-peak amplitude. The amplitude and waveshape of the output signal should give a good clue to the component fault.

   Disconnect the function generator and make any additional DC voltage checks you would like. Record the measurement point and data below.

   __________ ______________________________

   __________ ______________________________

   __________ ______________________________

Fault 2 - Input capacitor shorted

1. With circuit power off, remove the jumper shorting the emitter resistor. Place a jumper across C_1, the input capacitor.

2. Apply circuit power. Measure the amplifier mid-band gain and observe the output waveform. Is the signal saturation or cutoff clipping? Disconnect the function generator and make any DC voltage measurements you would like. Record measurement points and values in the blanks provided (next page).

DISCUSSION

Section I

1. Discuss what effect the input and output coupling capacitors and the emitter bypass capacitor have on the output frequency response.

2. Discuss what determined the dominant frequency in your BJT amplifier for both the lower and upper critical frequencies. What component values could you change to increase the amplifier bandwidth? Also, what one parameter change would increase the high frequency response?

Section II

Fault 1 - Emitter bypass capacitor shorted

Discuss the effect of this fault on the amplifier's gain and bias point shift. Is there any other component failure that could have the same effect on these two parameters?

Fault 2 - Input capacitor shorted

Did the DC measurements for this fault indicate any circuit problem? Discuss this with your idea of why the circuit behaved the way it did.

Quick Check

1. The coupling capacitors affect the lower critical frequency points for the small signal amplifier.

   True　　　　　　　　　　False

2. Which of the bias resistors will have the most effect on the overall gain of the amplifier? ____________

3. The output shunt network is always the most dominant network in a small-signal amplifier.

   True　　　　　　　　　　False

4. Bandwidth is the upper critical frequency minus the lower critical frequency.

   True　　　　　　　　　　False

41

FREQUENCY EFFECTS IN JFET AMPLIFIERS

INTRODUCTION

Practical JFET amplifiers have upper and lower frequency limitations on the signals they amplify. The frequency-limiting components in the circuit are the coupling and bypass capacitors and the capacitance in the JFET.

In this experiment you will calculate these values f_1 and f_2 and measure and record them. These measurements are a very useful check for amplifiers. In the troubleshooting section, you will simulate faults that will affect the corner frequencies and, through circuit measurements, you will see how the amplifier is affected.

In Section II you will observe the effects on our circuit if the output coupling capacitor and the JFET internal capacitors change values.

REFERENCE

Principles of Electronic Devices and Circuits - Chapter 10, Section 10.3

OBJECTIVES

In this experiment you will:

✓ Understand how circuit capacitors affect low frequency response

✓ Understand how JFET internal capacitance affects high frequency response

✓ Understand how the capacitors' changing values affect the circuit

EQUIPMENT AND MATERIALS

DC power supply
Function generator
Dual-trace oscilloscope
Circuit protoboard
N-channel JFET, MPF102 or equivalent
Resistors: 680 Ω, 3.3 kΩ, 40 kΩ, 1 MΩ
Capacitors: 1000 pF [3], 1 μF [2], 10 μF, 47 μF

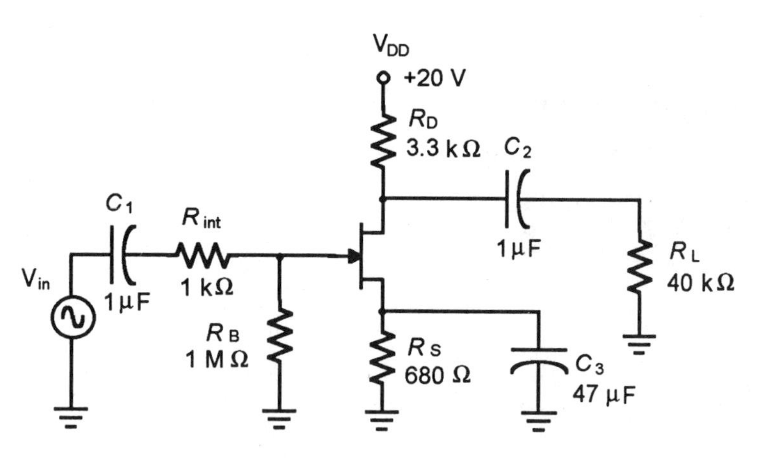

Figure 41.1

	Measured	Calculated
f_1		
f_2		
BW		
	$V_{in}(peak)$ = ________	
	$V_{out}(peak)$ = ________	

Table 41.1

1. Build the circuit of Figure 41.1 and apply DC power.

2. Connect one channel of your oscilloscope to the output of the signal generator, at the gate of the JFET, and the other channel to the amplifier across the load resistor. Set the signal generator to provide 200 $mV_{p\text{-}p}$ at 5 kHz. Measure and record in Table 41.1 the peak input and output signals.

3. Very slowly adjust the frequency down toward 0 Hz. When the v_{out} level changes to 0.707 of the peak V_{out} value (step 2), record the input frequency in Table 41.1. (*Hint:* During this procedure keep channel 1 V_{in} at 200 mV at all times.)

4. Adjust the function generator up to about 1 MHz. Slowly adjust upward until the V_{out} again falls to 0.707 times the peak V_{out} value (step 2).

5. Using the data of steps 3 and 4, calculate the bandwidth of the amplifier and record in Table 41.1.

6. For the circuit of Figure 41.1, calculate the low frequency cutoff point of the output coupling and bypass capacitors.

   Cutoff frequencies:

   Input = ________ Output = ________ Bypass = ________

7. Calculate and record in Table 41.1 the dominant low frequency cutoff. Remember, whichever capacitor cuts off the frequency at the highest point is the dominant low frequency cutoff.

8. To calculate the high dominant frequency cutoff, you calculate the effect of the internal capacitance of the JFET. Calculate and record the value of f_2 in Table 41.1.

9. Calculate the bandwidth of the amplifier, and record in Table 41.1. Compare your calculated data to the measured data.

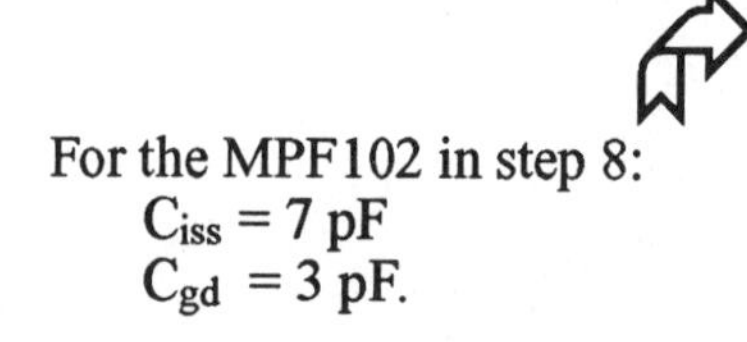

For the MPF102 in step 8:
C_{iss} = 7 pF
C_{gd} = 3 pF.

SECTION II TROUBLESHOOTING

Fault 1 - Drain bypass capacitor changes value

1. Remove the 47-μF source bypass capacitor and replace it with a 10-μF capacitor.
2. Calculate the new low frequency cutoff point: f_1 = ___________
3. Adjust the function generator to 5 kHz, and slowly adjust the input frequency down until V_{out} falls to 0.707 of the maximum V_{out}. Record this frequency: f_1 = ___________

Fault 2 - Incorrect transistor

1. For this portion of your troubleshooting, you will simulate an incorrect JFET transistor. Place three 1000-pF capacitors in your circuit, bypassing the JFET, as shown in Figure 41.2.
2. Apply DC circuit power.
3. In a real circuit this type of problem would be apparent only by a distorted output at high frequencies. As a technician you would have to check the high frequency cutoff experimentally as you have done previously.
4. Adjust the function generator again for 200 mV at 10 kHz, and slowly increase the input frequency. Measure and record the frequency where V_{out} is 0.707 times V_{out} maximum. This new frequency is:

   f_2 = ___________

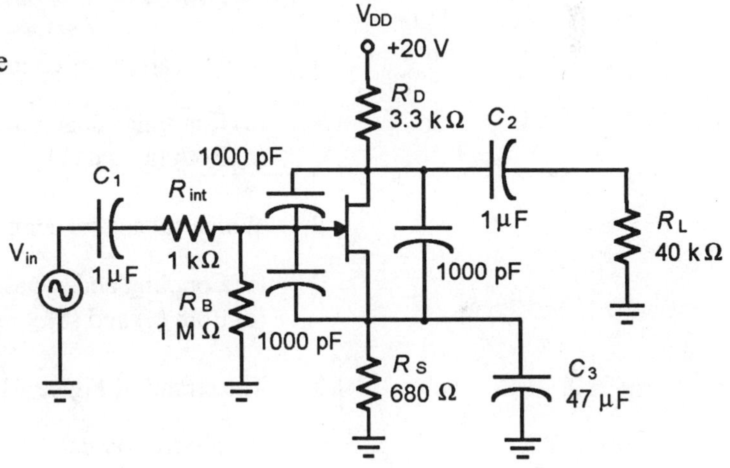

Figure 41.2

DISCUSSION

Section I

1. If the amplifier of Figure 41.1 were part of a music system amplifier, and if the lower corner frequency shifted up to 600 Hz, describe the effect this would have on the system.
2. If the amplifier of Figure 41.1 were used as part of a public address system used for voice announcements, and if the upper corner frequency shifted down to 1400 Hz, would this system be affected as much as would a stereo music amplifier? Explain why you chose your answer.
3. Today, many companies use automatic test equipment (ATE) to check the frequency response of amplifiers. ATE systems are usually controlled by

computerized test generators, DVMs, and oscilloscopes. These systems can make the frequency checks you have just made in this experiment in about one minute or less. In addition to speed, can you think of another advantage of ATE systems?

Section II

Fault 1 - Bypass capacitor changes value

In Fault 1 you discovered that a capacitor changing values can make a large difference in the amplifier's low frequency cutoff. Do you think this happens very often?

Fault 2 - Incorrect transistor

In Fault 2 you simulated a different value JFET in the amplifier. Do you think it is possible for this to happen?

Quick Check

1. Which capacitors control the low frequency response of a JFET amplifier?

   (a) Coupling capacitors (b) JFET capacitances
   (c) Both (a) and (b)

2. Which capacitors control the high frequency response of a JFET amplifier?

   (a) Coupling and bypass capacitors (b) JFET capacitances
   (c) Both (a) and (b)

3. The circuit of Figure 41.1 uses _____ bias.

   (a) current source (b) self
   (c) voltage divider (d) gate

4. The amplifier of Figure 41.1 is a common drain amplifier.

   True False

5. The JFET of the amplifier is an N-channel device.

   True False

42

BJT DIFFERENTIAL AMPLIFIERS

INTRODUCTION

The differential amplifier is a special amplifier that amplifies only the difference in two input voltages. This characteristic, amplification of a signal common to both inputs (common-mode rejection), is an important consideration in industrial electronics. The differential amplifier is used in instrumentation and operational amplifiers.

In this experiment, you will perform tests to determine the differential voltage gain (A_V), the common-mode voltage gain (A_{cm}), and the common-mode rejection ratio (CMRR).

In Section II of this experiment, you will observe the effects on the V_{out} of open and shorted transistors.

REFERENCE

Principles of Electronic Devices and Circuits - Chapter 10, Section 10.4

OBJECTIVES

In this experiment you will:

✓ Understand differential amplifier operation

✓ Determine common-mode A_V and CMRR for a differential amplifier

✓ Observe the effects of open and shorted transistors on the operation of a differential amplifier

EQUIPMENT AND MATERIALS

DC power supply
Function generator
Digital multimeter
Oscilloscope
Circuit protoboard
NPN transistor [2], 2N3904 or equivalent
Resistors: 100 Ω [2], 2.7 kΩ [2], 6.8 kΩ [3], 100 kΩ
Potentiometer: 10-kΩ ten-turn trimpot [2]

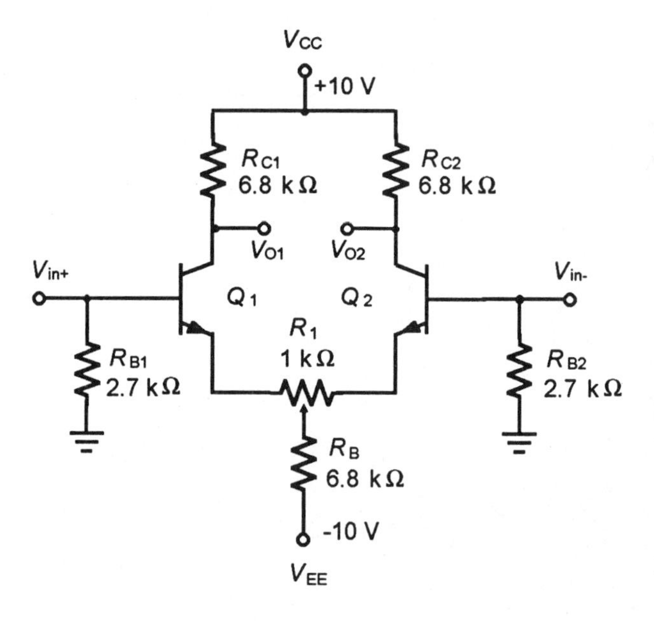

Figure 42.1

Output offset voltage:	V_{OO} = __________
Input offset voltage:	V_{IO} = __________
Differential input:	V_{ID} = __________
Differential output:	V_{OD} = __________
Differential gain:	V_D = __________
Common-mode input:	V_{ICM} = __________
Common-mode output:	V_{OCM} = __________
Common-mode gain:	A_{VCM} = __________
Common-mode rejection ratio:	CMRR = __________

Table 42.1

+10 V
R_1
100 kΩ
1 kΩ
T V_{in+}

Figure 42.2

1. Construct the circuit of Figure 42.1. Do not apply DC power to the circuit yet. Potentiometer R_1 is used to balance the amplifier by compensating for component and transistor differences in each half of the amplifier. Set the potentiometer to its mid-scale value so that each transistor sees the same value of resistance. Connect a 100-Ω resistor in parallel with the 2.7-kΩ base resistors (one for each base).

2. Apply DC power to your circuit. Connect a voltmeter between the two transistor outputs to read the differential voltage of the two transistor outputs.

   This voltage is the output offset voltage of the amplifier (V_{OD}).

   Record this voltage value in Table 42.1. You will use it later to determine the input offset voltage (V_{ID}).

3. Disconnect and remove the 100-Ω resistors installed in step 1. With the voltmeter still connected to read the amplifier differential output, adjust R_1 for a voltmeter output of zero volts. This compensates for the amplifier output and corrects it to zero.

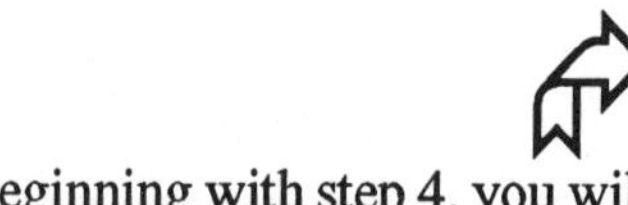

Beginning with step 4, you will measure the differential gain of your amplifier. Your input signal will be connected to the V_{in+} input. With no signal at the V_{in-} input the differential input signal is that of the V_{in+} input.

4. Connect the voltage divider of Figure 42.2. Set the potentiometer so that the wiper is at the ground end (zero output). Connect the wiper connection to the amplifier V_{in+} input, and adjust the potentiometer for a DC output of 15 mV. Record this value in Table 42.1 as V_{ID}.

5. Connect your voltmeter to read the amplifier differential output (between the two collectors). Record this voltage value (V_{OD}) in Table 42.1.

6. Calculate the differential voltage gain of your amplifier. Record your result in Table 42.1.

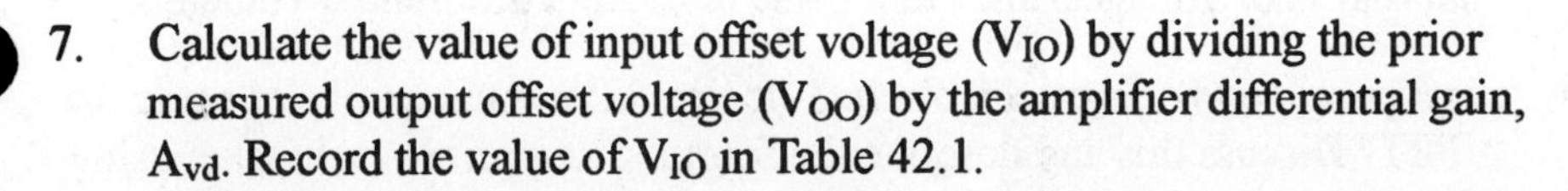

7. Calculate the value of input offset voltage (V_{IO}) by dividing the prior measured output offset voltage (V_{OO}) by the amplifier differential gain, A_{vd}. Record the value of V_{IO} in Table 42.1.

8. Connect a jumper from the V_{in+} to the V_{in-} input so that both inputs have the same input. Adjust the voltage divider potentiometer to provide its maximum output, approximately 100 mV. Measure and record the common-mode input voltage in Table 42.1 (V_{ICM}).

9. Use your voltmeter to measure the differential output of the amplifier. Record this value (V_{OCM}) in Table 42.1.

10. Calculate the common-mode gain (A_{VCM}) of the amplifier and record this value in Table 42.1.

11. Calculate the CMRR of the amplifier and record the value in Table 42.1.

Useful Formulas

Step 6:
$A_{vd} = V_{OD} / V_{ID}$

Step 12:
$CMRR = \frac{|A_{VD}|}{A_{VCM}}$

Beginning with step 8, you will measure the common-mode gain of your amplifier and determine the CMRR.

SECTION II TROUBLESHOOTING

Fault 1 - Q_1 C-E shorted

1. With circuit power off, place a shorting wire across the C-E of Q_1. Reapply DC power.

2. Measure and record the DC voltages for Q_1 AND Q_2 as indicated for Fault 1 in Table 42.2.

	Q_1 C-E shorted		Q_1 C-E open	
	Q_1	Q_2	Q_1	Q_2
V_E				
V_C				
V_B				

Table 42.2

Fault 2 - Q_1 C-E open

1. With circuit power off, remove the shorting wire and disconnect Q_1 from the circuit. Reapply DC power.

2. Measure and record the DC voltages for Q_1 AND Q_2 as indicated for Fault 2 in Table 42.2.

DISCUSSION

Section I

1. You should have measured a relatively large differential voltage gain for your amplifier. There are two features of the differential amplifier that contribute to this gain. One is that the gain of each transistor is the collector resistance divided by $2r'_e$: $A_v = \frac{R_C}{2r'_e}$. Describe and discuss the other feature that contributes to the gain.

2. For your differential amplifier, assume a V_{ID} of 5 mVDC, with V_{in+} the more positive. Describe, for the amplifier, the current paths, the individual transistor voltages, and from these the output differential voltage.

3. Is it possible to construct a discrete-device differential amplifier using the JFET? Discuss this and describe the form the amplifier would take, using schematic diagrams in your description. Would you, for example, use a BJT current source bias?

Section II

Fault 1 - Q_1 C-E shorted

From your measured data, describe the operating state of each transistor for this fault condition, and discuss the reason for each transistor's state. In this fault, for example, is Q_1 in cutoff, saturated, or somewhere in between?

Fault 2 - Q_1 C-E open

From your measured data, describe the operating state of each transistor and discuss the reason for each state.

Quick Check

1. The higher the CMRR value, the better.

   True   False

2. Why is direct coupling desirable in a differential amplifier?

3. In the circuit of Figure 42.1, the base of each transistor is zero-referenced.

   True   False

4. The theoretical differential gain of the circuit of Figure 42.1 is _____.

   (a) 53   (b) 86.5
   (c) 173   (d) 200

5. In the circuit of Figure 42.1, the input marked V_{in+} has the designation because a positive input signal causes the collector of Q_1 to be positive.

   True   False

43

BASIC OP-AMP PARAMETERS

INTRODUCTION

The operational amplifier, abbreviated op-amp, is an amplifier with a very high gain, a high input impedance, and a low output impedance. The op-amp has many applications in electronics. Some of these include active filters, comparators, level detection, and instrumentation amplifiers.

In this experiment you will perform tests to measure and let you see some of the basic characteristics of an IC op-amp.

REFERENCE

Principles of Electronic Devices and Circuits - Chapter 11, Sections 11.1 and 11.2

OBJECTIVES

In this experiment you will:

✓ Measure the input offset voltage of an op-amp

✓ Determine and understand the relationship between gain and bandwidth

✓ Through measurement observe and determine the op-amp response to a common-mode signal

EQUIPMENT AND MATERIALS

DC power supply
Function generator
Digital multimeter
Oscilloscope
Circuit protoboard
Operational amplifier, 741 or equivalent
Resistors: 47 **Ω, 4.7 kΩ, 1 kΩ, 2.2 kΩ, 22 kΩ, 100 kΩ**

SECTION I FUNCTIONAL EXPERIMENT

Input Offset Voltage

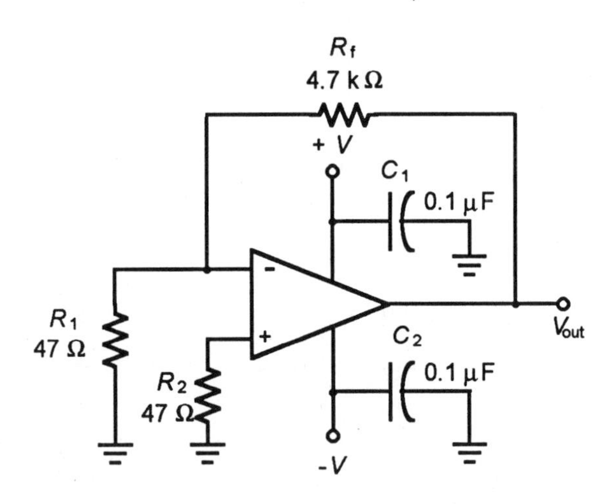

Figure 43.1

Input offset voltage cannot be measured directly. You will make your measurement indirectly by measuring the output due to the input offset voltage and calculating the magnitude of input offset voltage by the following formula:

$$V_{io} = \frac{V_o}{1 + \frac{R_f}{R_i}}$$

where:

V_{io} = input offset voltage

V_o = amplifier output

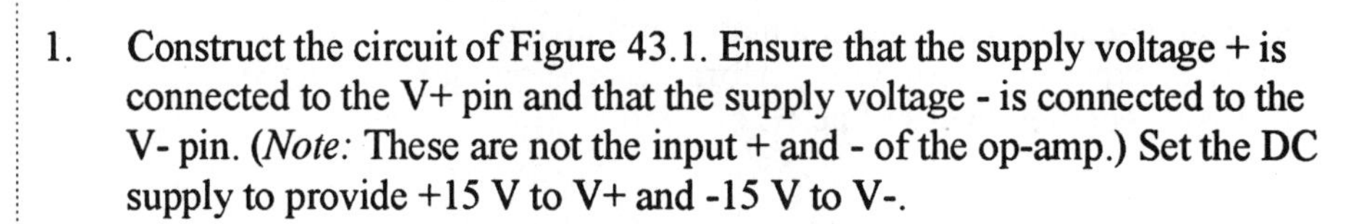

1. Construct the circuit of Figure 43.1. Ensure that the supply voltage + is connected to the V+ pin and that the supply voltage - is connected to the V- pin. (*Note:* These are not the input + and - of the op-amp.) Set the DC supply to provide +15 V to V+ and -15 V to V-.

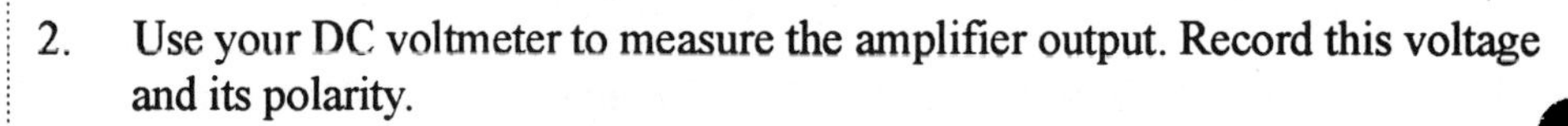

2. Use your DC voltmeter to measure the amplifier output. Record this voltage and its polarity.

   V_o = ___________

The maximum value you should have measured in step 2 is approximately 70 mV. If your measured value is greater, turn off circuit power and check your circuit connections.

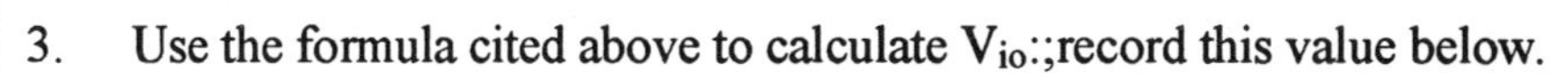

3. Use the formula cited above to calculate V_{io}:;record this value below.

   V_{io} = ___________

   Is this value within the op-amp specifications?

Gain Bandwidth

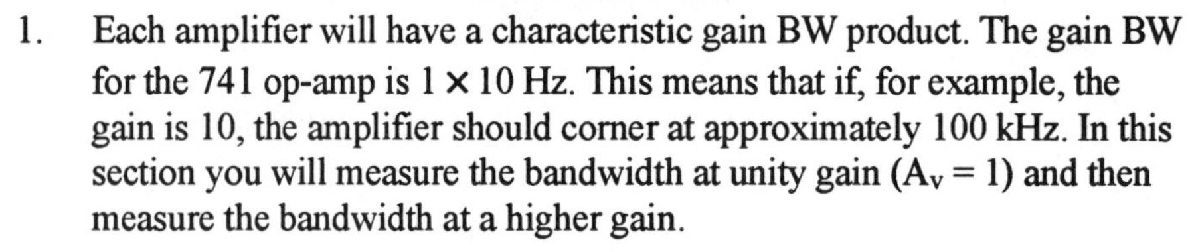

1. Each amplifier will have a characteristic gain BW product. The gain BW for the 741 op-amp is 1 × 10 Hz. This means that if, for example, the gain is 10, the amplifier should corner at approximately 100 kHz. In this section you will measure the bandwidth at unity gain (A_v = 1) and then measure the bandwidth at a higher gain.

2. Construct the circuit of Figure 43.2a.

3. Apply DC power (+ and - 15 volt). Set your function generator to provide a 1-$V_{p\text{-}p}$ sine wave at 1 kHz. Because the 741 can operate to DC, you can take 0 Hz for the lower corner frequency.

4. Connect your oscilloscope to measure V_{out} of the op-amp. Throughout the step periodically ensure that the input signal is at exactly 1 $V_{p\text{-}p}$. Measure and record the amplifier peak-to-peak output voltage and calculate the gain of the amplifier.

   V_{out} (1 kHz) = ________ A_v = ______________

Increase the generator frequency while monitoring the amplifier output. Measure and record the frequency where the output is 0.707 of the 1-kHz output amplitude. This is the upper corner frequency and should occur at approximately 1 MHz.

f_2 (upper corner) = ___________

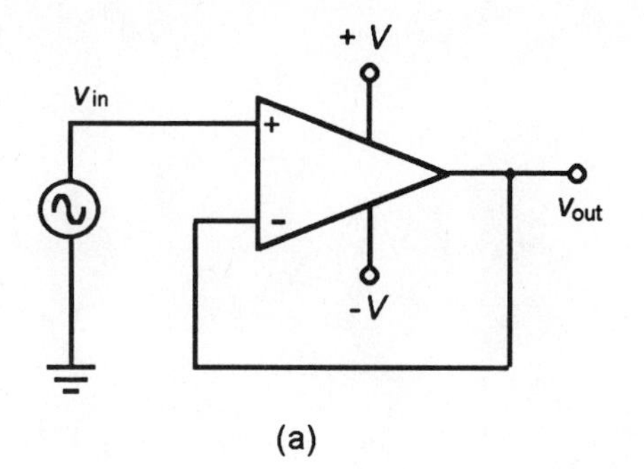

5. Compute the op-amp gain BW product as the product of the 1-kHz gain and the upper corner frequency.

   gain BW = ___________

6. Turn off the DC supply and modify your circuit to that of Figure 43.2b.

7. Reapply DC power and repeat the measurements of step 4. Record the 1-kHz output, the circuit gain, and the upper corner frequency (f_2).

   V_{out} (1 kHz) = ________

   A_v = _______________ f_2 = _______________

8. Compute the gain BW product again. You should find that this value is in close agreement with the value of step 4.

   gain BW = ___________

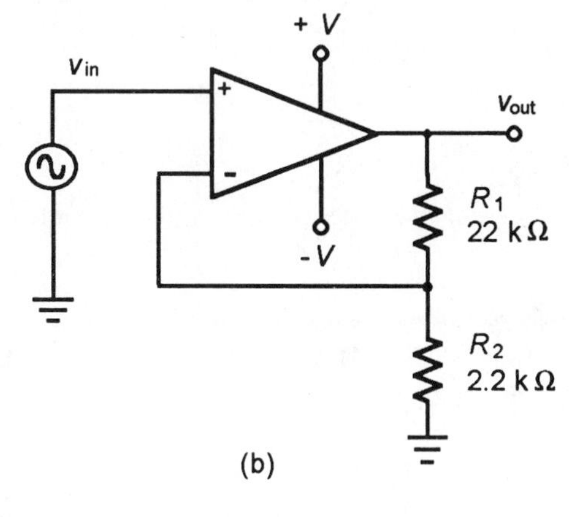

Figure 43.2

Common-Mode Rejection Ratio

1. The common-mode rejection ratio (CMRR) is a measure of the amplifier's ability to reject a common-mode signal input. The formula you will use to calculate the CMRR is

$$\text{CMRR} = \frac{\text{amplifier gain}}{\text{common–mode gain}}$$

2. Connect the circuit of Figure 43.3a. You will use this circuit to obtain the op-amp differential gain.

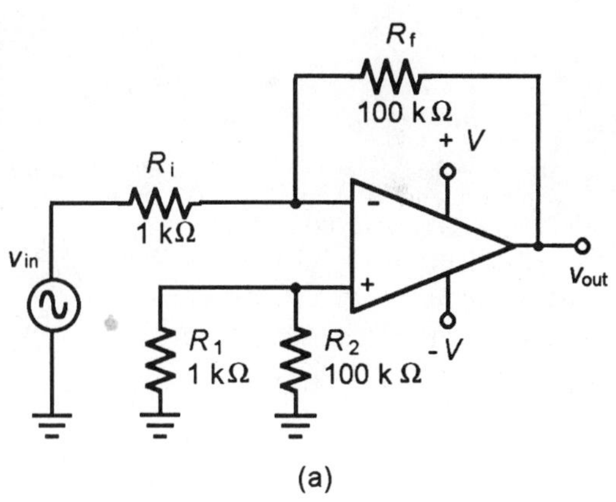

3. Apply DC power to your circuit. Set the function generator to provide an input sinewave of 1 $V_{p\text{-}p}$ at 1 kHz. Measure and record the peak-to-peak amplitude of the signal input and op-amp output.

   $V_{in(p\text{-}p)}$ = ______________ $V_{out(p\text{-}p)}$ = ___________

   Calculate the op-amp gain.

   A_v = ____________ (amp gain)

4. Turn off DC power. Modify your circuit to that of Figure 43.3 b. Reapply DC power. Set the function generator to provide a 1-$V_{p\text{-}p}$ sine wave at 1 kHz. Measure and record the peak-to-peak input and output amplitudes.

   $V_{in(p\text{-}p)}$ = ______________ $V_{out(p\text{-}p)}$ = ___________

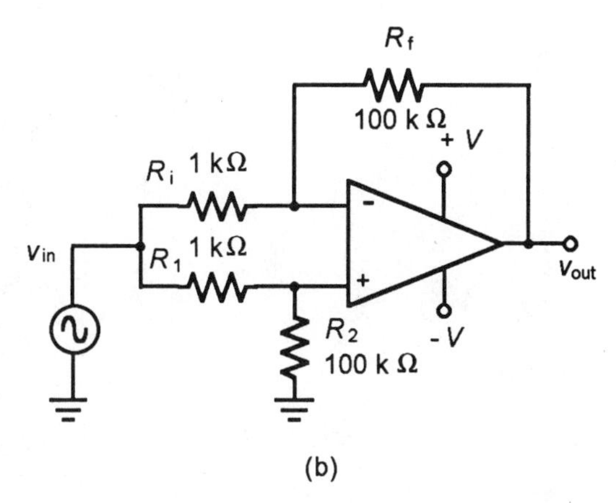

Figure 43.3

Calculate the op-amp gain.

A_V = ___________ (common-mode gain)

5. Calculate the CMRR using the formula at the beginning of this experiment.

   CMRR = ___________

DISCUSSION

1. Considering the input offset voltage you measured for your 741 op-amp, describe the output signal you would expect for that amplifier operating with a gain of 10 and an input sinewave signal of 200 $mV_{p\text{-}p}$ at 1 kHz.

2. Given the gain BW product that you measured for your 741 op-amp, what gain would you expect to find for your amplifier if the corner frequency were 50 kHz?

   Discuss why the knowledge of an amplifier gain BW product is useful to the user.

3. What kinds of applications can you suggest for an amplifier with a high CMRR? Select one application and describe.

Quick Check

1. The operational amplifier has high gain and low output impedance.

   True False

2. Closed-loop bandwidth is a function of the voltage gain.

   True False

3. The operational amplifier has three inputs: inverting, noninverting, and null.

   True False

4. An op-amp has a differential gain of 25 and a common-mode gain of 0.02. What is the CMRR?

   CMRR = ___________

5. What is the unity-gain bandwidth for a 741 op-amp?

   BW = ___________

44

OP-AMP SLEW RATE AND CMRR

INTRODUCTION

The slew rate of an op-amp tells how fast the output voltage can change with respect to a change at the input. The unit of measurement of slew rate is volts per unit of time, typically V/μs. **The slowest rate of change of output occurs at unity gain; therefore it is under this condition that the parameter is most often stated.**

Common-mode rejection (CMRR) is defined as an op-amp's ability to minimize the effects of unwanted input signals. These signals may be the result of power supply fluctuations or electromagnetic interference (EMI). CMRR is the ratio of differential-mode gain (A) to common-mode gain (A_{cm}) and is expressed in dB.

In this experiment you will construct an op-amp circuit to measure slew rate. Two different measurements will be made. The second measurement, the more difficult, will also demonstrate the power bandwidth characteristic of a op-amp. You will also set up a circuit to measure the op-amp response to a common-mode input and from this derive the amplifier CMRR.

REFERENCE

Principles of Electronic Devices and Circuits - Chapter 11, Section 11.3

OBJECTIVES

Upon completing this experiment, you will be able to:

✓ Determine the slew rate of an op-amp

✓ Determine the CMRR of an op-amp

EQUIPMENT AND MATERIALS

DC power supply
Function generator
Dual-trace oscilloscope
Digital multimeter
Circuit protoboard
Operational amplifier, 741 or equivalent
Resistors: 1 kΩ, 1 MΩ [2]

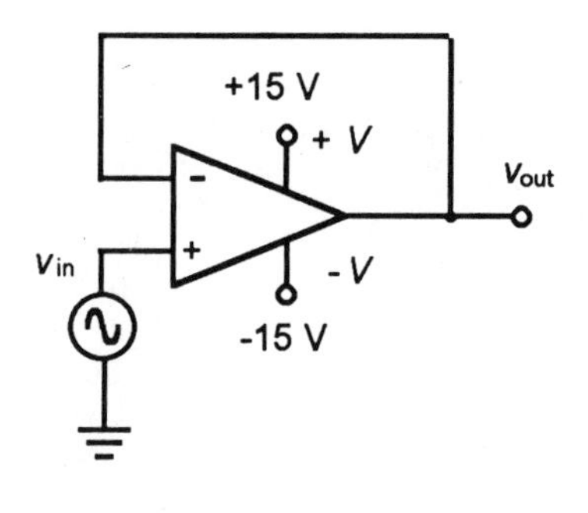

Figure 44.1

SECTION I FUNCTIONAL EXPERIMENT

Op-Amp Slew Rate

1. Connect the circuit shown in Figure 44.1. The op-amp is configured as a unity-gain, noninverting voltage follower. Connect the oscilloscope as shown, and set the controls to the following positions:

   Time base: 5 µs/div
   Channel 1: 2 V/div
   Channel 2: 2 V/div

   Positive slope
   Internal trigger on Channnel 1

2. Apply DC power and a 25-kHz, 6-$V_{p\text{-}p}$ square wave to the noninverting input. You should observe two waveforms similar to those in Figure 44.2.

3. In Graph 44.1, add time and amplitude units and sketch the waveforms.

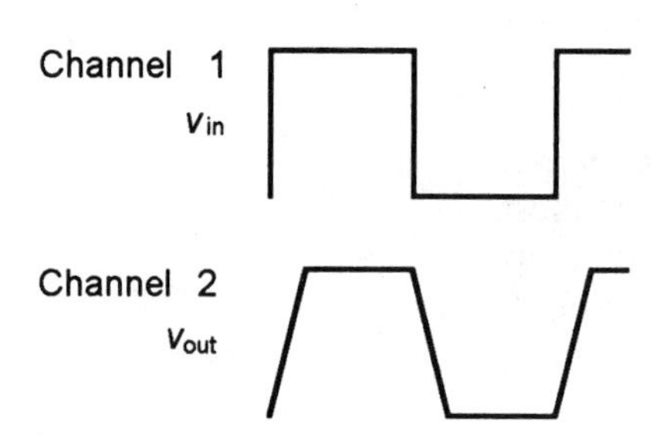

Figure 44.2

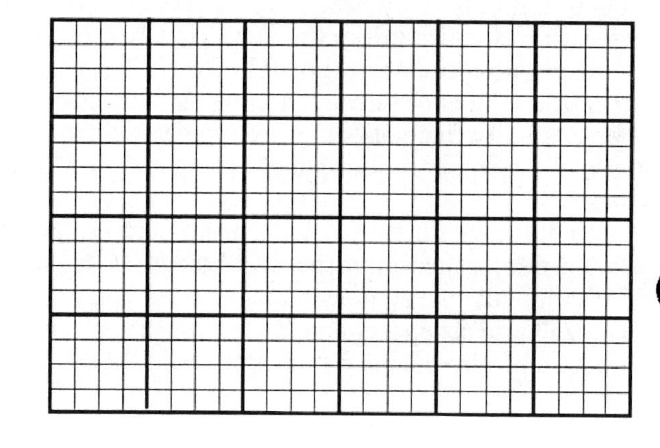

Graph 44.1

4. Measure the peak-to-peak voltage of the output waveform and record this below as ΔV.

   ΔV = ___________

5. Measure the amount of time it takes the waveform to change from one peak to the other (this can be on either the positive-going or the negative-going edge). Record this value as Δt below.

   Δt = ___________

6. Calculate the slew rate of the op-amp using the following formula:

   slew rate = ΔV/Δt

   slew rate = ___________

7. The slew rate of an op-amp limits its performance with a sinusoidal signal also. If the slope of the sinusoidal signal is greater than the op-amp slew rate distortion or power bandwidth. The relationship between the frequency where the signal starts to distort and the amplifier slew rate is given by

$$SR = 2\pi f_m V_p$$

8. Change the function generator output to a sinewave signal. Maintain the output at 6 V_{p-p}. Starting at a frequency of 1 kHz, observe the op-amp output while increasing the frequency. Adjust the generator frequency to the value where you can just begin to detect distortion of the sine wave. This is f_m. Record this frequency.

   f_m = ___________

As the frequency is increased, you will note that the output signal will become triangular shaped.

9. Using the formula of step 7, calculate the op-amp slew rate, and record your result.

   slew rate = ___________

10. Although the power bandwidth measurement is more difficult to make, it illustrates the large signal limitation that the slew rate imposes on an op-amp.

11. Turn off power from the function generator; then remove DC power from the circuit. Disassemble the circuit.

Op-Amp CMRR

1. Connect the circuit in Figure 44.3.

2. Calculate the differential-mode gain (A) using the following formula:

$$A = R_f/R_1$$

   A = ___________

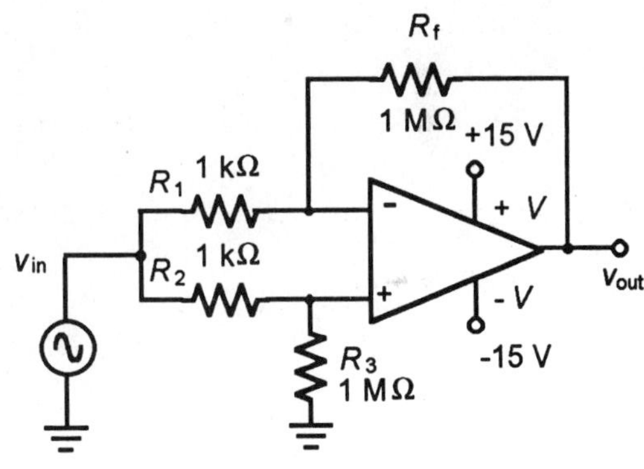

Figure 44.3

3. Apply DC power and a 6-V_{p-p}, 60-Hz sine wave to the input. This voltage will be used to represent EMI from a nearby piece of machinery.

4. Observe the input and output signals on the oscilloscope. You should observe a 6-V_{p-p} input signal and a small signal at the output.

5. Measure the AC voltage (v_{in}) at pin 2 (or pin 3) of the op-amp. Measure the AC voltage (v_{out}) at pin 6 of the op-amp. Use the formula v_{out}/v_{in} to find the common-mode gain (A_{cm}). Record these values below.

   v_{in} = ___________ v_{out} = ___________ A_{cm} = ___________

6. Calculate the common-mode rejection ratio, in dB, using this formula:

$$CMRR = \log \frac{A}{A_{cm}}$$

   CMRR = ___________

7. Turn off the power from the function generator; then remove DC power from the circuit.

DISCUSSION

1. The op-amp parameter, slew rate, can have an impact on how a particular op-amp will perform in different situations. Describe what effect frequency variations might have on slew rate.

2. Compare the data sheet parameters of a µA741 to those of an LM308 op-amp. Which op-amp you think would be better suited for use in an area susceptible to high electromagnetic interference? Justify your answer.

Quick Check

1. The slew rate of an op-amp is independent of the frequency that will be applied.

   True | False

2. If a certain op-amp has a voltage swing from +5.5 V to -7.5 V in 0.25 µs, the slew rate is equal to _____.

   (a) 8 V/µs | (b) 3.25 V/µs
   (c) 52 V/µs | (d) 0.5 V/µs

3. The unit in which the common-mode rejection ratio is stated is _____.

   (a) dB | (b) V/µs
   (c) Ω | (d) no unit

4. A certain op-amp has a CMRR of 76 dB. The ratio of A to A_{CM} is _____.

   (a) 6310 | (b) 3.8
   (c) 1520 | (d) 76

45

NONINVERTING VOLTAGE AMPLIFIER

INTRODUCTION

The noninverting voltage amplifier has the characteristics of an ideal voltage amplifier: exceptionally high input impedance and very low output impedance. Additionally, this amplifier, like any other based on a high-gain op-amp, has a voltage gain and stability that are dependent upon the external circuit resistors and independent of amplifier variations.

In this experiment you will construct a series-parallel negative feedback op-amp noninverting voltage amplifier circuit. You will verify and observe the effects of feedback resistors in setting the voltage gain, and you will demonstrate that negative feedback reduces the output impedance of the op-amp.

REFERENCES

Principles of Electronic Devices and Circuits - Chapter 12, Sections 12.1 and 12.2

OBJECTIVES

In this experiment you will:

✓ Demonstrate the operation of the noninverting amplifier

✓ Verify voltage gain control by feedback resistors

✓ Demonstrate the reduced Z_{out} of the op-amp circuit

✓ Show the effects of faulty resistors

EQUIPMENT AND MATERIALS

DC power supply
Function generator
Dual-trace oscilloscope
Digital multimeter
Operational amplifier, 741 or equivalent
Resistors: 1 kΩ [2], 15 kΩ

SECTION I FUNCTIONAL EXPERIMENT

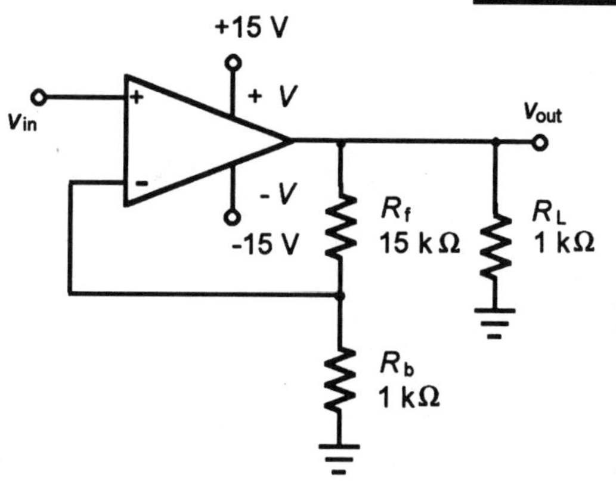

Figure 45.1

1. Construct the circuit in Figure 45.1.

2. Apply DC power. Connect the function generator to v_{in}. Set the generator to supply a 1-kHz sinewave at 200 $mV_{p\text{-}p}$. Using your oscilloscope, verify that the output is in phase with the input.

   Phase shift = ___________ degrees

   With the oscilloscope, measure the peak-to-peak voltage levels of v_{in} and v_{out}. Record these values in Table 45.1.

3. From your measured data, calculate the amplifier gain and record in Table 45.1.

   Calculate the expected gain of the amplifier based on your resistance values. Record this value in Table 45.1.

4. Your measured value of gain should agree with the calculated value, within resistor and equipment tolerances. If you find a greater deviation, recheck your measurements.

5. Select a new value of R_f to set the amplifier gain in the range of 6 to 25. Record the new value of R_f in Table 45.1 with your predicted gain.

6. Turn off AC and DC sources. Install the new R_f resistance you selected. Reapply DC power and signal input. Repeat the gain measurement of steps 2 and 3. Record the peak-to-peak values of v_{in} and v_{out} in Table 45.1. Calculate the amplifier gain from the measured data and record in Table 45.1.

Beginning with step 7, the procedure demonstrates that negative feedback reduces the effective op-amp output impedance. The specified Z_o of the 741 op-amp is 75 Ω.

7. With both AC and DC sources off, install a 10-kΩ resistor for R_f. Reapply DC power. Set the function generator to supply a 200 $mV_{p\text{-}p}$ input signal. Measure and record in Table 45.1 the peak-to-peak input and output voltage levels.

R_f = 15 kΩ	R_f = ________	R_f = 10 kΩ, R_L = 1 kΩ	R_f = 10 kΩ, R_L = 75Ω
Measured	Measured	Measured	Measured
V_{in} = ________	V_{in} = ________	V_{in} = ________	V_{in} = ________
V_{out} = ________	V_{out} = ________	V_{out} = ________	V_{out} = ________
A_V = ________	A_V = ________		
Calculated	Calculated		
A_V = ________	A_V = ________		

Table 45.1

8. Turn DC and AC sources off. Change R_L to a 75-Ω resistor. Reapply DC power and the amplifier signal input. Ensure that the peak-to-peak input signal is exactly the same as used in step 8. Measure and record in Table 45.1 the peak-to-peak output voltage level. This measurement won't give the output impedance value, however; if V_{out} is one-half that of step 8, then $Z_{out} = 75\ \Omega$.

   Your measured V_{out} should be nearly that of step 8 (approximately 10% to 20% less), demonstrating that the amplifier output impedance is significantly less than 75 Ω and has been reduced by the negative feedback.

SECTION II TROUBLESHOOTING

Fault 1 - Feedback resistor (R_f) open

1. Turn off all sources. Replace R_L with a 1-kΩ resistor. Remove the feedback resistor (R_f) and replace with a 2.2-MΩ resistor.

   Note: This simulation of a failure won't perfectly emulate the failure of an open R_f, but it will serve to give you a good idea of what to expect with that failure mode.

2. Apply DC power and a 1-kHz, 200-$mV_{p\text{-}p}$ input signal. Using your oscilloscope, observe the amplifier output, noting in particular the waveshape and peak-to-peak amplitude of the signal.

Fault 2 - Feedback resistor (R_b) open

Turn off DC and AC sources. Restore R_f to a 10-kΩ resistor. Disconnect the ground end of R_b. Apply DC power and AC input signal. Measure and record below the peak-to-peak input and output voltage levels.

V_{in} = ______________ V_{out} = ______________

DISCUSSION

Section I

1. Describe a discrete device amplifier (2 stages maximum) that would approximate the noninverting op-amp voltage amplifier.

2. Compare and contrast your discrete device amplifier to the noninverting voltage amplifier.

Section II

Fault 1 - Feedback resistor (R_f) open

Describe the output waveform you observed for this fault. Can you think of an op-amp failure mode that would produce this same waveform? Discuss op-amp failure modes and the probable output waveform result.

Fault 2 - Feedback resistor (R_b) open

Describe the result of this fault. Discuss why the amplifier should produce thi result with just one faulted resistor.

Quick Check

1. The noninverting amplifier has a very stable voltage gain.

   True   False

2. An open feedback resistor does not affect gain.

   True   False

3. The voltage gain of a noninverting op-amp is controlled by the relationship between R_f and R_b.

   True   False

4. The input signal is in phase with the output signal in a noninverted series-parallel negative feedback circuit.

   True   False

5. Negative feedback increases Z_{in} and Z_{out} of an op-amp circuit.

   True   False

6. A nonverting voltage amplifier has an R_f of 27 kΩ and R_b of 4.7 kΩ. The amplifier gain is _____.

   (a) 5.7   (b) 6.7
   (c) 8.7   (d) 10.7

46

INVERTING VOLTAGE AMPLIFIERS

INTRODUCTION

The op-amp inverting voltage amplifier is a very popular amplifier. It displays most of the characteristics of the noninverting amplifier except the extremely high input impedance. In addition, at low gains, it will have a smaller gain bandwidth product.

In this experiment you will build and test an inverting op-amp voltage amplifier and examine the characteristics of input impedance and gain bandwidth product.

The troubleshooting section will let you see the effects of simulated component failures on your amplifier.

REFERENCE

Principles of Electronic Devices and Circuits - Chapter 12, Section 12.3

OBJECTIVES

In this experiment you will:

✓ Demonstrate the characteristics of the inverting amplifier

✓ Determine the input impedance and gain bandwidth product of an inverting amplifier

✓ Demonstrate circuit problems caused by faulty resistors

EQUIPMENT AND MATERIALS

DC power supply
Function generator
Oscilloscope
Operational amplifier, 741 or equivalent
Resistors: 1 kΩ, 2.2 kΩ, 15 kΩ

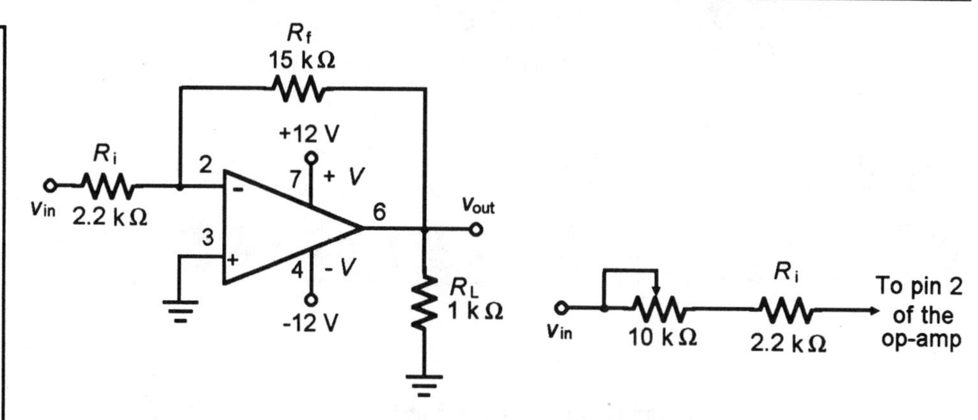

Figure 46.1 Figure 46.2

Amplifier Gain (R_f = 15 kΩ)
Measured Values
V_{in} = ______
V_{out} = ______
A_V = ______
Calculated Value
A_V = ______

Amplifier Gain (R_f = ______ kΩ)
Measured Values
V_{in} = ______
V_{out} = ______
A_V = ______
Calculated Value
A_V = ______

Input Z
Zin = ______

Gain-BW
BW = ______
Gain-BW = ______

Table 46.1

1. Construct the circuit of Figure 46.1.

2. Connect power to the circuit. Apply a 500-m$V_{p\text{-}p}$, 1-kHz sinewave signal to the input v_{in} of the amplifier.

3. Measure and record in Table 46.1 the input and output peak-to-peak voltages using your oscilloscope. Use the data from step 3 to determine the actual A_V; record it in Table 6.1.

4. Calculate the expected gain for your amplifier and record in Table 46.1.

   Note: If your measured and calculated values for A_V are not close, check your calculations and your circuit measurements.

5. Remove power from the amplifier. Select a new value of R_f in the range of 2.2 kΩ to 39 kΩ. Record your R_f value in the second box of Table 46.1.

6. Reapply DC power and input signal of 500 m$V_{p\text{-}p}$ at 1 kHz. Measure the peak-to-peak input and output voltages, and use the data to determine the actual amplifier gain. Record these values in the second box of Table 46.1.

7. Calculate the expected gain for your amplifier and record in Table 46.1.

8. Disconnect the AC signal source; turn off the DC source. Connect a 10-kΩ potentiometer in series with R_i as shown in Figure 46.2. Change R_f back to the 15-kΩ resistor.

Beginning with step 8, you will measure the input impedance of your amplifier.

9. Adjust the potentiometer to be zero ohms. Reapply DC power and reconnect the AC signal source. Set the input signal to obtain the same amplifier output recorded in step 3. Note this value below. Adjust the potentiometer until the input signal is exactly one-half the noted value.

   V_{out} = ______

10. Turn off the AC and DC sources. Disconnect the potentiometer and measure its resistance. This is the input impedance value of your amplifier. Record in Table 46.1.

11. Restore the input circuit connections, omitting the potentiometer. In the next procedure steps you will measure the amplifier gain and bandwidth to determine the gain bandwidth product.

12. Turn on DC power and the AC signal source. Measure the amplifier peak-to-peak input and output voltages. Note these below. Calculate the amplifier gain and note below.

    V_{in} = __________ V_{out} = __________ A_v = __________

13. While ensuring that V_{in} stays constant, increase the generator frequency to the amplifier upper corner frequency. Measure and record this frequency.

    f_2 = __________

    Since the amplifier should operate to DC, f_2 is the bandwidth.

14. Calculate the gain bandwidth product. You should find that your value is less than the expected result of 1 MHz. For the gain of your amplifier, your gain bandwidth product is expected to be between 10% and 15% less. Lower gains produce even smaller gain bandwidth products.

SECTION II TROUBLESHOOTING

Fault 1 - Feedback resistor (R_f) open

1. Turn off DC and AC sources. Remove the 15-kΩ R_f and replace it with a 2.2-MΩ resistor.

This won't simulate a completely opened resistor, but will simulate the failure clearly enough for this experiment.

2. Apply DC power and the AC signal input of 500 mV_{p-p}. With your oscilloscope connected to the amplifier output, observe the output waveshape. Note the peak-to-peak value below.

   $V_{out\ (p-p)}$ = __________

3. Measure and record below the following voltages.

   $V_{in\ (p-p)}$ = ______ V_{+DC} = __________ V_{-DC} = __________

Fault 2 - Input resistor (R_i) open

1. Turn off DC and AC sources. Remove the 2.2-MΩ resistor and install the original 15-kΩ R_f. Remove the 2.2-kΩ R_i resistor and replace with the 2.2-MΩ resistor.

Again, this won't completely simulate an open resistor, but it is sufficient for demonstrating this fault.

2. Apply DC power and the AC signal input of 500 mV_{p-p}. Measure and record the circuit values listed below.

   V_{in} (p-p) = __________ V_{out} (p-p) = __________

   $V_{+DC} = V_{-DC}$ = ______

DISCUSSION

Section I

1. Compare and contrast the op-amp inverting amplifier to a BJT common emitter amplifier by listing as many advantages and disadvantages as you can.

2. Given an audio amplifier application requiring a voltage gain of 75 in the frequency band of 30 Hz to 20 kHz, discuss how well or poorly an inverting amplifier based on the 741 would meet this application.

Section II

Fault 1 - Feedback resistor (R_f) open

1. Describe the output waveform observed for this fault and discuss the reason for that waveform.

2. Discuss how well the measurements made point to the external circuitry as the fault rather than the op-amp.

Fault 2 - Input resistor (R_i) open

Do the data measured for this fault clearly indicate that the op-amp is okay, with the fault having to be in the external circuitry? Discuss this concept, considering what op-amp failures could produce the same circuit fault measurements.

Quick Check

1. The output signal of the inverting amplifier is 90° out of phase.

   True False

2. If the value of the input resistor decreases, the gain will increase.

   True False

3. If the value of the feedback resistor increases, the gain will increase.

   True False

4. For the amplifier of Figure 46.1, if $R_f = 27\ k\Omega$ and $R_i = 3.9\ k\Omega$, the amplifier gain would be _____.

   (a) 27 (b) 10.6
   (c) 7.9 (d) 6.9

47

OP-AMP CURRENT AMPLIFIERS

INTRODUCTION

The ideal current amplifier has zero input impedance, infinite output impedance, and a constant current gain. The ideal current amplifier can be approached using an op-amp with an inverting current feedback. This current amplifier form provides a linear gain response to the point where the op-amp reaches its output voltage saturation limit.

In this experiment you will construct a current amplifier and make measurements to see the basic characteristics. Then you will set up a simulated circuit to obtain concepts involved in a circuit application.

The troubleshooting section will aid you in developing skills for effective troubleshooting of op-amp circuits.

REFERENCE

Principles of Electronic Devices and Circuits - Chapter 12, Section 12.4

OBJECTIVES

In this experiment you will:

✓ Determine the operating characteristics of the current amplifier

✓ Understand the operation of a current amplifier in a circuit illustrating an application

✓ Develop skills in troubleshooting an op-amp circuit

EQUIPMENT AND MATERIALS

DC power supply
Digital multimeter [2]
Circuit protoboard
Operational amplifier, 741 or equivalent
Resistors: 560 Ω, 1 kΩ, 2.2 kΩ, 3.9 kΩ, 10 kΩ [2], 22 kΩ
Potentiometer: 10-kΩ ten-turn trimpot [2]
Capacitor: 0.01 uF [2]

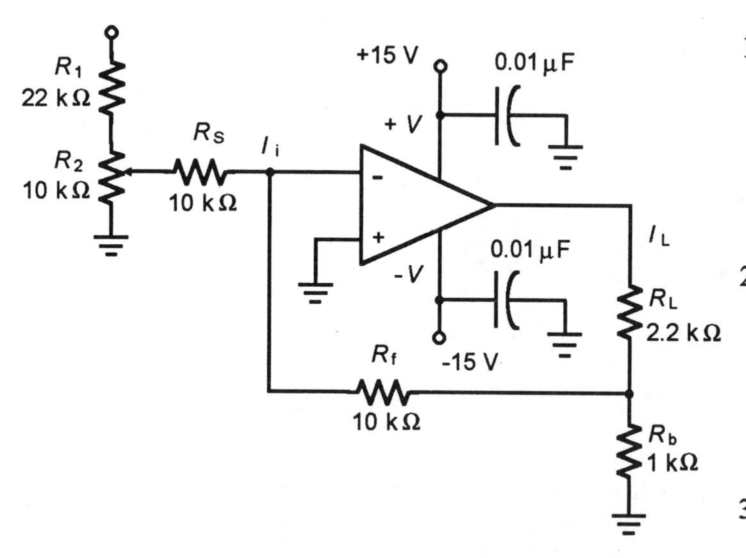

Figure 47.1

1. Construct the circuit of Figure 47.1. Connect an ammeter in the circuit between R_s and the junction of R_f. Connect a second ammeter between the op-amp output and R_L. Network R_1, R_2, and R_S provides the current source for your circuit.

2. Apply DC power to the circuit. While reading the input current ammeter, adjust R_2 to obtain an input current of 150 µA. Record your set value and the output current value in the *Current Gain* box of Table 47.1.

3. From the data of step 2 calculate the amplifier current gain (A_i) and record in Table 47.1.

4. If the current amplifier has a high output impedance< as a good amplifier should, changing the size of R_L should not significantly affect the load current. In the next procedure steps you will test the amplifier with different R_L values.

5. Turn off the DC supply. Remove the 2.2-kΩ load resistor, and install a 560-Ω load. Reapply DC power. Ensure that the input current is still 150 µA. Measure and record the output current (I_{out}) in Table 47.1 in the box labeled $R_L = 560\ \Omega$.

6. Repeat step 5, this time using a 3.9 kΩ load resistor. Record the output current in the box labeled $R_L = 3.9$ kΩ in Table 47.1. You should have found that changing the load resistance through a range of almost 4:1 had little effect on the output current.

7. Turn off the DC power. Disconnect the output current ammeter and complete the output circuit connections without the ammeter. Reapply DC power, and ensure that the input current is still 150 µA. Measure the voltage from the op-amp output to ground. You should find a reading of between 7.5 V and 8.0 V. As long as the size of the load resistance doesn't require more than the saturated op-amp output, the current amplifier will be linear.

8. To see one application of a current amplifier, assume that you have a 1-mA ammeter with a meter resistance of 1.8 kΩ, and you must be able to measure a current of 250 µA. You can use a current amplifier with a gain of 4 to scale the current up to the meter full-scale range.

Since $A_i = \frac{R_f}{R_b} + 1$, the $\frac{R_f}{R_b}$ ratio needs to be 3. So when $R_f = 10$ kΩ, R_b should be 3.0 kΩ.

Current Gain
I_{in} = ________
I_{out} = ________
A_i = ________
Load Variations ($R_L = 560\ \Omega$)
I_{in} = ________
I_{out} = ________
Load Variations ($R_L = 3.9$ kΩ)
I_{in} = ________
I_{out} = ________
V_{out} = ________

Table 47.1

9. Turn off the DC supply. If you have access to a 1-mA meter movement, install it in place of R_L. If not, you can use a 1.8-kΩ resistor and an ammeter to measure the current. Install a 10-kΩ potentiometer which is set to provide 3 kΩ in place of the 1-kΩ R_b.

10. Reapply DC power. Adjust R_2 for an input current of 250 μA and measure the output current. You may need to adjust the potentiometer of R_b for an exact scaling to produce the 1-mA output.

SECTION II TROUBLESHOOTING

Fault 1 - Circuit feedback problem

In this fault, it is assumed that the op-amp feedback circuitry is designed to provide a current gain of 11. Furthermore, the current amplifier is spread through several circuit boards in the system. You will fault the circuit by providing excessive feedback, thereby reducing the gain.

1. Turn off the DC power. Replace R_f with a 100-kΩ resistor and R_b with a 50-kΩ resistor. This simulates an excessive feedback.

2. Turn on the power. Adjust R_2 for an input current level of 0.1 mA. Measure the input and output current levels. Calculate the gain.

3. Turn off power. Replace R_f with a 250-kΩ potentiometer.

4. While monitoring the input and output currents, adjust R_f until a current ratio of approximately 11:1 is reached.

   If you can achieve current gain called for in the specifications, the fault lies in the feedback circuitry. If not, the problem lies in the op-amp circuitry. If you have a discrete component op-amp, you'll have to troubleshoot each BJT or FET network independently.

DISCUSSION

Section I

1. Discuss your current response to changing load resistance. For example, does your amplifier have a constant I_{out} independent of R_L values (within limits)?

2. What effect does changing the ratio of R_f to R_b have on the operation of the circuit? For example, if R_b increases, what changes does this have on the output current?

Section II

Fault 1 - Feedback circuit problem

Discuss the concepts of the troubleshooting example. Consider in particular the situation where parts of the feedback circuitry might be on two different circuit boards in a system. Give advantages and disadvantages of this troubleshooting approach.

Quick Check

1. The gain of a current amplifier can be expressed by $A_i = R_f/R_b$.

   True False

2. An inverting current feedback op-amp that has an output current of 120 mA and an input current of 120 μA has a gain of approximately _____.

   (a) 10:1 (b) 12:1
   (c) 1:10 (d) 1000:1

3. The ideal current amplifier will have a very high input impedance.

   True False

4. A good current amplifier will have a very low output impedance.

   True False

48

VOLTAGE-TO-CURRENT CONVERTERS

INTRODUCTION

The voltage-to-current converter uses noninverting current feedback to control the operation of the amplifier. The noninverting feedback increases input impedance and output impedance, so that the amplifier approaches the ideal voltage-to-current converter.

This experiment will examine the basic voltage-to-current converter. You will construct a floating-load noninverting voltage-to-current converter, and you will make measurements of this basic converter. Then you will build and test a circuit illustrating an application of the voltage-to-current converter.

REFERENCE

Principles of Electronic Devices and Circuits - Chapter 12, Section 12.5

OBJECTIVES

In this experiment you will:

✓ Understand the operation of the op-amp voltage-to-current converter

✓ Examine an application of the voltage-to-current converter

EQUIPMENT AND MATERIALS

DC power supply
Digital multimeter [2]
1-0-1 mA meter movement
Circuit protoboard
Resistors: 1 kΩ, 2.2 kΩ, 22 kΩ
Potentiometer: 10-kΩ ten-turn trimpot
Capacitors: 0.01 μF [2]

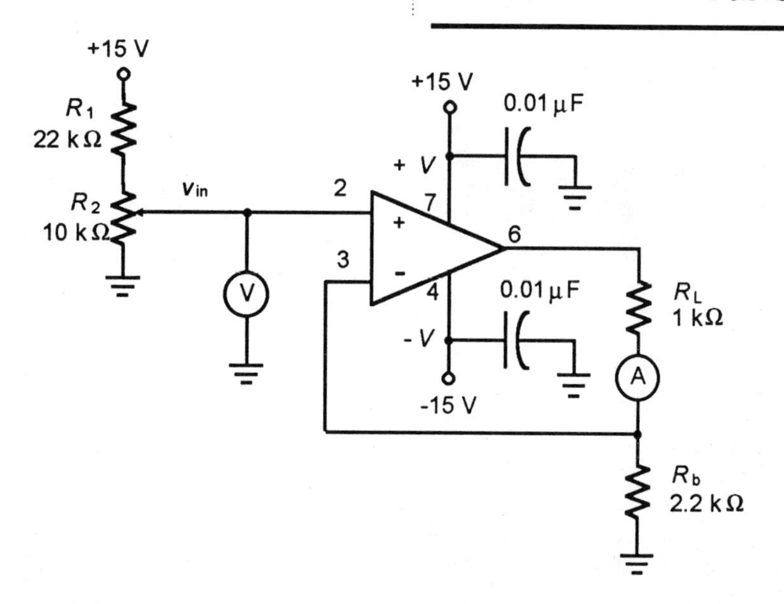

Figure 48.1

Converter	
Vin	iout
________	________
________	________
Conversion Scaling	
V/mA = ________	
Voltmeter	
Vin	iout
0.5 V	________
0.75 V	________

Table 48.1

1. Construct the circuit in Figure 48.1.

2. Apply DC power. Adjust divider potentiometer R_2 for an input voltage of 1.1 V. Measure the output current. Record the input voltage and output current in the *Converter* box of Table 48.1.

3. Adjust the divider potentiometer for an input voltage of 2.2 V. Measure the input voltage and output current. Also record these values in the *Converter* box of Table 48.1.

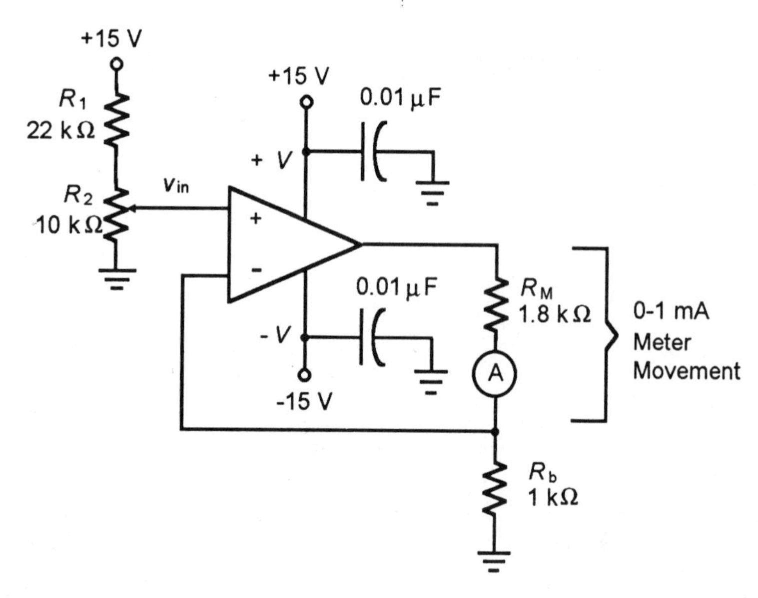

Figure 48.2

4. From the data of steps 2 and 3, calculate the voltage-to-current ratio, in V/mA, for your converter. Record this value in the *Conversion Scaling* box of Table 48.1.

In the following procedure steps, you will modify your converter to simulate a high input impedance analog voltmeter. A 0-1 mA meter movement will display 0-1 V full scale.

Note: If a 1-mA meter movement is not available, it can be simulated as shown in Figure 48.1. The scaling is the same: 1 mA = 1 V.

5. Turn off the DC power. Modify your circuit to that of Figure 48.2. Use a 1-mA meter movement, if available. If not, use your digital multimeter, set to 1-mA range as shown in Figure 48.2.

6. Apply DC power. Adjust the divider potentiometer for an input of 0.5 V. Your scale factor is

   $I_M = V_{in} / R_b = 1\text{ V} / 1\text{ mA}$ (depending on the accuracy of R_b)

   Measure and record in the *Voltmeter* box of Table 48.1 the output current as voltage: 1 mA = 1 V.

7. Adjust the divider potentiometer for an input of 0.75 V. Measure and also record the output current as voltage in the *Voltmeter* box of Table 48.1.

DISCUSSION

1. As the input voltage increased, what happened to the output current? Do your measurements indicate a voltage-to-current conversion? Is the circuit inverting or noninverting?

2. Could the output of the op-amp in Figure 48.1 be connected to a BJT whose emitter is connected to a motor? Explain how the BJT would be turned on, thereby driving the motor.

Quick Check

1. The operation of the voltage-to-current converter is linear.

   True False

2. In a noninverting voltage-to-current converter, the output is in phase with the input.

   True False

3. For the circuit of Figure 48.1, if v_{in} were 10 V, the output current would be _____.

   (a) 2.8 mA (b) 1.36 mA
   (c) 2.3 mA (d) 4.6 mA

4. For the circuit of Figure 48.1, to set the conversion scale to 1 V/0.5 mA, the value of R_b should be _____.

   (a) 1.0 kΩ (b) 1.5 kΩ
   (c) 2.0 kΩ (d) 3.0 kΩ

49

SUMMING AMPLIFIERS

INTRODUCTION

The purpose of a summing amplifier is to provide an output that is proportional to the sum of several inputs. The output of the summing can reflect AC and DC input values. Summing amplifiers can be used to convert digital to analog levels, produce an output equal to (L + R) FM intelligence, and mix audio signals.

In this experiment you will construct two different summing amplifiers. One lets you explore the basic summing amplifier, and the other illustrates a simple mixing function.

The troubleshooting section will let you observe the effect of resistor failures on the summing amplifier circuit parameters.

REFERENCE

Principles of Electronic Devices and Circuits - Chapter 12, Section 12.5

OBJECTIVES

In this experiment you will:

✓ Understand the operating characteristics of the summing amplifier

✓ Explore the function of mixing an AC and a DC signal

✓ Understand the circuit effect caused by a component failure

EQUIPMENT AND MATERIALS

DC power supply
Function generator
Oscilloscope
Digital multimeter
Circuit protoboard
Operational amplifier, 741 or equivalent
Resistors: 2.2 kΩ [2], 6.8 kΩ, 47 kΩ
Potentiometer:, 10-kΩ ten-turn trimpot

SECTION I FUNCTIONAL EXPERIMENT

1. Construct the circuit in Figure 49.1.

2. Temporarily remove R_2. Apply DC power to your amplifier. Connect a function generator set to supply an 800 Hz sine wave at 500 $mV_{p\text{-}p}$ to input V_{in1}. Ensure that your signal has no DC offset.

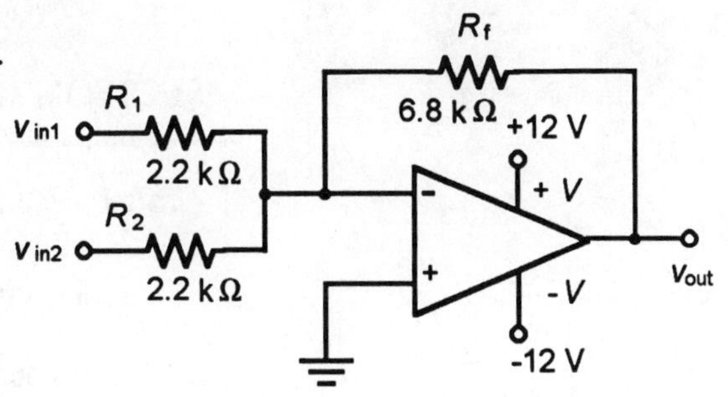

Figure 49.1

3. Use your oscilloscope to measure the peak-to-peak input (V_{in1}) and output voltage (V_{out}) levels. Record the values in the column labeled *R_2 Open* of Table 49.1.

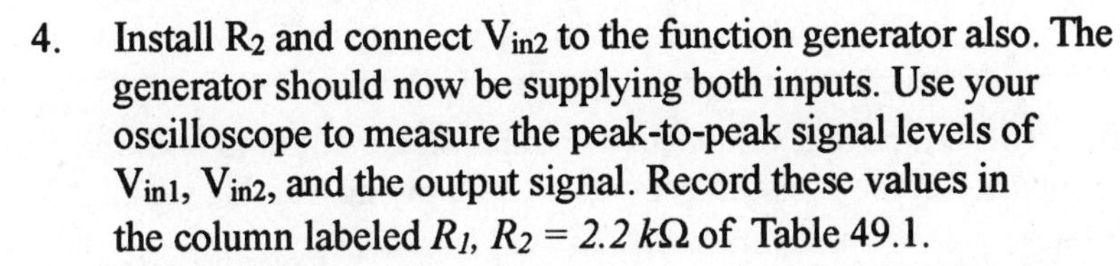

4. Install R_2 and connect V_{in2} to the function generator also. The generator should now be supplying both inputs. Use your oscilloscope to measure the peak-to-peak signal levels of V_{in1}, V_{in2}, and the output signal. Record these values in the column labeled *R_1, R_2 = 2.2 kΩ* of Table 49.1.

   What happened to the amplifier output measured in this step compared to step 2?

5. Turn off the AC and DC sources. Remove the 2.2-kΩ R_2 and replace it with a 3.3-kΩ resistor. Apply DC power and the AC signal of step 2. Use your oscilloscope and measure the peak-to-peak voltage levels of V_{in1}, V_{in2}, and V_{out}. Record these values in the column labeled *R_2 = 3.3 kΩ* of Table 49.1.

6. Change the circuit input connections so that the function generator is supplying V_{in1} only. Connect the voltage divider of Figure 49.2 and connect the voltage divider output to V_{in2}. Adjust the divider to provide a 0.5-V output to V_{in2}.

You have now seen the summing amplifier add input voltages. Beginning with step 6, the procedure steps illustrate a mixing circuit using the summing amplifier.

7. Set your oscilloscope to DC coupling and observe the output signal. Is there a DC offset in the output signal?

   Measure and record below the peak positive voltage level and the peak negative voltage level of the amplifier output.

   $V_{out(pk+)}$ = ____________ $V_{out(pk-)}$ = ____________

   Measure and record the DC offset: V_{DC} = ____________

	R2 open	R1, R2 = 2.2 kΩ	R2 = 3.3 kΩ
Vin1			
Vin2	--		
Vout			
Av			

Table 49.1

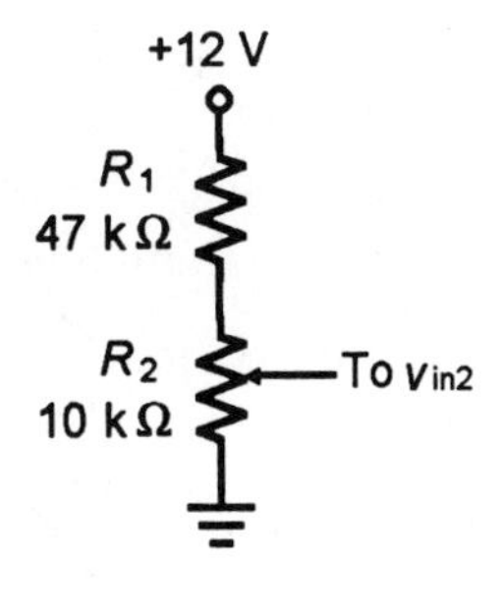

Figure 49.2

Notice that your summing amplifier effectively mixed the AC and DC input signals, providing a composite AC and DC output. It will do exactly that same mixing with different input signal levels and different input frequencies.

SECTION II TROUBLESHOOTING

Fault 1 - Feedback resistor open

1. Remove power from your circuit. Remove the feedback resistor and replace it with a 2.2-MΩ resistor.

2. Can you predict what effect an open R_f will have on the circuit operation?

   ____

3. Reapply circuit DC power. Connect your function generator to both the V_{in1} and V_{in2} inputs. Set the function generator to supply an 800-Hz signal at 500 mV_{p-p}.

4. Connect your oscilloscope to the amplifier output. What kind of waveshape do you observe?

   ____

5. Measure and record below the peak-to-peak output voltage.

   $V_{out(p-p)}$ = ___________

6. Measure the input signal at the op-amp input (pin 2). Is this signal essentially 0?

   Pin 2 input = ___________

7. Measure the op-amp supply voltage values (V+ and V-). Are these normal?

   Power supply voltages = ___________

DISCUSSION

Section I

1. Suppose you are given the circuit of Figure 49.1 supplied with two 1-kHz signals of exactly the same amplitude, but exactly 180 degrees out of phase (V_{in1} with a signal at 0 degrees and V_{in2} with a signal of 180 degrees). Discuss what you would expect to see at the output of the summing amplifier. Include a description of the amplifier operation that would produce the output you expect.

2. Given the circuit of Figure 49.1, what sort of output signal would you expect to see if one input were a 1-kHz signal and the other was a 2-kHz signal?

Section II

Fault 1 - Feedback resistor open

1. From all of the measurements made for this fault, can you rule out the probability of an op-amp failure? Discuss this and cite the measurement(s) that support your answer.

2. Do the measurements made for this fault clearly indicate a feedback resistor failure? Discuss the measurement(s) made that support your conclusion.

Quick Check

1. The output of a summing amplifier is proportional to the sum of the inputs.

   True False

2. V_{out} can be described as the sum of the input voltages times the ratio, $R_f : R_{in}$.

   True False

3. If R_f opens, the output _____.

   (a) could saturate
   (b) increases in proportion to V_{in}
   (c) increases the current gain
   (d) could go to zero

4. The summing amplifier can mix AC and DC voltages.

   True False

50

RC OSCILLATORS

INTRODUCTION

RC oscillators serve to provide very good sources of sinewave signals from very low frequencies to high frequencies in the range of 1 MHz. These oscillators are characterized by a frequency-determining RC network connected in a regenerative feedback loop and a degenerative feedback loop to control the overall gain to meet the Barkhausen criteria for oscillation.

This experiment will let you explore the popular Wien-bridge oscillator and also the twin-T oscillator. The Wien-bridge oscillator uses a series and a parallel RC circuit as bandpass filters connected to the noninverting input. A portion of the output signal is fed back to the inverting input and is used to control the gain of the amplifier.

The twin-T filter oscillator is sometimes called a *notch filter* because it notches out or attenuates frequencies near f_r.

REFERENCE

Principles of Electronic Devices and Circuits - Chapter 13, Sections 13.3 and 13.44

OBJECTIVES

In this experiment you will:

✓ Understand the basic RC oscillator fundamentals

✓ Demonstrate the operation of the Wien-bridge oscillator

✓ Demonstrate the operation of the twin-T oscillator

EQUIPMENT AND MATERIALS

DC power supply
Oscilloscope
Operational amplifier, 741 or equivalent
Small-signal diode [2], 1N914 or similar
Red LED, TIL221 or similar
Resistors: 100 Ω [2], 10 kΩ, 12 kΩ [4], 50 kΩ [2]
Potentiometer: 50-kΩ ten-turn trimpot
Capacitors: 0.01 μF [2], 10 μF [2], 200 μF

Wien-Bridge Oscillator

1. Construct the circuit in Figure 50.1. Set R_5 to 25 kΩ, and temporarily short R_4 and the Z diodes by placing a jumper in parallel.

2. Calculate and record the expected f_c.

   f_c = ___________

3. Turn on circuit power. Monitor the output of the oscillator with your oscilloscope. It is necessary for your oscillator operation that the closed-loop gain be very close to 3, and this requires R_5 to be twice the value of R_3. If R_5 is larger than $2R_3$, the gain will be greater than 3 and the oscillator will produce a squared waveform. If R_5 is less than $2R_3$, the gain will be less than 3 and there will be no oscillation.

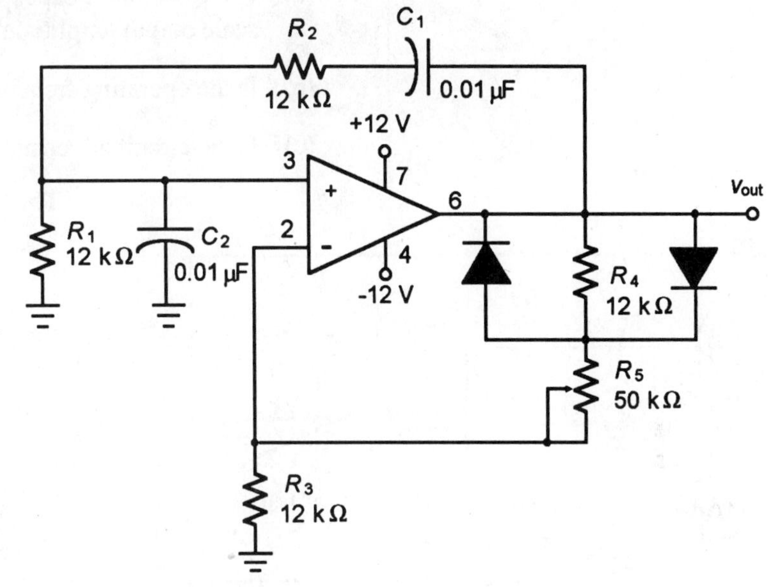

Figure 50.1

4. Slowly adjust R_5 to the point of a stable sinusoidal output waveform. Measure and record below the peak-to-peak oscillator output signal.

   V_{out} (peak-to-peak) = ___________

   Measure and record the frequency at the output signal.

   f_c = ___________

   Does the oscillator appear to be a symmetrical sine wave? _______

5. Slowly readjust R_5 to increase its resistance and note below the results on the output waveform.

   ____

   Slowly readjust R_5 to decrease its resistance and note below the effect on the output waveform.

   ____

6. Readjust R_5 to produce a stable sinusoidal waveform. (You should have found that a very small change in R_5 made a significant change in the oscillator output.)

   R_5 = ___________

7. Replace R_5 in the circuit. Remove the jumper shorting R_4 and the diode network. Adjust R_5 to obtain a stable sinewave form of approximately 4 to 5 $V_{p\text{-}p}$.

Could you do this with the diode network shorted? ___________

8. Slowly adjust R_5 to increase and decrease its resistance while observing the output waveform. Note below the differences in circuit operation. Consider such things as:

   (a) Is the oscillator capable of maintaining oscillation at less than full-scale output amplitude?

   (b) Is the operating frequency the same?

   (c) Is the circuit as sensitive to changes in feedback (R_5 :R_3 ratio)?

The Twin-T Oscillator

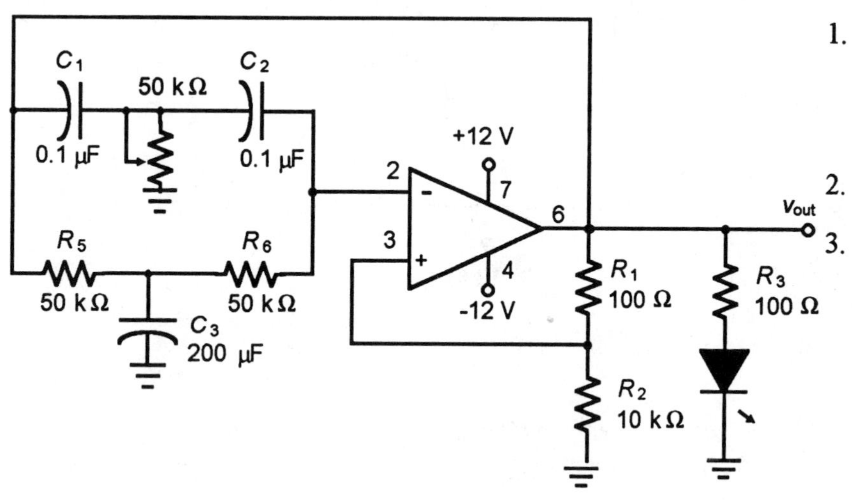

Figure 50.2

1. Calculate f_c for the circuit in Figure 50.2.

   f_c = ___________

2. Construct the circuit in Figure 50.2.

3. Turn on the power. Use your oscilloscope to monitor V_{out}. While observing the LED, adjust potentiometer R_4 until the LED no longer blinks. This is the notch of the twin-T oscillator. Readjust R_4 for a stable sinewave output. Measure and compare the oscillator frequency to that calculated in step 1.

4. Remove power from the circuit. Replace C_1 and C_2 with 0.01-μF capacitors. Recalculate f_c.

   f_c = ___________

5. Return the potentiometer R_4 to its greatest resistance value (50 kΩ).

6. Reapply circuit power. Adjust R_4 to obtain a stable, undistorted sinewave output. Measure the frequency of the output and compare to your frequency calculation of step 4.

   Note: You should find pretty close agreement in the two values.

   f_c = ___________ (measured)

7. Monitor the circuit output and slowly adjust R_4 (near its midpoint) to a point of minimum of no output. This the notch point of the twin-T. Recall that at the notch frequency minimum regenerative feedback is obtained, and at this frequency the phase shift through the twin-T filter is 0°. This will occur at an R_4 setting of 1/2R: the 50 k**Ω of R_5** and R_6.

8. Remove the potentiometer and measure the resistance value. Does it measure R/2? ___________

DISCUSSION

1. Discuss the effect of adjusting the potentiometer on the output of the Wien-bridge oscillator. Include in your discussion why the output would go to saturation (clipping or no output).
2. Discuss the effect of disabling the diode network of the Wien-bridge oscillator. Did the diode network add stability to the oscillator?
3. What effect did changing the twin-T capacitors C_1 and C_2 have on the operation of the circuit?

Quick Check

1. Another name for the twin-T filter oscillator is _____________.
2. The Wien-bridge oscillator is used for oscillation below 1 MHz.

   True False
3. The twin-T oscillator is part of the negative feedback circuit and therefore increases gain.

   True False
4. Gain is dependent on feedback signal strength.

   True False

51

COLPITTS AND CLAPP OSCILLATORS

INTRODUCTION

The Colpitts and the Clapp oscillators are used wherever frequencies greater than 1 MHz are needed. The Clapp oscillator is a variation of the Colpitts oscillator, having an extra capacitor in the tank circuit, which is the primary frequency-determining capacitor, providing a more stable oscillator.

In this experiment you will construct a Colpitts and Clapp oscillator, calculate the output frequency, and observe and measure the output frequencies.

The troubleshooting section will let you see the effect of an open capacitor on the oscillator operation.

REFERENCE

Principles of Electronic Devices and Circuits - Chapter 13, Section 13.5

OBJECTIVES

In this experiment you will:

✓ Determine the operating characteristics of the Colpitts oscillator

✓ Determine the operating characteristics of the Clapp oscillator

EQUIPMENT AND MATERIALS

DC power supply
Oscilloscope
Frequency counter
Circuit protoboard
NPN transistor, 2N3904 or equivalent
Resistors: 4.7 kΩ, 120 kΩ, 240 kΩ
Capacitors: 20 pF, 100 pF [2], 0.01 μF [2], 0.022 μF
Inductors: 10 mH, 1 mH

SECTION I **FUNCTIONAL EXPERIMENT**

The Colpitts Oscillator

1. Construct the circuit in Figure 51.1.

2. Connect power to the circuit. Observe the output frequency with the oscilloscope.

   Note: Use the ×10 probe for this measurement.

3. Using the frequency counter, measure the output frequency.

   f_c = ________ (measured)

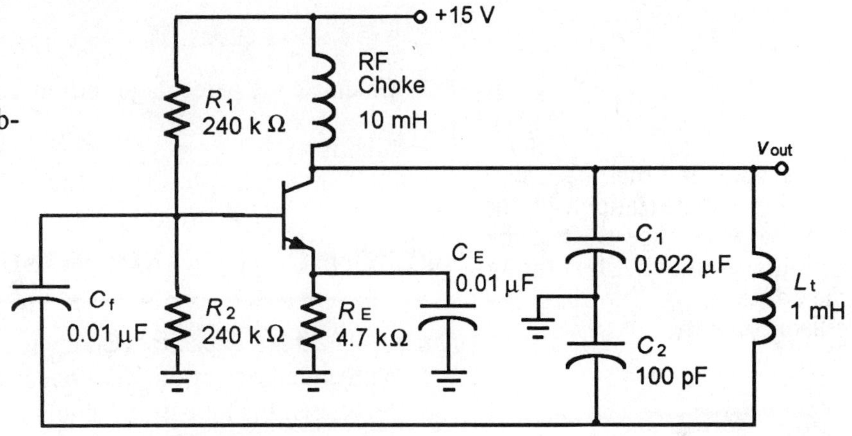

Figure 51.1

4. Calculate the expected resonant frequency for your oscillator. How well does this compare to the measured frequency?

   f_c = ___________(calculated)

5. Turn the DC power off.

6. Add a 20-pF capacitor from the base of the transistor to ground. Connect this capacitor right at the base of the transistor.

7. Reapply circuit power. Measure the output frequency of your oscillator and record below.

   f_c = ___________

   Calculate the percentage of frequency shift from the original frequency measured.

   % shift = ___________

The difference between the Colpitts and Clapp oscillators is their frequency sensitivity to changes in the input capacity of the amplifier. Beginning with step 6, you will simulate a change and measure the corresponding frequency shifts for the Colpitts. In a later part of this experiment, you will do the same for a Clapp oscillator.

The Clapp Oscillator

1. Disconnect power. Insert a 100-pF capacitor between the output and L_t to create the Clapp oscillator of Figure 51.2. Disconnect the 20-pF capacitor in the transistor base.

2. Using the oscilloscope and the frequency counter, observe and measure the output signal.

   f_c = ___________

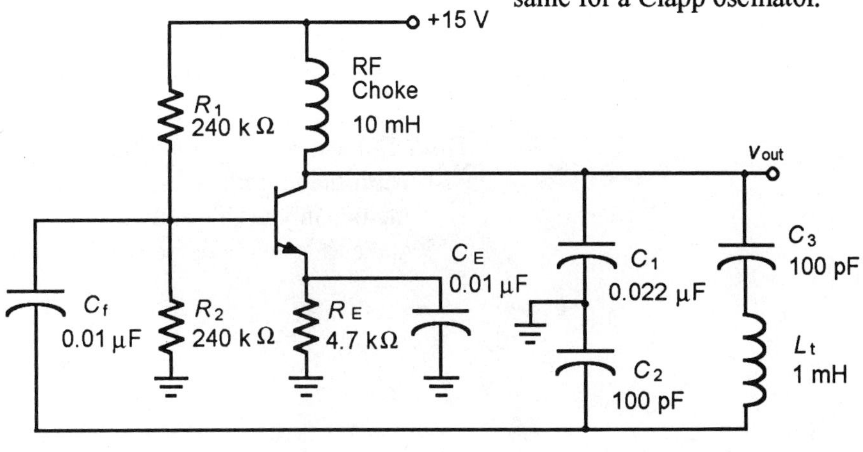

Figure 51.2

3. Turn DC power off. Add a 20-pF capacitor between the transistor base and ground.

4. Reapply circuit power. Using your oscilloscope, observe the output signal. With the frequency meter, measure the output frequency and record it below.

   f_c = __________

   Calculate the percentage shift in frequency caused by the additional capacity.

   % shift = __________

You should find that the Clapp oscillator experienced about half the percentage shift of the Colpitts oscillator, illustrating the advantage of the Clapp oscillator stability.

SECTION II TROUBLESHOOTING

Fault 1 - Feedback capacitor (C_f) open

1. With circuit power off, disconnect and remove the 20pF capacitor shunting the base of the transistor. Simulate an open feedback capacitor by disconnecting one end of C_f.

2. Using the oscilloscope, monitor the output signal. What do you observe?

DISCUSSION

Section I

1. Discuss the reason, using frequency calculations, why the 20-pF capacitor affected the Colpitts oscillator to a greater extent than the Clapp oscillator.

2. For the Clapp oscillator, at what frequency did the capacitive reactance of C_{eq} equal the inductive reactance of L_t?

3. Discuss the feedback signal to the base of the transistor. Is the feedback regenerative or degenerative?

Section II

Fault 1 - Feedback capacitor (C_f) open

Discuss the effect of an open feedback capacitor on your oscillator. Include in your discussion the additional measurements you need to make to isolate a defective feedback capacitor.

Quick Check

1. The Colpitts oscillator is typically used where frequencies between 1 MHz and 500 MHz are needed.

   True False

2. The Clapp oscillator is more stable than the Colpitts oscillator.

   True False

3. The Colpitts oscillator uses an RC network as feedback.

   True False

4. The Clapp oscillator is a modification of the Colpitts oscillator.

   True False

52
HARTLEY OSCILLATORS

INTRODUCTION

The Hartley oscillator is another type of LC oscillator. Like the Colpitts and the Clapp oscillators, the Hartley oscillator requires feedback and uses a resonant LC network to determine the frequency of operation. In the Hartley oscillator, feedback is developed by the inductive voltage divider of L_1 and L_2.

This experiment will let you build and test a Hartley oscillator and check your measured values against your calculated values.

In the troubleshooting section, you will simulate a circuit failure and make measurements to see the effect on circuit operation.

REFERENCE

Principles of Electronic Devices and Circuits - Chapter 13, Section 13.5

OBJECTIVES

In this experiment you will:

✓ Determine the characteristics of the Hartley oscillator

✓ Determine the effect of a circuit failure on the circuit operation

EQUIPMENT AND MATERIALS

DC power supply
Oscilloscope
Frequency counter
Circuit protoboard
NPN transistor, 2N3904 or equivalent
Resistors: 2.2 kΩ, 10 kΩ [2]
Capacitors: 20 pF, 47 pF, 100 pF, 0.01 μF [2]
Inductors: 15 μH, 1 mH, 10 mH

SECTION I FUNCTIONAL EXPERIMENT

1. Construct the circuit in Figure 52.1.

2. Using the oscilloscope and the frequency counter, measure and observe the waveform at the output of the oscillator.

   Note: Use the ×10 probe for oscilloscope observations.

   f_c = ___________

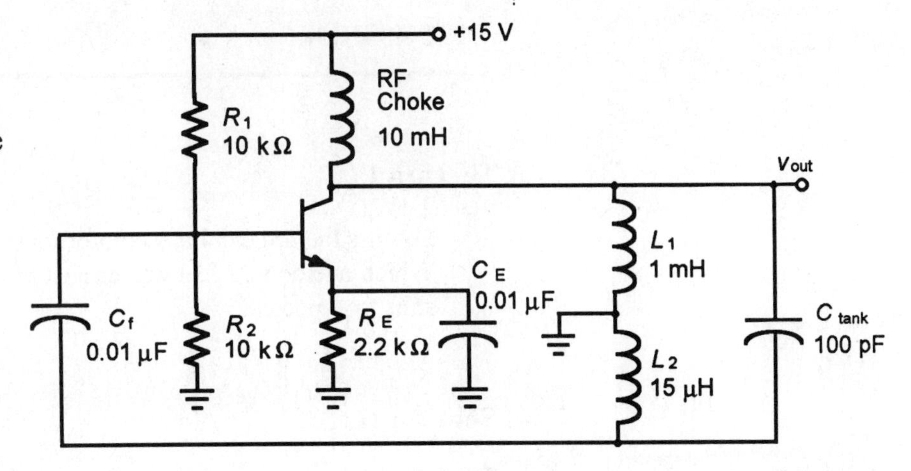

Figure 52.1

3. Calculate the resonant frequency for the circuit in Figure 52.1.

   f_c = ___________

4. Remove power from the circuit. Replace the 100-pF capacitor with a 47-pF capacitor. What effect do you predict this will have on the operation of the circuit?

   ____

5. Reconnect power and measure and observe the output signal.

   f_c = ___________

SECTION II TROUBLESHOOTING

Fault 1 - Tank capacitor open

1. Remove power from the circuit. Disconnect the tank capacitor from the circuit.

2. Reconnect power. Observe and measure the output frequency.

   f_c = ___________

Fault 2 - Reduced feedback signal

1. Disconnect power from your circuit. Reconnect the tank capacitor. Connect an 82-kΩ resistor in series with C_f, between C_f and the junction of R_1 and R_2.

   Note: This is not a normal failure mode for the circuit; however, it will let you observe the effect of a low amplifier input impedance in reducing the effective feedback of the oscillator tank.

2. Reconnect power and measure the output signal.

f_c = ___________

DISCUSSION

Section I

1. Discuss the three delta oscillator forms (Colpitts, Clapp, and Hartley). Which one do you feel was easiest to work with and gave the closest results to expected?

Section II

Fault 1 - Tank capacitor open

Describe the effect of an open tank capacitor on the operation of your oscillator.

Note: An inductor will self-resonate due to the distributed capacity of the windings. Could this have had an effect on your observed results?

Quick Check

1. The Hartley oscillator is an example of an LC oscillator.

   True   False

2. An open C_{fb} has no effect on the operation of the oscillator.

   True   False

3. The feedback ratio of a Hartley oscillator is C_1/C_2.

   True   False

4. The term for introducing a small amount of energy back into an electronic oscillator is _____.

   (a) a current mirror   (b) feedback
   (c) reactance   (d) resonance

5. The feedback to the transistor base is degenerative.

   True   False

53 RELAXATION OSCILLATORS

INTRODUCTION

An important application of RC networks is in the relaxation oscillator, where the RC timing controls the circuit, producing a square wave. This type of oscillator is often called a *multivibrator*.

In this experiment you will build an op-amp relaxation oscillator and, through circuit measurements, see the basic characteristics of the relaxation oscillator.

The troubleshooting section will let you simulate a circuit failure and see the effect on the circuit operation.

REFERENCE

Principles of Electronic Devices and Circuits - Chapter 13, Section 6

OBJECTIVES

In this experiment you will:

✓ Understand the characteristics of a relaxation oscillator

✓ Determine the effects of changing the capacitive value on the output waveform

✓ Understand the effect of a circuit failure on the circuit operation

EQUIPMENT AND MATERIALS

DC power supply
Oscilloscope
Frequency counter
Circuit protoboard
Operational amplifier, 741 or equivalent
Resistors: 2.2 kΩ, 18 kΩ, 2.2 M
Potentiometer: 5-kΩ ten-turn trimpot
Capacitors: 0.01 μF, 0.47 μF [2], 0.1 μF

SECTION I FUNCTIONAL EXPERIMENT

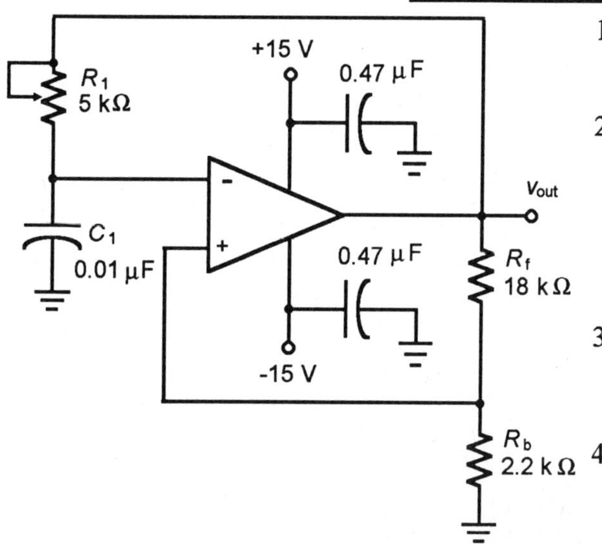

Figure 53.1

1. Construct the circuit in Figure 53.1. Adjust the potentiometer for a resistance of 5 kΩ.

2. Turn on the power. Measure and observe the waveform and frequency at the output.

   V_{out} = ________________

   f_o = ________________ (measured)

3. Determine the output frequency for your oscillator.

   f_o = ________________ (calculated)

4. Connect channel 2 of your oscilloscope to the top of capacitor C_1 (at the inverting input of the op-amp). With channel 1 connected to the oscillator output, observe the capacitor waveform versus the output diagrams of Figure 13.35 in your text.

   Can you relate the UTP and LTP to the output waveform? ________

   ____

5. Disconnect power and adjust the potentiometer for a resistance of 2.5 kΩ. What effect do you predict this will have on the operation of the circuit?

   ____

   ____

6. Turn on power. Measure and observe the waveform and frequency at the output.

   f_o = ________________ (measured)

7. Turn power off. Replace C_1 with the 1μF capacitor.

8. Recalculate the output frequency.

   f_o = ________________ (calculated)

9. Reconnect power. Monitor and measure the output frequency.

   f_o = ________________ (measured)

SECTION II TROUBLESHOOTING

Fault 1 - Potentiometer R_1 open

1. Turn off the power. To simulate an open (very high resistance) failure of the potentiometer, disconnect the potentiometer (R_1) and replace it with a

2.2-MΩ resistor. What do you predict the effect will be on the operation of the oscillator?

2. Turn on the power. Monitor the output signal. Is there an output square wave?

Fault 2 - Capacitor C_1 open

1. Turn off power. Replace the potentiometer in the circuit and remove the 2.2-MΩ resistor. Discnnnect the ground lead of C_1.

2. Turn on the power. Monitor the output signal and note below the output you observe.

____

DISCUSSION

Section I

1. Discuss the effect of the feedback value on the frequency of operation of the relaxation oscillator. For example, if in the circuit of Figure 53.1, R_b were changed to 4.7 kΩ, what effect would this have on the output frequency?

2. Describe how you would modify or what additional circuitry you would add to the circuit of Figure 53.1 to make the output TTL compatible, that is, to obtain an output square wave that changes from 0 V to +5 V.

Section II

Fault 1 - Potentiometer R_1 open

In step 1 of this section you simulated a very large resistance failure of the potentiometer. Describe the effect on the circuit if R_1 opened completely.

Fault 2 - Capacitor C_1 open

Could an op-amp failure have resulted in the same measured output from your oscillator? Discuss what additional measurements you would make to isolate the failure to capacitor C_1.

Quick Check

1. A relaxation oscillator produces a sawtooth wave form.

   True   False

2. The term *trip point* refers to the voltage the capacitor must exceed to produce changes in output.

True False

3. The output frequency depends on the charge and discharge time of the resistors.

True False

4. If the R_1 value were 3 kΩ and the C_1 value were 0.47 μF, what would the output frequency be for the circuit in Figure 53.1?

(a) 2.5 kHz (b) 25 kHz
(c) 250 Hz (d) 250 k Hz

54

DIFFERENTIATORS AND INTEGRATORS

INTRODUCTION

The *differentiator* is an electronic circuit that performs the mathematical function of differentiation. The output voltage is proportional to the slope (rate of change) of the input voltage. The *integrator* performs the mathematical function of integration.

Differentiator and integrator circuits find application in industrial electronic control loops. Additionally, they can be used to change signal waveforms. The differentiator will produce a rectangular waveform from a triangular input. The integrator will produce a triangular waveform from a rectangular input.

In this experiment you will build and explore the op-amp differentiator and integrator. You will also see the waveform-changing function of these circuits.

The troubleshooting section will let you see the effect of a simulated circuit failure on circuit operation.

REFERENCE

Principles of Electronic Devices and Circuits - Chapter 13, Section 13.7

OBJECTIVES

Through this experiment you will:

✓ Understand the operation of the differentiator

✓ Understand the operation of the integrator

EQUIPMENT AND MATERIALS

DC power supply
Function generator
Dual-trace oscilloscope
Circuit protoboard
Operational amplifier, 741 or equivalent
Resistors: 1 kΩ, 10 kΩ
Capacitors: 0.01 μF, 0.47 μF [2]

SECTION I FUNCTIONAL EXPERIMENT

The Differentiator

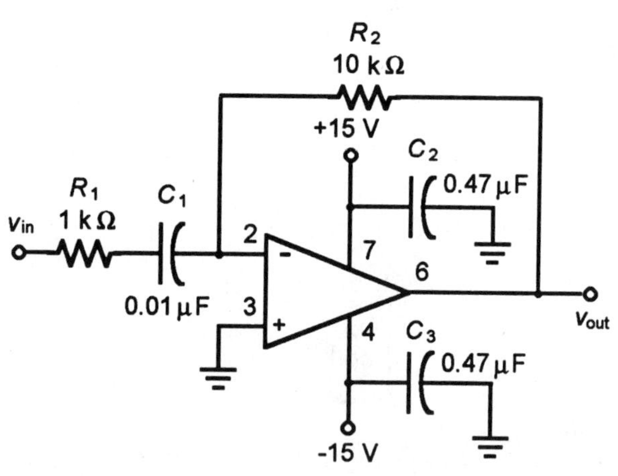

Figure 54.1

1. Construct the circuit in Figure 54.1.

2. Connect your function generator set to squarewave output and adjust to 2 $V_{p\text{-}p}$ at 1 kHz. Apply DC power to the circuit.

3. Connect your oscilloscope in dual-trace mode to monitor both v_{in} and v_{out}. Observe the signal waveforms. The output of the differentiator is given by:

$$v_{out} = C_1 R_2 \left(\frac{\Delta v_{in}}{\Delta t} \right)$$

Notice that at the leading and trailing edges of the input signal there is an output from the differentiator. Also when the input is constant, even though it is not zero, there is no output.

Sketch the oscilloscope display in Graph 54.1, noting the peak values of the signals.

4. Switch the function generator to obtain a triangular waveform of 2 $V_{p\text{-}p}$ at 1 kHz. Observe the oscilloscope display. Notice that as long as the input signal is changing at a fixed rate, the output is a nearly constant value.

Sketch the oscilloscope display in Graph 54.2, noting the signal peak values.

5. To illustrate the differentiator function, set your function generator to obtain a sinewave output of 2 $V_{p\text{-}p}$ at 1 kHz. Notice that your output is also a sinusoidal waveform. It is, however, phase shifted by 90 degrees to the input. This is a cosine wave, since the derivative of the sine is the cosine.

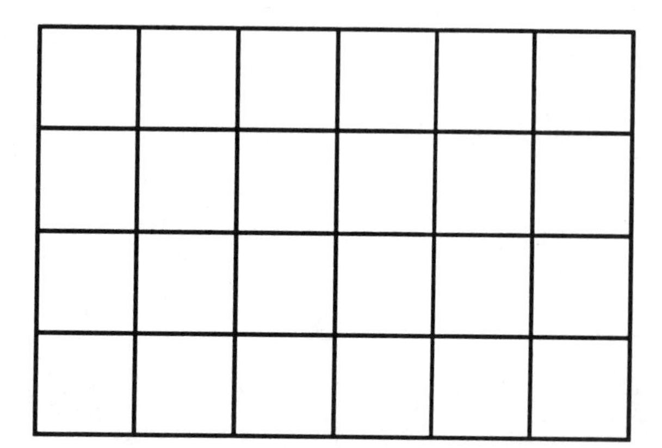

Graph 54.1

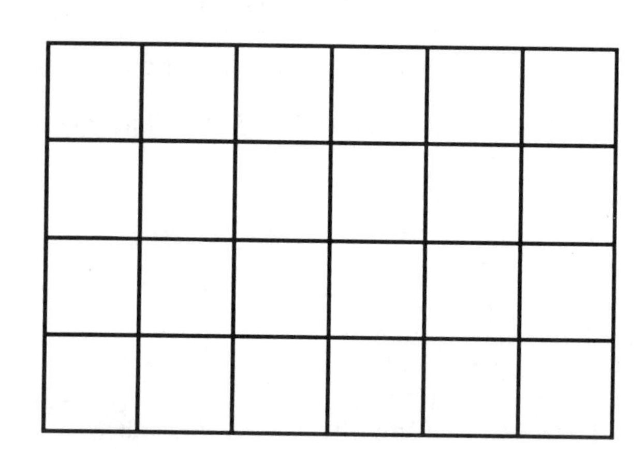

Graph 54.2

The Integrator

1. Construct the circuit in Figure 54.2.

2. Apply DC power to your circuit. Set the function generator to provide a squarewave signal of 2 $V_{p\text{-}p}$ at 10 kHz.

3. Connect your oscilloscope to monitor both the input and output signals.

As long as the input signal is a constant value, either positive or negative, the output of the integrator will be a voltage increasing at a rate determined by R_1C_1 in the opposite polarity of V_{in}.

The output of the integrator is given by:

$$vout = -\frac{1}{R_1C_1}(vin\Delta t)$$

Sketch the oscilloscope display in Graph 54.3, noting the peak values of the signals.

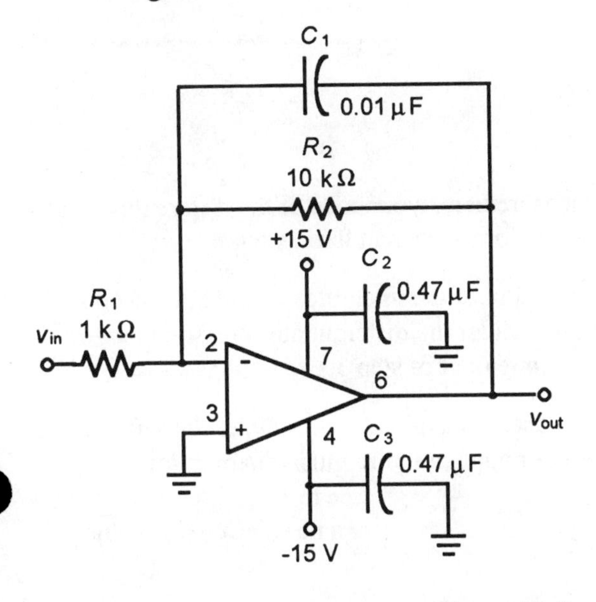

Figure 54.2

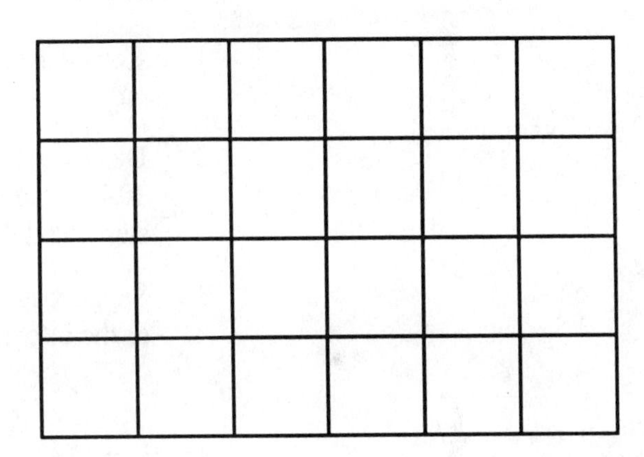

Graph 54.3

SECTION II TROUBLESHOOTING

Fault 1 - Integrator R_2 open

1. Remove power from the circuit. Remove R_2 from the circuit.

2. Reapply power to the circuit. Make two sets of measurements, the first with the function generator disconnected and V_{in} jumpered to ground. Repeat this measurement two or three times, watching the amplifier output on your oscilloscope carefully as you switch the DC power on. After a short interval, switch the DC power off, and briefly short capacitor C_1 to discharge before reapplying power.

   For the second measurement connect your function generator and apply a 2-$V_{p\text{-}p}$, 10- kHz squarewave input. Measure the input and output signals. Record your observations for each measurement.

   First Measurement ______________________________

   Second Measurement ______________________________

Fault 2 - Integrator C_1 open

1. Remove power from the circuit. Replace R_2.

2. Remove C_1 from the circuit. Before reconnecting power, can you predict what the output might be?

3. Reapply DC power and the squarewave input. Monitor the input and output signals. Note your observations.

____

____

DISCUSSION

Section I

1. Considering the various measurements you made of the differentiator, discuss the one you found most informative and interesting.

2. Discuss what you think the output of the integrator would have been with a 10-kHz sinewave applied. (Consider the differentiator measurement of step 4 and the integrator measurement of step 3.)

3. Suggest an automotive or household application for either the differentiator or the integrator. Describe your application in block diagram form.

Section II

Fault 1 - Integrator R_2 open

Do you think your circuit would have displayed the results you observed if you had been using an "ideal" op-amp? An ideal op-amp has an input offset voltage of zero and a perfectly balanced input bias current. Discuss this concept, indicating why you chose your response to the question.

Fault 2 - Integrator C_1 open

Discuss your measurement results for this fault. Did your observed results point clearly to C_1 being the problem?

Quick Check

1. A differentiator produces a spiked output if the input waveform is a square wave.

   True False

2. An integrator produces a square wave output from a spiked input.

   True False

3. The differentiator response to the input pulse width is dependent on the RC time constant.

True False

4. The integrator response to a squarewave input is a spiked output.

True False

55

OP-AMP DIODE CIRCUITS

INTRODUCTION

The use of an op-amp in diode circuits provides nearly "ideal" circuit results. The active rectifier, limiter, and clamper circuits all have the diode in the feedback loop, effectively eliminating the forward voltage drop of the diode in the circuit output.

In this experiment you will build the active halfway rectifier, limiter, and clamper. In addition, you will examine the each circuit in operation, making measurements to see the circuit characteristics.

REFERENCE

Principles of Electronic Devices and Circuits - Chapter 14, Section 14.1

OBJECTIVES

In this experiment you will:

✓ Understand the operation of the half-wave rectifier

✓ Observe the operation of an active limiter

✓ Work with an active clamper

EQUIPMENT AND MATERIALS

Function generator
DC power supply
Dual-trace oscilloscope
Frequency counter
Circuit protoboard
Diode, 1N914 or similar
Operational amplifier, 741 or equivalent
Resistors: 1 kΩ [2], 2.2 kΩ, 10 kΩ, 12 kΩ, 100 kΩ, 220 kΩ
Potentiometer: 10-kΩ ten-turn trimpot
Capacitors: 0.01 μF [2], 0.33 μF

The Half-Wave Rectifier

The half-wave rectifier you will be testing in this first part of the experiment not only provides half wave rectification; it also provides a signal gain. You should expect to see an output waveform that eliminates the diode voltage drop and is larger than the input signal supplied.

1. Construct the circuit in Figure 55.1.
2. Connect your oscilloscope to monitor the input signal and the output signal across the load resistor.
3. Set the generator to provide a 300-m$V_{p\text{-}p}$ sinewave at 500 Hz. Apply DC power to the circuit.
4. Sketch the input and output waveforms in Graph 55.1, noting the signal peak values.
5. Adjust the oscilloscope vertical trace controls so that the input and output waveforms overlap each other. Is there any indication of diode forward drop in the output waveform? ___________
6. Turn off the DC power. Reverse the polarity of the diode. Sketch the expected output waveform in Graph 55.2.
7. Reapply DC power and the input signal of step 3. Sketch the input and output waveforms in Graph 55.3, noting the signal peak values.

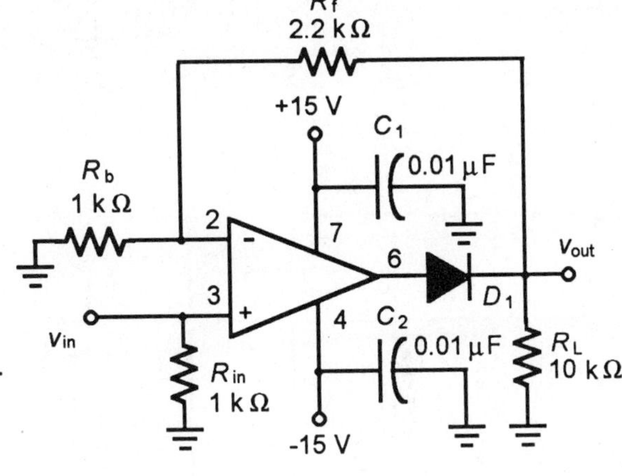

Figure 55.1

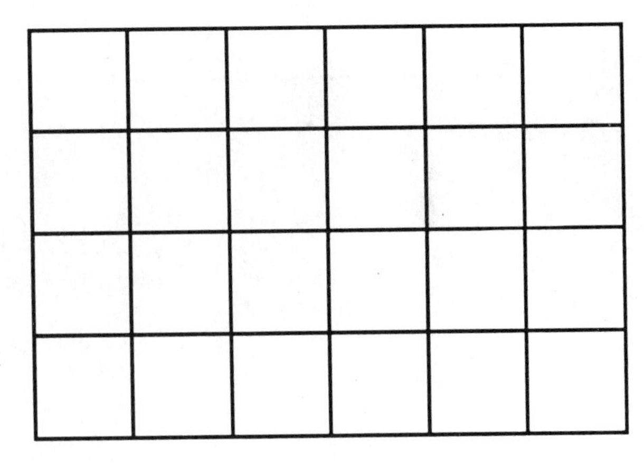

Graph 55.1

The Active Limiter

1. Construct the circuit in Figure 55.2. Initially adjust voltage divider resistor R_2 for zero output.
2. Set the function generator to sinewave output of 2 $V_{p\text{-}p}$ at 1 kHz and connect the function generator to the limiter input junction. Turn on DC power.

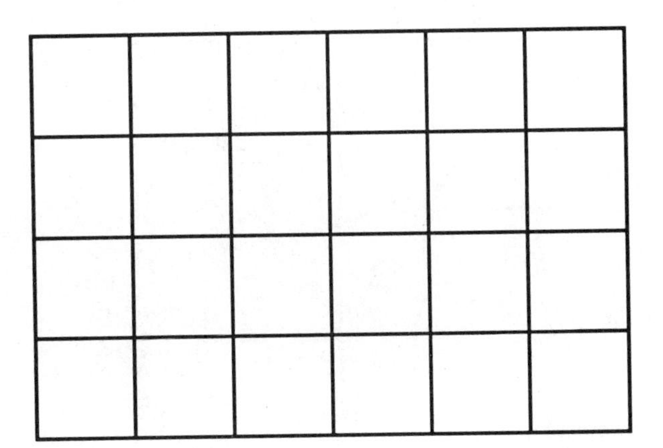

Graph 55.2

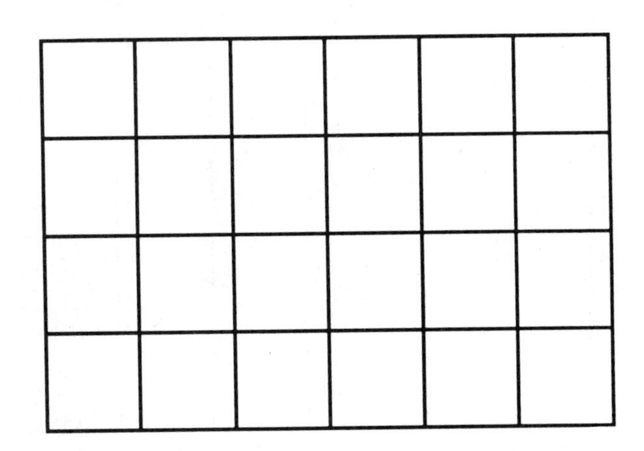

Graph 55.3

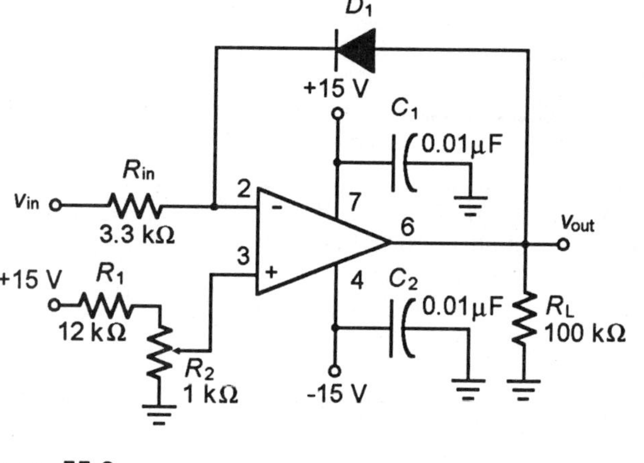
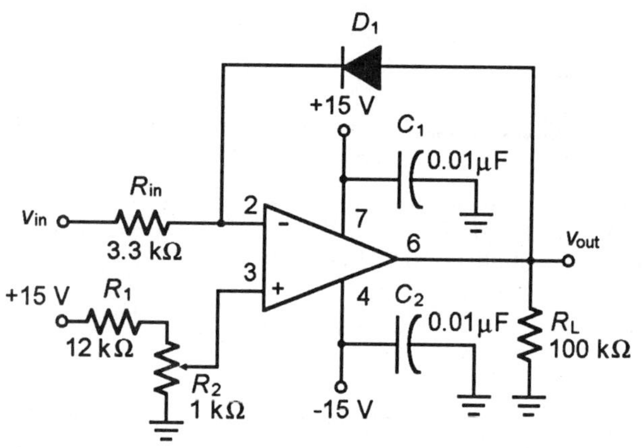

Figure 55.2

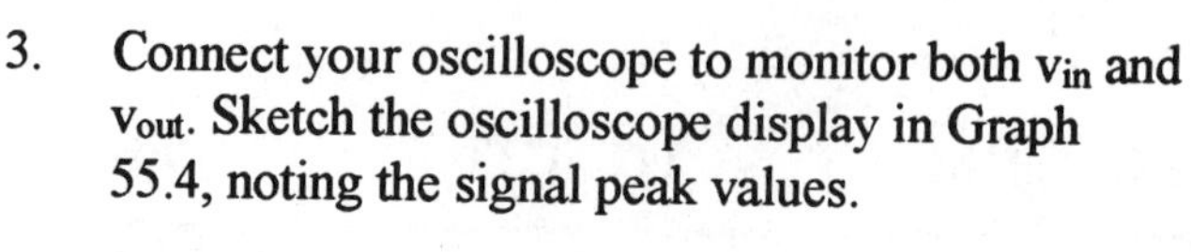

3. Connect your oscilloscope to monitor both v_{in} and v_{out}. Sketch the oscilloscope display in Graph 55.4, noting the signal peak values.

4. Adjust the voltage divider R_2 for a voltage of 0.75 V to the op-amp noninverting input. Sketch the output waveform in Graph 55.5, noting the 0-V reference levels and peak signal values.

5. Vary the setting of R_2 while observing the output waveform. Describe the effect of a larger and smaller voltage to the op-amp noninverting input.

____

____

Graph 55.4

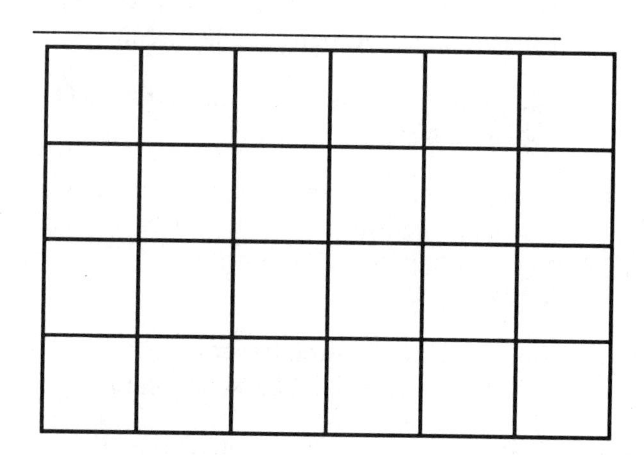

Graph 55.5

The Active Clamper

1. Construct the circuit in Figure 55.3. Connect your function generator to the clamper input and apply DC power.

2. Set the function generator to provide a 2-$V_{p\text{-}p}$ sine wave at 1 kHz.

3. Connect your oscilloscope to monitor v_{in} and v_{out} (v_{out} DC coupled).

4. Sketch the oscilloscope display in Graph 55.6, noting the 0-V reference and peak values of the signals.

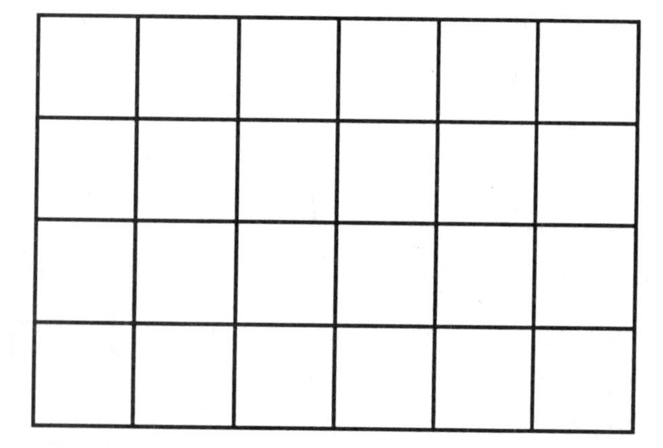

Graph 55.6

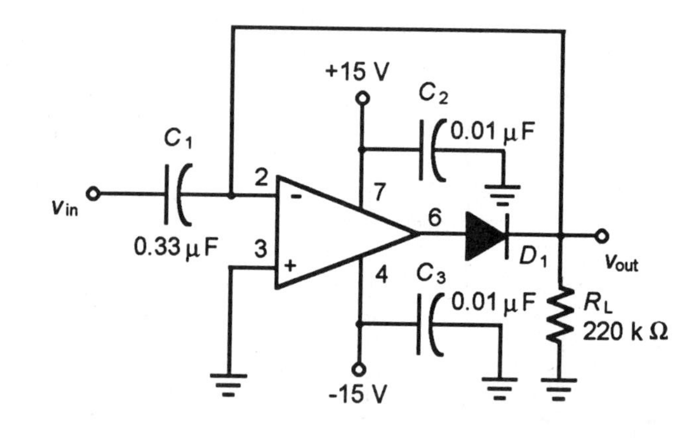

Figure 55.3

5. Turn off circuit power. Connect the diode in reversed direction (cathode to the op-amp).

6. Reapply DC power and input signal of 2 V_{P-P} at 1 kHz. With your oscilloscope connected as in step 3, observe the circuit input and output waveforms.

   Is the output signal positively or negatively clamped?

   ____

   Is the signal clamped to the 0-V reference, or is there a small overshoot due to the diode voltage drop?

   ____

Graph 55.7

7. Sketch the oscilloscope display in Graph 55.7, noting the 0-V reference and signal peak values.

DISCUSSION

The Half-Wave Rectifier

Contrast the active diode rectifier to the discrete device rectifier. Consider what the expected output of the discrete device rectifier would be with a 2-V_{P-P} input signal, taking into account the diode forward voltage.

The Active Limiter

Discuss the effect of the positive DC bias on the active limiter, and describe the circuit operation to produce the output you observed with the DC voltage set to 0.75 V.

The Active Clamper

1. Why is the effect of the diode forward voltage drop not seen in the clamping levels in the active clamper? Describe the circuit operation to produce the output waveforms you observed.

2. Contrast the active clamper to a discrete device clamper. Consider the input signal magnitudes, the complexity of circuitry, and the two circuit results. For example, are there signal levels where you would not recommend the active clamper?

Quick Check

1. The term *nonlinear* means that something is not moving in a straight line.

   True False

2. The output limit of the limiter is established by the DC level at the non-inverting input.

   True False

3. The clamper changes the reference of the output signal.

   True False

4. Diode polarity has no effect on the polarity of the limiter output signal.

   True False

56

SCHMITT TRIGGER CIRCUITS

INTRODUCTION

Schmitt trigger circuits are comparators that use positive feedback to accentuate the switch transition. Positive feedback aids in reducing spurious triggering. The Schmitt trigger is useful in industrial and control applications.

In this experiment you will construct an inverting and a noninverting Schmitt trigger and observe the operation of both of these circuits.

REFERENCE

Principles of Electronic Devices and Circuits-Chapter 14, Section 14.2

OBJECTIVES

In this experiment you will:

✓ Study the operating characteristics of the inverting Schmitt trigger

✓ Observe the operating characteristics of the noninverting Schmitt trigger

EQUIPMENT AND MATERIALS

DC power supply
Function generator
Dual-trace oscilloscope
Circuit protoboard
Small-signal diode [2], 1N914 or similar
Operational amplifier, 741 or equivalent
Resistors: 1 kΩ, 4.7 kΩ, 10 kΩ, 12 kΩ, 15 kΩ, 68 kΩ, 82 kΩ

SECTION I FUNCTIONAL EXPERIMENT

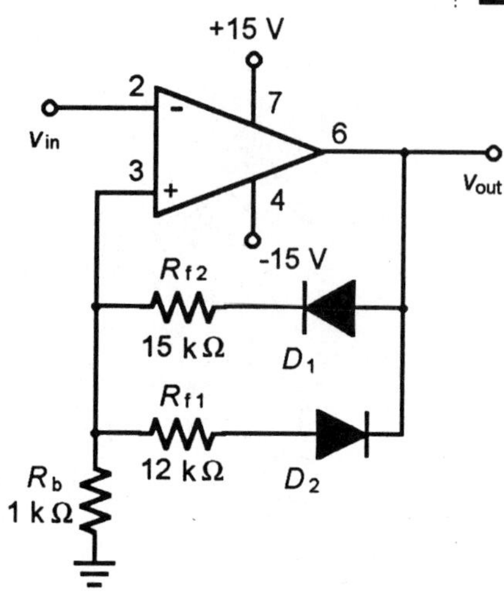

Figure 56.1

The Inverting Schmitt Trigger

1. Construct the circuit in Figure 56.1. Apply power to the circuit. Adjust the function generator for a 10-V_{p-p} sine wave at 400 Hz.

2. Monitor the input and output signals and sketch the waveforms in Graph 56.1, noting the voltage level of the input signal when your Schmitt trigger changes state.

3. Remove power from the circuit. Replace R_{f2} with a 10-kΩ resistor. Predict the UTP and LTP.

   UTP = ______________ LTP = ______________

4. Turn on the power, and monitor the input and output waveforms. Sketch the oscilloscope display in Graph 56.2, noting the input signal voltage when the Schmitt trigger changes state.

Graph 56.1

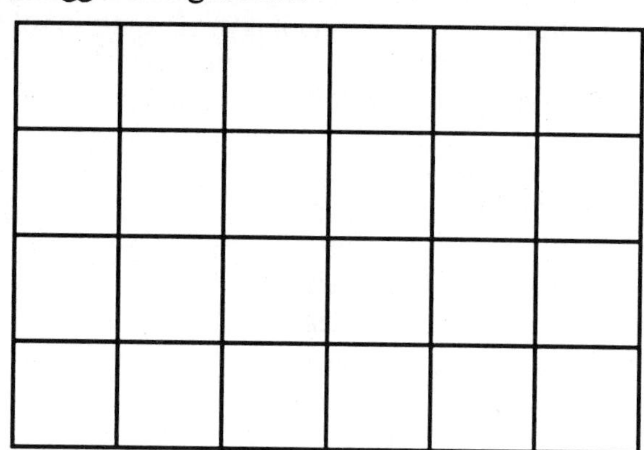

Graph 56.1

The Noninverting Schmitt Trigger.

1. Construct the circuit in Figure 56.2. Apply DC power. Adjust the function generator for a 10-V_{p-p} sine wave at 1 kHz, and connect it to the circuit input.

2. Monitor the input and output of the Schmitt trigger. Sketch the oscilloscope waveforms in Graph 56.3, noting the input signal voltage at the point where the Schmitt trigger output switches.

3. Remove power from the circuit. Replace R_{in} with a 4.7-kΩ resistor. Predict the effect on the output waveform. Recalculate the new UTP and LTP.

   UPT = ______________ LTP = ______________

4. Reapply power and monitor the input and output signals. Sketch your oscilloscope display in Graph 56.4, noting the input voltage level at the point where the Schmitt trigger output switches.

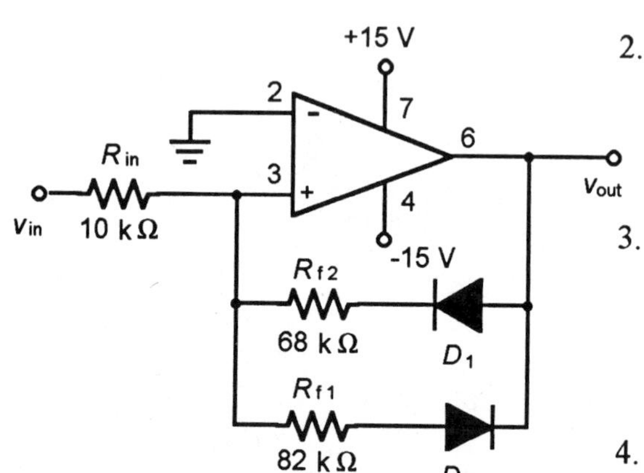

Figure 56.2

Graph 56.3

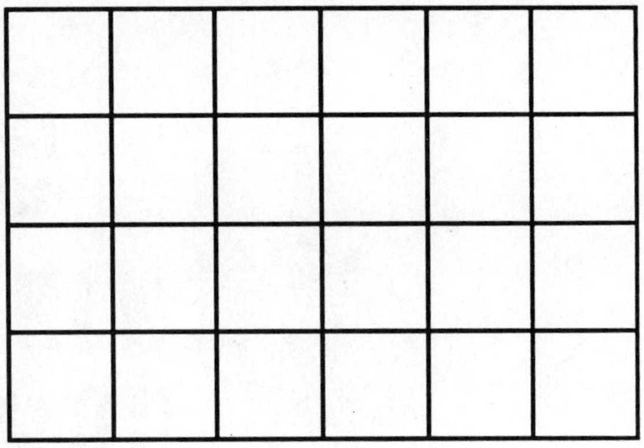

Graph 56.4

DISCUSSION

The Inverting Schmitt Trigger

1. Your inverting Schmitt trigger had a built in hysteresis. Define *hysteresis* and illustrate your definition using the measured values of step 3.
2. Explain how you might use a Schmitt trigger with a temperature transducer whose output is 700 mV at 70°F and 750 mV at 75°F, to control temperature in a house.

The Noninverting Schmitt Trigger

Sketch the hysteresis curve for your Schmitt trigger circuit of Figure 56.2 using the data of step 3. Then determine the new value of UTP and LTP if diode D_1 and R_{f2} were disconnected. Sketch the new hysteresis curve for that circuit.

Quick Check

1. The Schmitt trigger is a wave-shaping circuit.

   True False

2. The term LTP stands for _____.

   (a) Lower Trigger Point (b) Lower Transition Point
   (c) Last Transition Point (d) Last Trigger Point

3. The Schmitt trigger produces a squarewave output for any wave form at the input.

   True False

4. What is the UTP and LTP for the circuit of Figure 56.1 with $R_b = 2.2\ k\Omega$?

   UTP = ____________ LTP = ____________

57
WINDOW COMPARATORS

INTRODUCTION

The window comparator is a special form of comparator. The output of the window comparator is active only when the input signal is within a given voltage range. Hence, the name *window comparator*.

In this experiment you will build an op-amp window comparator. Then you will make measurements to let you see the operating characteristics of this comparator.

The troubleshooting section will let you simulate a circuit fault and make measurements to see the effect of the fault.

REFERENCE

Principles of Electronic Devices and Circuits - Chapter 14, Section 14.3

OBJECTIVES

In this experiment you will:

✓ Understand the operating characteristics of the window comparator

✓ Determine the effect of a failure on the circuit parameters

EQUIPMENT AND MATERIALS

Function generator
DC power supply
Digital multimeter
Circuit protoboard
Small-signal diode [2], 1N914 or similar
Operational amplifier, 741 or equivalent
Resistors: 1 kΩ [2], 2.2 kΩ, 3.3 kΩ

SECTION I FUNCTIONAL EXPERIMENT

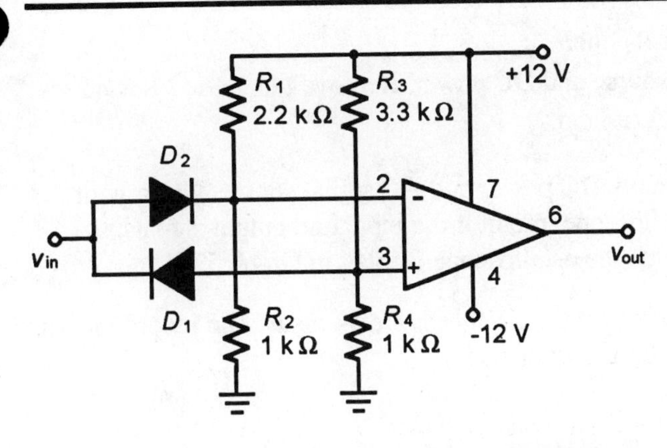

Figure 57.1

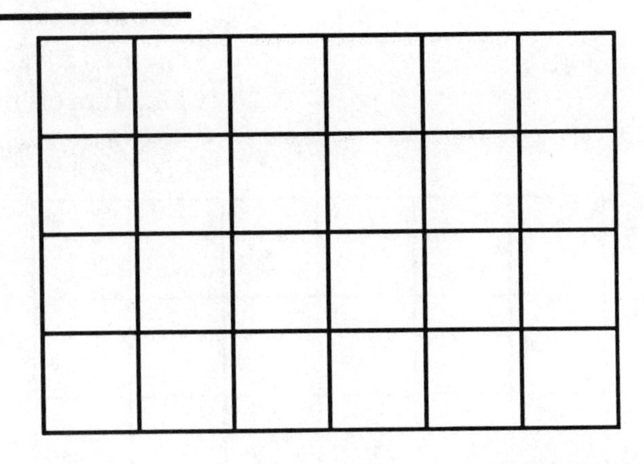

Graph 57.1

1. Build the circuit of Figure 57.1. Apply DC power. Connect your function generator, set to provide a 400-Hz triangular wave at 10 V_{p-p}.

2. Connect your oscilloscope, in dual-trace mode, to monitor the input and output signals. Sketch the oscilloscope display in Graph 57.1.

3. Measure and record the voltage level of the input signal at the point where the output goes high (v_{in1}) and the input signal voltage at the point where the output switches from high to low (v_{in2}).

   V_{in1} = ______________ V_{in2} = ______________

4. Using the data of step 3, calculate the input window voltage range, $V_W = V_{in1} - V_{in2}$.

   V_W = ______________

5. Disconnect your function generator. Using your voltmeter, measure the DC voltage at the anode of D_1 to ground. This is v_{n1}. Record it below.

   V_{n1} = ______________

6. Measure and record the voltage from the cathode of D_2 to ground.

   V_{n2} = ______________

7. From your data of steps 5 and 6, calculate the lower and upper trip points for your circuit.

   V_{LTP} = ______________ V_{UTP} = ______________

   How well do your calculated data compare to the measured values of step 2?

   ____

SECTION II TROUBLESHOOTING

Fault 1 - Divider resistor R_3 open

1. Turn off the signal source and DC power. Remove the 3.3-kΩ R_3 and replace it with a 1-MΩ resistor.

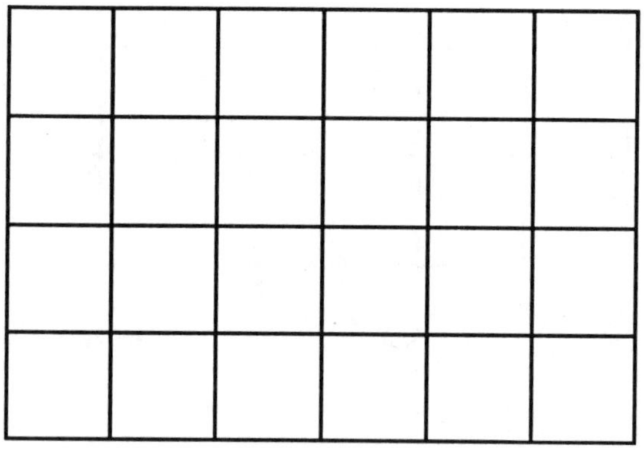

Graph 57.2

2. Reapply DC power and AC signal source. Using your oscilloscope, monitor the input and output signals. Sketch the oscilloscope display in Graph 57.2.

3. Disconnect the signal source. Measure and record the following DC voltages:

   $V_{+(pin7)}$ = ____________

   $V_{-(pin4)}$ = ____________

   V_{n1} = ____________

   V_{n2} = ____________

DISCUSSION

Section I

1. Suppose you are required to monitor a 4.5-V power supply and indicate a problem if the supply output shifts by more than 0.5 V. Discuss the use of a window comparator for this application.

2. Can you suggest another application for a window comparator? Describe your application and draw a block diagram to go with your description.

Section II

Fault 1 - Divider resistor R_3 open

Do your measurements for this fault prove the failure of R_3? In your response, cite your measured data to illustrate your conclusions.

Quick Check

1. A window comparator indicates when an input is between two limits.

   True False

2. The output voltage goes high when the input voltage reaches LTP.

   True False

3. The output voltage goes low when the input voltage reaches UTP.

True False

4. For the circuit of Figure 57.1, what is the new V_{LTP} if $R_3 = 3.9\ k\Omega$?

V_{LTP} = ___________

58

ACTIVE LOW-PASS FILTERS

INTRODUCTION

Active filters are popular because they produce better filtering at a lower cost. The active component, the op-amp, makes up for losses the passive filters usually produce. Active filters are smaller, have less weight, and are less expensive than the LC filters. The coils in the LC filters are expensive and large.

In Section I of this experiment you will calculate and measure the A_v in the mid-band and at f_c of the op-amp Butterworth active filter. You will also contrast the rolloff differences above f_c between a first- and second-order filter.

In Section II you will see how a shorted C_1 affects the filter output.

REFERENCE

Principles of Electronic Devices and Circuits - Chapter 15, Section 15.3

OBJECTIVES

Through this experiment you will:

✓ Understand the operation of the low-pass active filter

✓ Be able to contrast first- and second-order active low-pass filters

✓ Observe the effects on the filter output for a common failure

EQUIPMENT AND MATERIALS

DC power supply
Dual-trace oscilloscope
Function generator
Circuit protoboard
Operational amplifier, 741 or equivalent
Resistors: 33 kΩ, 22 kΩ, 10 kΩ [2]
Capacitors: 0.022 μF, 0.01 μF

SECTION I FUNCTIONAL EXPERIMENT

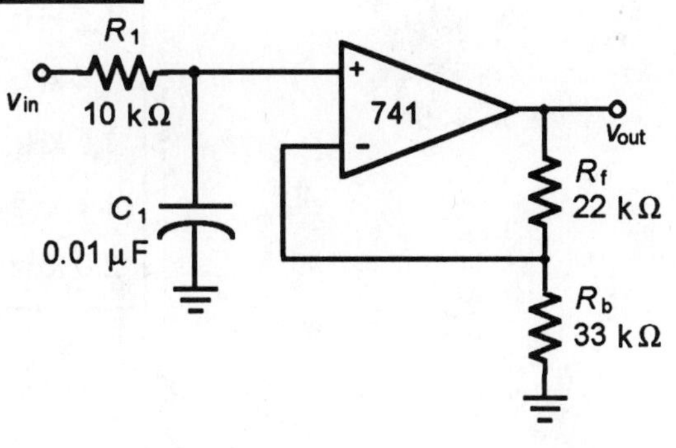

Figure 58.1

1. Build the circuit of Figure 58.1.

2. Connect one channel of your oscilloscope to V_{in} and one channel to V_{out}. V_{in} must be monitored and adjusted when necessary to maintain the set value.

3. Connect your signal generator to V_{in}. Set the generator to provide a sinewave signal of 500 $mV_{p\text{-}p}$ at 800 Hz. Measure and record V_{in} and V_{out}. This is the mid-band A_V.

   V_{in} = ________ V_{out} = ________ A_V = ________

4. Adjust the signal generator to increase frequency until v_{out} drops off to 0.707 of the v_{out} recorded in step 3. Record this frequency.

   f_c = ________

5. Calculate and record the A_V, f_c, v_{out} (at mid-band), and v_{out} (at f_c) of Figure 58.1.

   A_V = ________

   f_c = ________

   $V_{out(mid)}$ = ________

   $V_{out(fc)}$ = ________

freq.	V_{out}
1.0 kHz	________
1.2 kHz	________
1.4 kHz	________
1.6 kHz	________
1.8 kHz	________
2.0 kHz	________

Table 58.1

V_{out}: 1.6, 1.2, 0.8, 0.4

0.8 1.0 1.2 1.4 1.6 1.8 2.0

Frequency (kHz)

Graph 58.1

6. Adjust the signal generator to the frequencies listed in Table 58.1. Measure v_{out} and record for each of the listed frequencies.

7. Plot your data of step 6 in Graph 58.1.

8. Build the circuit of Figure 58.2. Apply DC power.

9. Connect your signal generator to the circuit input and adjust the generator to provide a 500-Hz sinewave at 500 $mV_{p\text{-}p}$. Connect your oscilloscope for the dual-trace mode in order to monitor the circuit input and output.

10. Measure and record below the input and output voltage levels.

    $V_{in(p\text{-}p)}$ = ________ $V_{out(p\text{-}p)}$ = ________

11. Repeat the measurements of step 6, using Table 58.2 for recording your data.

12. Plot the data of step 11 in Graph 58.2.

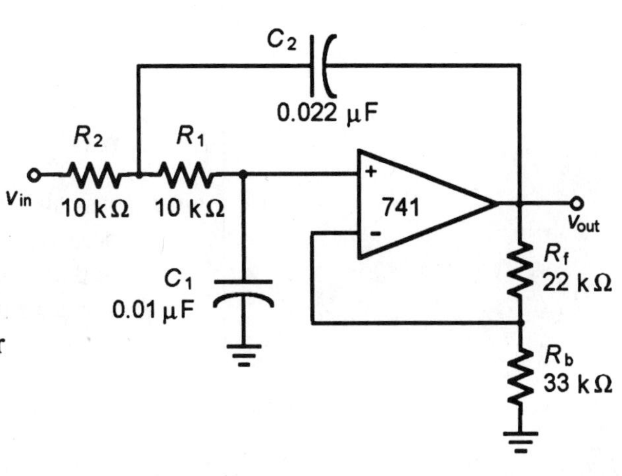

Figure 58.2

freq.	Vout
1.0 kHz	________
1.2 kHz	________
1.4 kHz	________
1.6 kHz	________
1.8 kHz	________
2.0 kHz	________

Table 58.2

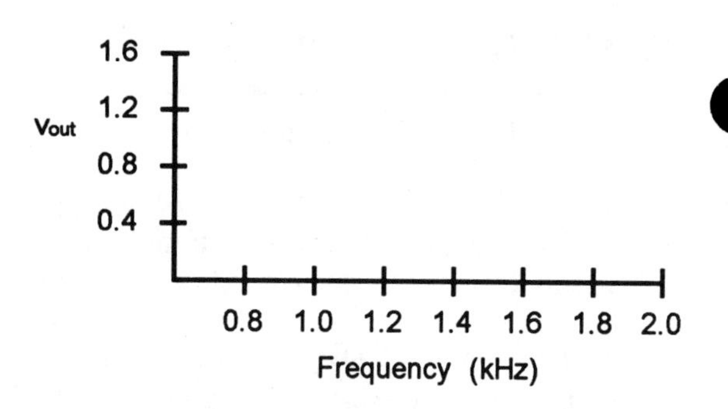

Graph 58.2

SECTION II TROUBLESHOOTING

Fault 1 - C_1 shorted

1. Turn DC power off. In your circuit of Figure 58.2, place a shorting wire across C_1. Reapply DC power.

2. Apply an input of 500 Hz at 500 $mV_{p\text{-}p}$. Measure and record the circuit output.

   V_{out} = ____________

DISCUSSION

Section I

1. You should have found that the A_v (mid-band) of Figure 58.2 is approximately 1.60. Discuss this gain value; is there anything significant about this figure to you?

2. You should have found that in the second order active low-pass filter of Figure 58.2 the output voltage fell off very quickly. Discuss the rolloff rate vs. the rolloff rate for the first-order filter of Figure 58.1.

Section II

Fault 1 - C_1 shorted

Describe your circuit result for this failure. Could this same result have been due to an op-amp failure? What do you think would be the effect if C_1 changed values?

Quick Check

1. A low-pass filter actually cuts off _____ frequencies.

   (a) high　　　　(b) low

2. The active low-pass filter is actually more expensive than the LC filter.

   True　　　　False

3. The rolloff for a second-order active LP filter is _____.

   (a) 20 dB per decade　　　　(b) 40 dB per decade
   (c) 60 dB per decade

4. For the circuit of Figure 58.1, what is the corner frequency if $C_1 = 0.047\ \mu F$ and $R_1 = 5.6\ k\Omega$?

   $f_c =$ ____________

59

ACTIVE HIGH-PASS FILTERS

INTRODUCTION

Active high-pass filters have the same advantages as the active low-pass filters discussed in Experiment 58. Active filters are smaller, they weigh less, and they are less expensive than their passive-filter counterparts. In addition, active filters have fewer losses because of the active components they use. The major difference between high- and low-pass active filters is that low-pass filters pass low frequencies up to a cutoff frequency, and high-pass filters cut off frequencies at some lower frequency and pass all frequencies above that point.

In Section I of this experiment you will calculate and measure the A_v in the midband and at f_c of the op-amp Butterworth active filter. You will also contrast the rolloff differences above f_c between a first- and sccond-order filter.

In Section II you will see how an open C_1 and a shorted resistor will affect the circuit output.

REFERENCE

Principles of Electronic Devices and Circuits - Chapter 15, Section 15.3

OBJECTIVES

In this experiment you will:

✓ Study the operation of the active high-pass filter

✓ Contrast first- and second-order active high-pass filters

✓ Observe the effects on the circuit output of some common failures

EQUIPMENT AND MATERIALS

DC power supply
Dual-trace oscilloscope
Function generator
Circuit protoboard
Operational amplifier, 741 or equivalent
Capacitor, 0.01 μF [2]
Resistors: 5 kΩ, 27 kΩ, 30 kΩ, 47 kΩ

SECTION I FUNCTIONAL EXPERIMENT

1. Build the circuit of Figure 59.1. Apply DC power.

2. Connect your oscilloscope with one channel to V_{in} and one channel to v_{out}. Level v_{in} must be monitored and adjusted in order to maintain a constant level of input voltage.

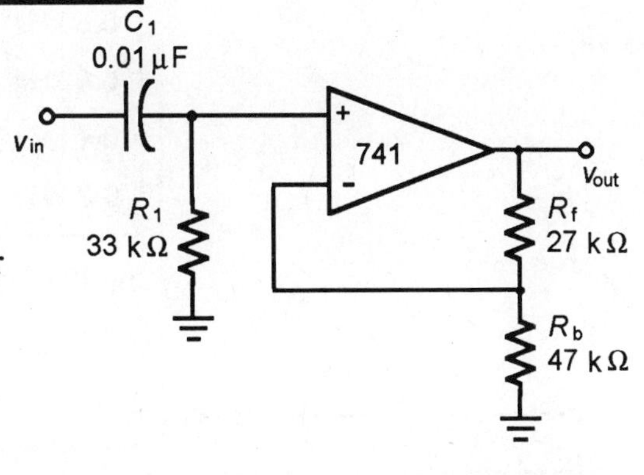

Figure 59.1

3. Connect your function generator to v_{in}, and adjust the generator to provide a sinewave signal of 500 mV_{p-p} at 1 kHz. Measure v_{in} and v_{out}, and record below. This is the mid-band A_v.

   v_{in} = ________________ v_{out} = ________________

   A_v = ________________

4. Adjust the function generator to decrease frequency until the circuit output drops off to 0.707 of the v_{out} recorded in step 3. Record this frequency.

   f_c = ____________

5. Calculate and record the A_v, f_c, $v_{out(mid)}$, and $v_{out(fc)}$ of Figure 59.1.

   A_v = ________________ f_c = ________________

   $v_{out(mid)}$ = ______________ $v_{out(fc)}$ = ______________

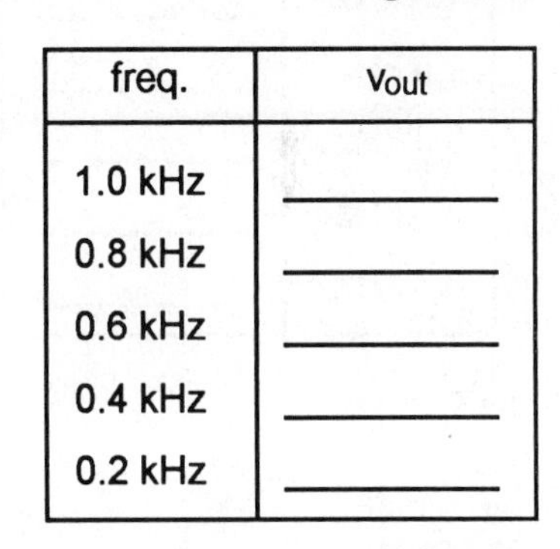

freq.	Vout
1.0 kHz	________
0.8 kHz	________
0.6 kHz	________
0.4 kHz	________
0.2 kHz	________

Table 59.1

6. Adjust the function generator to the frequencies listed in Table 59.1. Measure the filter output and record in the table.

7. Plot your data of step 6 in Graph 59.1.

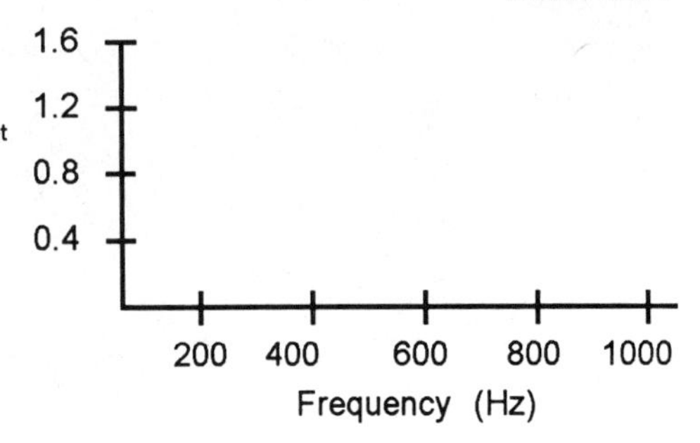

Graph 59.1

8. Build the circuit of Figure 59.2. Apply DC power.

9. Connect your function generator to the circuit input. Adjust the generator to provide a 1.5-kHz sine wave at 500 mV_{p-p}. Connect your oscilloscope in dual-trace mode to monitor the filter input and output signals.

10. Measure and record the passband input and output voltages for your filter.

    $V_{in(p-p)}$ = ________ $V_{out(p-p)}$ = ________

11. Repeat measurements of step 6 for this filter, but use Table 59.2 for your data.

12. Plot the data of step 11 in Graph 59.2.

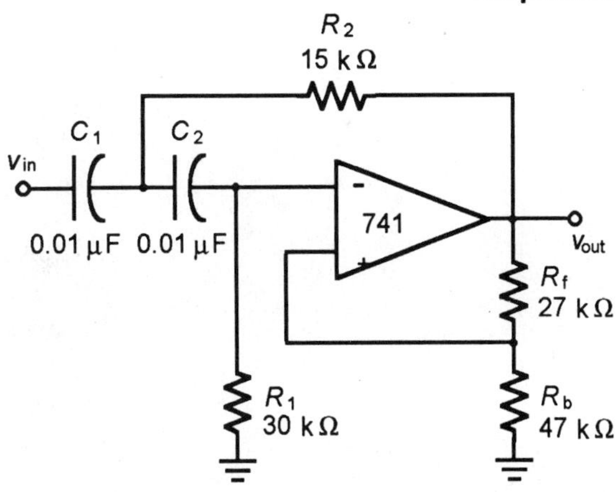

Figure 59.2

freq.	Vout
1.0 kHz	________
0.8 kHz	________
0.6 kHz	________
0.4 kHz	________
0.2 kHz	________

Table 59.2

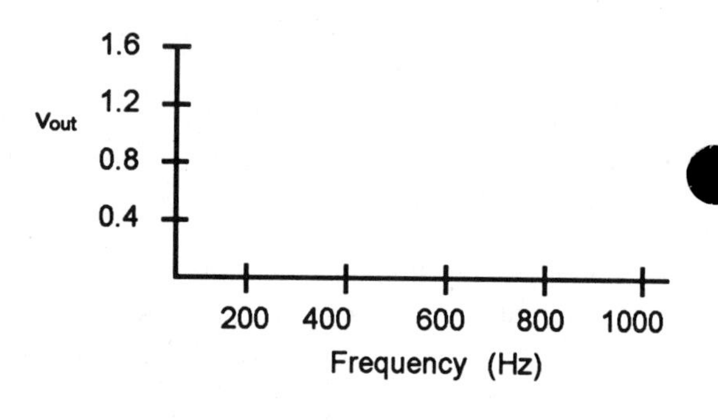

Graph 59.2

freq.	Vout
1.5 kHz	________
1.0 kHz	________
0.8 kHz	________
0.6 kHz	________
0.4 kHz	________

Table 59.3

SECTION II TROUBLESHOOTING

Fault 1- C_1 shorted

1. Turn off the circuit power. For the circuit of Figure 59.2, place a jumper wire in parallel with capacitor C_1. Reapply circuit power. Adjust your function generator to provide a 1.5-kHz, 500-m$V_{p\text{-}p}$ input signal.

2. With your oscilloscope connected to monitor both the filter input and output signals, measure and record the filter output for the frequencies listed in Table 59.3.

Fault 2 - C_1 open

1. Turn off the DC power. Remove the jumper shorting C_1 and lift the end of C_1 from the junction of C_2 and R_2. Reapply circuit power. Adjust your generator to provide an input signal of 800 Hz at 500 m$V_{p\text{-}p}$.

2. Measure and record below the filter output.

   V_{out} = ____________

   What kind of output waveform do you observe?

DISCUSSION

Section I

1. For the filter of Figure 59.1, you should have found the passband gain to be approximately 1.6. Discuss the significance of this gain value in terms of the filter characteristics.

2. Compare the two filter response graphs. Describe the attenuation slopes you measured and discuss the slopes you obtained versus the ideal (predicted) slope for a first-order filter and second order filter.

Section II

Fault 1 - C_1 shorted

From the data measured, can you eliminate the op-amp as a problem and determine the fault in the filter network? Discuss this, citing your measured data that illustrate your answer.

Fault 2 - C_1 open

Does the measurement of the filter output give sufficient information to suggest a filter network problem? Discuss this failure mode and indicate what additional measurements you would make to isolate the fault.

Quick Check

1. A high-pass filter actually cuts off the _____ frequencies.

   (a) high (b) low

2. The active high-pass filter is more expensive than the LC filter.

   True False

3. The rolloff for a first-order active high-pass filter is _____ per decade.

   (a) 20 dB (b) 40 dB
   (c) 60 dB

4. For the filter of Figure 59.1, what would the corner frequency be if $C_1 = 0.001\ \mu F$ and $R_1 = 27\ k\Omega$?

   $f_c =$ ___________

60

ACTIVE BAND-PASS FILTERS

INTRODUCTION

Band-pass filters fill the application need for a circuit that will pass a range of frequencies and reject (attenuate) frequencies above and below that range. Active filters provide nearly ideal characteristics for filters in the frequency range of a few Hz to approximately 150 kHz. For higher frequencies, passive LC filters are likely the filter of choice.

In this experiment you will build and test two band-pass filter forms. The first is a single-stage filter, and the second is a two-stage filter composed of a low-pass section and a high-pass section. The troubleshooting section will let you see the effect of a component fault on the circuit operation.

REFERENCE

Principles of Electronic Devices and Circuits - Chapter 15, Section 15.3

OBJECTIVES

Through this experiment you will:

✓ Understand how a specific band of frequencies can be passed by the active band-pass filters

✓ Determine the actual band of frequencies that are passed by the band-pass filter

✓ Observe the effects of shorted capacitors on a band-pass circuit

EQUIPMENT AND MATERIALS

Dual-trace oscilloscope
DC power supply
Circuit protoboard
Function generator
Operational amplifier [2], 741 or equivalent
Capacitors: 20 pF, 0.0033 μF, 0.0047 μF [2], 0.022 μF [2]
Resistors: 330 Ω, 1 kΩ, 10 kΩ [2], 15 kΩ, 22 kΩ, 27 kΩ, 33 kΩ [2], 47 kΩ, 100 kΩ

SECTION I FUNCTIONAL EXPERIMENT

Single-Stage Band-Pass Filter

1. Build the circuit of Figure 60.1. Apply DC power.

2. Connect the signal generator to v_{in} and adjust for a sinewave signal of 80 kHz at 100 $mV_{p\text{-}p}$.

   Note: You may need a voltage divider to reduce the generator signal to v_{in}.

3. Connect your oscilloscope to monitor both V_{in} and V_{out}. Maintain the input signal at 100 $mV_{p\text{-}p}$. Measure and record the circuit output with the 80-kHz input.

   V_{out} at 80 kHz = ___________

4. Adjust the signal generator to each of the frequencies listed in Table 60.1, and measure and record the filter peak-to-peak output at each frequency.

5. Set your signal generator to 90 kHz and 100 $mV_{p\text{-}p}$. Measure and record below the filter output.

   v_{out} at 90 kHz = ___________

6. Adjust your signal generator to each of the frequencies listed in Table 60.2; measure and record the filter output for each frequency.

7. For the circuit of Figure 60.1, calculate and record the cutoff frequencies below.

   f_1 = _______________ f_2 = _______________

   f_3 = _______________ f_4 = _______________

   Calculate and record the bandwidth (BW) for the circuit of Figure 60.1.

   BW = ___________

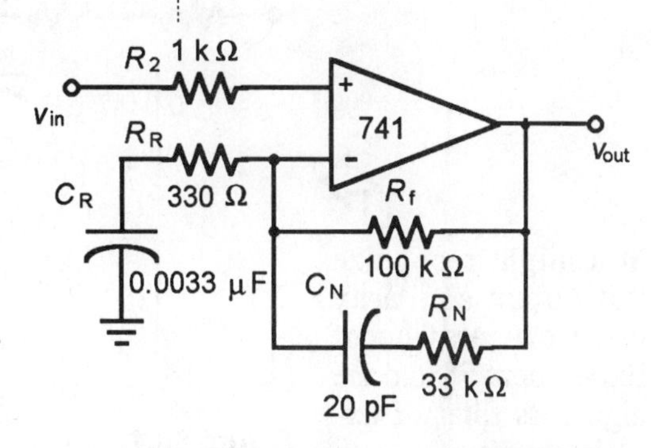

Figure 60.1

Freq.	$V_{out(p\text{-}p)}$
60 kHz	________
55 kHz	________
50 kHz	________
45 kHz	________
40 kHz	________
35 kHz	________
30 kHz	________
25 kHz	________
20 kHz	________

Table 60.1

Freq.	$V_{out(p\text{-}p)}$
100 kHz	________
120 kHz	________
130 kHz	________
140 kHz	________
150 kHz	________
160 kHz	________
170 kHz	________
180 kHz	________

Table 60.2

Two Stage Band-Pass Filter

1. Build the circuit of Figure 60.2 (next page). Apply DC power to your circuit.

2. With your oscilloscope connected to monitor both V_{in} and V_{out}, measure the filter output and record below.

   V_{out} at 1.5 kHz = ___________

3. Decrease the generator frequency until the filter output drops to 0.707 of the value measured in step 2. Record this frequency below. This is the lower corner frequency.

   f_c = ___________

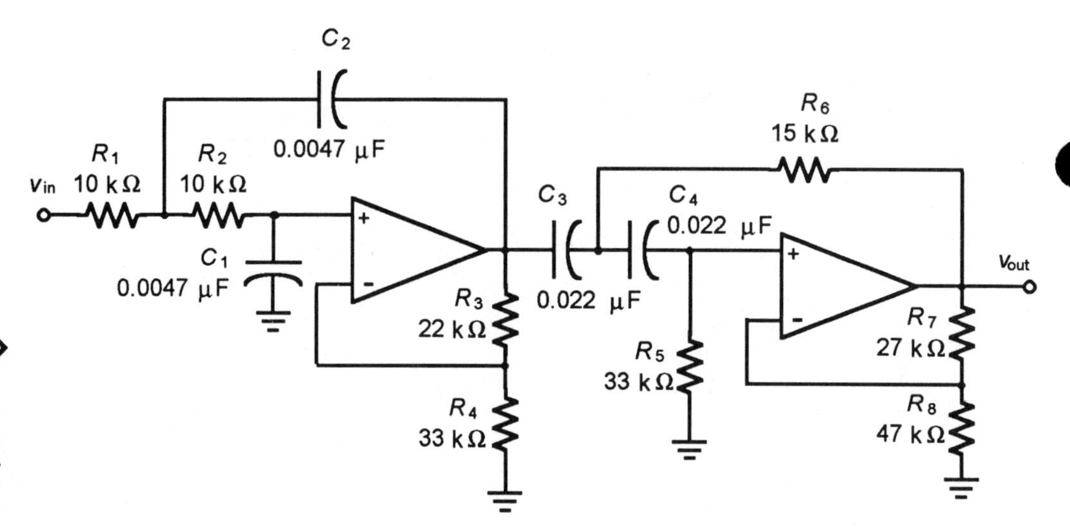

You might recognize this circuit similarity to the low-pass filter of Experiment 58 and the high-pass filter of Experiment 59.

Figure 60.2

4. Increase your generator frequency until the filter output again drops to 0.707 of the value measured in step 2. This is the upper corner frequency. Record this frequency below.

   f_c = ___________

6. From your data of steps 3 and 4, calculate the bandwidth of the filter.

   BW = ___________

SECTION II TROUBLESHOOTING

Fault 1 - C_R shorted

1. Use the circuit of Figure 60.1. With circuit power off, use a jumper wire to short C_R. Reapply DC power. With your signal generator connected to V_{in}, apply a sinewave signal of 80 kHz at 100 $mV_{p\text{-}p}$.

2. Measure and record below the filter output.

   V_{out} (80 kHz) = ___________

3. Check the filter output at several of the frequencies of Section I, steps 4 and 6. Record your frequencies and measured filter outputs in Table 60.3.

Freq.	$V_{out(p\text{-}p)}$
______ kHz	______
______ kHz	______
______ kHz	______
______ kHz	______
______ kHz	______
______ kHz	______

Table 60.3

DISCUSSION

Section I

1. Describe, for the circuit of Figure 60.1, which component you would change to lower the filter f_2 point. Include in your description the type of change. Would it increase or decrease the schematic value given?

2. For the two-stage band-pass circuit of Figure 60.2, describe the circuit operation. For example, is the lower corner frequency less than or greater than the upper corner frequency, and which section controls the lower corner frequency?

3. Given a requirement to pass a band of frequencies from 1 kHz to 12 kHz with maximum attenuation on each side of the lower and upper corner frequencies, which of the filter forms you've explored in this experiment would you select. Discuss your answer and indicate why you selected a given filter.

SECTION II

Fault 1 - C_R shorted

Describe what information about the circuit failure the measured data gave you. Discuss also how well your measured data pointed to, or exonerated, the op-amp as a possible failed component.

Quick Check

1. The cascaded active band-pass filter is make up of what two circuits?

   a.____

   b. ____

2. It is important for a band-pass filter to be able to reject unwanted frequencies efficiently as it is to pass the desired frequencies.

   True　　　　　　　　　　False

3. What is the first check you would make if there was no output from the circuit of Figure 60.1?

   ____

4. To have a Butterworth filter response the low-pass and high-pass filter sections of the filter of Figure 60.2 must have gains of _____.

   (a) 1.414　　　　　　(b) 1.60
   (c) 1.83　　　　　　(d) 1.95

61

ACTIVE BAND-REJECT FILTERS

INTRODUCTION

Active band-reject filters are used where a specific band of frequencies are desired to be attenuated, while passing the frequencies above and below the stop-band. Often you may see a combination low-pass and high-pass active filter. In this experiment, the filters are in parallel and their outputs are fed into a summing amplifier. One difference between the band-reject and the band-pass filter is that, in the band-reject filter, the low-pass cutoff frequency must be lower than the high-pass filter cutoff frequency.

In this experiment you will measure the f_c for both the low-pass and high-pass filters as well as the bandwidth (BW). You will also predict these parameter. In Section II you will observe the effects on the v_{out} when an op-amp is inoperative. You will work through a troubleshooting sequence for these faults.

REFERENCE

Principles of Electronic Devices and Circuits - Chapter 15, Section 15.3

OBJECTIVES

In this experiment you will:

✓ Understand the operation of an active band-reject filter

✓ Observe the effect on the v_{out} when the low-pass and high-pass filters are paralleled

✓ Determine, through measurement, the effect of an op-amp failure on a circuit

EQUIPMENT AND MATERIALS

DC power supply
Dual-trace oscilloscope
Function generator
Circuit protoboard

Operational amplifier [3], 741 or equivalent
Capacitor, 0.01 µF [2]
Resistors: 10 kΩ [3], 5.1 kΩ, 22 kΩ [2], 27 kΩ, 33 kΩ, 47 kΩ, 51 kΩ, 100 kΩ

SECTION I FUNCTIONAL EXPERIMENT

1. Build the circuit of Figure 61.1. R_{10} is used to minimize the input offset of U_3 due to differences of input bias currents.

2. Insert a signal voltage of 500 mV_p at 200 Hz.

3. Measure and record the v_{out}.

   v_{out} at 200 Hz = ____________

4. Increase the function generator frequency; measure and record the v_{out} for each listed frequency listed in Table 61.1.

5. Now adjust the function generator to 4.0 kHz at 500 $mV_{p\text{-}p}$.

6. Measure and record v_{out}.

   v_{out} at 4.0 kHz = ____________

7. Decrease the function generator frequency and measure v_{out} at each frequency listed in Table 61.2. Record v_{out} at these frequencies in that table.

8. From your data of steps 4 and 7, calculate and record the measured bandwidth.

   BW = ____________

9. Plot the data of steps 4 and 7 in Graph 61.1 (next page).

Freq.	$V_{out(p\text{-}p)}$
400 kHz	________
600 kHz	________
800 kHz	________
1000 kHz	________
1200 kHz	________

Table 61.1

Freq.	$V_{out(p\text{-}p)}$
3.5 kHz	________
3.0 kHz	________
2.8 kHz	________
2.6 kHz	________
2.4 kHz	________
2.2 kHz	________
2.0 kHz	________
1.8 kHz	________

Table 61.2

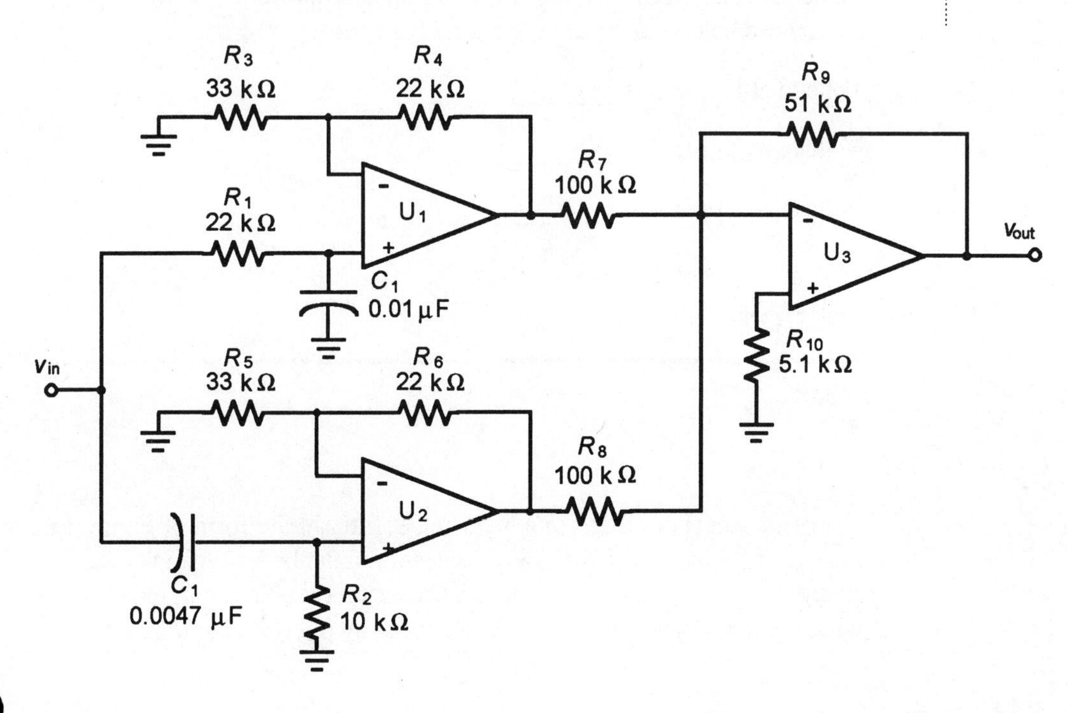

Figure 61.1

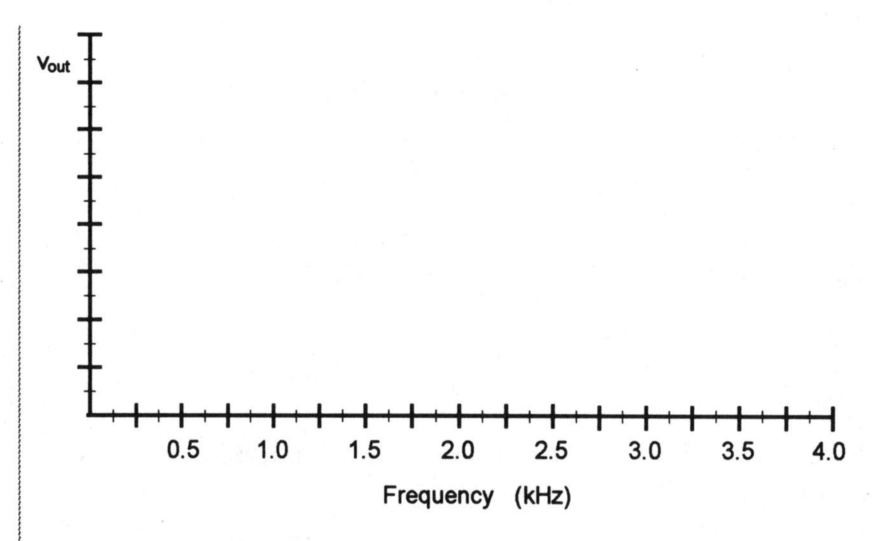

Graph 61.1

SECTION II TROUBLESHOOTING

Fault 1 - Op-amp U_1 inoperable

1. Turn off the circuit power. Disconnect the DC supply connections from pins 4 and 7 of op-amp U_1.
2. Reapply circuit power. With your oscilloscope connected to monitor both v_{in} and v_{out} of the circuit, check several of the low-frequency and several of the high-frequency data points measured in Section I, steps 4 and 7.
3. Your check of the data points of steps 4 and 7 should indicate a problem in the low-pass section. Set your signal generator to 800 Hz. Measure and record below the following signal voltages:

   Voltage at U_1 pin 3 = __________

   Voltage at U_1 pin 2 = __________

   Voltage at U_1 pin 6 = __________

DISCUSSION

Section I

Your text points out that, for a band-reject filter in the form of Figure 61.1, the cutoff frequency of the low-pass section must be less than the cutoff frequency of the high-pass section. Discuss this requirement, indicating what you feel would be the result of not following this requirement.

Section II

Fault 1 - Op-amp U_1 inoperative

Following the measurements that identified the band-reject filter problem in the low-pass section, you made additional measurements in step 3. Discuss these measurements and their values, and explain how these measured values point to an op-amp failure.

Quick Check

1. It is important that the low-pass filter have a higher cutoff frequency than the high-pass filter.

   True False

2. Are the low-pass and high-pass filters in this experiment Butterworth filters?

   Yes No

3. What components determine the cutoff frequencies of the low-pass filter in Figure 61.1? ____________________

   The high-pass filter? ____________________

4. How many poles are there in the low-pass filter section in the circuit of Figure 61.1? ___________

62

VCVS ACTIVE FILTERS

INTRODUCTION

Voltage-controlled voltage source (VCVS) active filters have advantages such as simple design, low output impedance, small spread of component values, and a high Q. One main disadvantage, however, is that they are sensitive to component value changes due to age. This causes the cutoff frequencies to drift and makes the filter unstable.

In this experiment you will build a low-pass VCVS Butterworth active filter. You will measure the frequency cutoff (f_c) and gain of the circuit. You will also build a high-pass VCVS active Butterworth filter, measure the cutoff frequency (f_c) and response curves, and predict the filter values.

In the troubleshooting section you will observe the effects on the high-pass filter f_c when C_1 opens.

REFERENCE

Principles of Electronic Devices and Circuits - Chapter 15, Section 15.4

OBJECTIVES

Through this experiment you will:

✓ Understand the low- and high-pass VCVS active filters

✓ Determine the effect on v_{out} rolloff of a VCVS high-pass filter when C_1 opens

EQUIPMENT AND MATERIALS

Dual-trace oscilloscope
DC power supply
Function generator
Circuit protoboard
Operational amplifier, 741 or equivalent
Resistors: 2.7 kΩ [2], 10 kΩ [2], 56 kΩ, 100 kΩ
Capacitors: 0.01 μF [2], 0.022 μF [2]

SECTION I FUNCTIONAL EXPERIMENT

1. Build the low-pass filter circuit of Figure 62.1.

2. Connect the function generator to provide an input of 1-V_{p-p} sinewave signal at 1 kHz. Connect your oscilloscope to monitor v_{in} and v_{out}. Measure and record the filter output at the frequencies listed in Table 62.1. (Remember to maintain v_{in} at 1 V_{p-p}.)

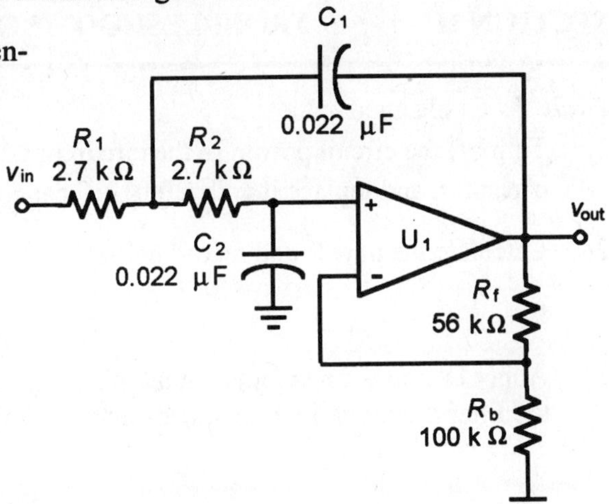

Figure 62.1

Freq.	$V_{out(p-p)}$
1.0 kHz	________
2.0 kHz	________
2.2 kHz	________
2.4 kHz	________
2.6 kHz	________
2.8 kHz	________
3.0 kHz	________
3.2 kHz	________

Table 62.1

3. From your measured data determine the passband gain (A_v) and the cutoff frequency of your filter.

A_v = ______________ f_c = ______________

4. Build the high-pass filter circuit of Figure 62.2. Apply DC power.

5. Connect the function generator to provide a 1 V_{p-p} sinewave signal at 4 kHz. Connect your oscilloscope to monitor both v_{in} and v_{out}. Measure and record the filter output at the frequencies listed in Table 62.2.

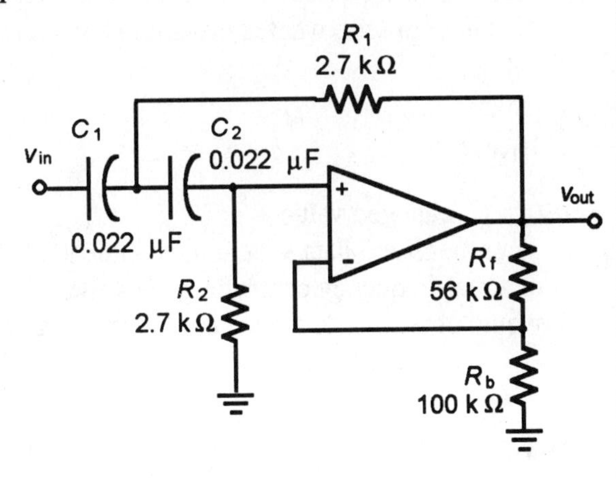

Figure 62.2

Freq.	$V_{out(p-p)}$
4.0 kHz	________
3.0 kHz	________
2.8 kHz	________
2.6 kHz	________
2.4 kHz	________
2.2 kHz	________
2.0 kHz	________
1.8 kHz	________

Table 62.2

6. From your data determine the passband gain (A_v) and the filter cutoff frequency.

A_v = ______________ f_c = ______________

7. Calculate the gain and cutoff frequency for the high-pass filter of Figure 62.2.

A_v = ______________ f_c = ______________

8. How well do your measured data compare to your predicted data?

____

Freq.	$V_{out(p-p)}$
4.0 kHz	________
3.0 kHz	________
2.8 kHz	________
2.3 kHz	________
2.0 kHz	________
1.0 kHz	________
0.5 kHz	________

Table 62.3

SECTION II TROUBLESHOOTING

Fault 1 - C_1 changed value

1. Turn off the circuit power. In the circuit of 62.2 remove C_1, a 0.022-µF capacitor, and replace it with a 0.01-µF capacitor.
2. Calculate the new f_c and record below.

   f_c = ____________

3. Apply DC power and input signal of 1 $V_{p\text{-}p}$ at 4 kHz. Measure and record the filter output at the frequencies listed in Table 62.3.

DISCUSSION

Section I

Discuss your impressions of the VCVS filter. Is it as simple and predictable as you expected? From your measured data and the data of Table 15.4 in your text, what is the filter characteristic?

Section II

Fault 1 - C_1 changed value

Your measured data should have indicated that the fault was in the external frequency controlling networks. What additional measurements would you make to isolate the failed component?

Quick Check

1. The VCVS active filter is a unity gain filter.

   True False

2. The VCVS filter can be a _____ filter.

   (a) Butterworth (b) Bessel
   (c) Chebyshev (d) All of these

3. The VCVS filter is restricted to only a two-pole filter.

   True False

4. The VCVS active filter has the advantage of being very tolerant of component aging.

   True False

5. For a Butterworth VCVS active filter all resistors and capacitors in the frequency-determining network are the same value.

   True False

6. The VCVS active filter is a _____ circuit.

   (a) high-Q (b) low-Q

63

INSTRUMENTATION AMPLIFIERS

INTRODUCTION

Instrumentation amplifiers have their largest application in industrial electronics. These applications involve working with low output level transducers whose signals may travel long distances. The instrumentation amplifier must provide signal gain with a large common-mode noise rejection. Most instrumentation amplifiers used now are normally integrated circuit devices.

In this experiment you will build an instrumentation amplifier using three op-amps. You will explore the functional characteristics of the instrumentation amplifier through measurements made on the operating circuit.

REFERENCE

Principles of Electronic Devices and Circuits - Chapter 16, Section 16.2

OBJECTIVES

Through this experiment you will:

✓ Increase your knowledge of common mode rejection

✓ Be able to calculate the output of an instrumentation amplifier

✓ Show how an instrumentation amplifier amplifies the signal and rejects the common-mode signal

EQUIPMENT AND MATERIALS

DC Power SupplY
Dual-trace oscilloscope
Digital multimeter
Function generator
Circuit protoboard
Operational amplifier [3], 741 or equivalent
Resistors: 2.2 kΩ, 10 kΩ [2], 22 kΩ, 100 kΩ [5]
Potentiometers: 200-kΩ and 10-kΩ [2] ten-turn trimpots

SECTION I FUNCTIONAL EXPERIMENT

1. Build the circuit of Figure 63.1. Do not apply DC power to the op-amps at this time. Connect each input voltage divider (Figure 63.2) to +15 V and adjust the divider to obtain an output of 5.000 V as closely as you can.

2. Apply 15 V to the op-amps. Connect divider A to both input A and input B. Using your voltmeter to measure v_{out} adjust R_7 for a minimum output–as closely to 0 as you can. Reconnect the inputs: divider A to input A and divider B to input B.

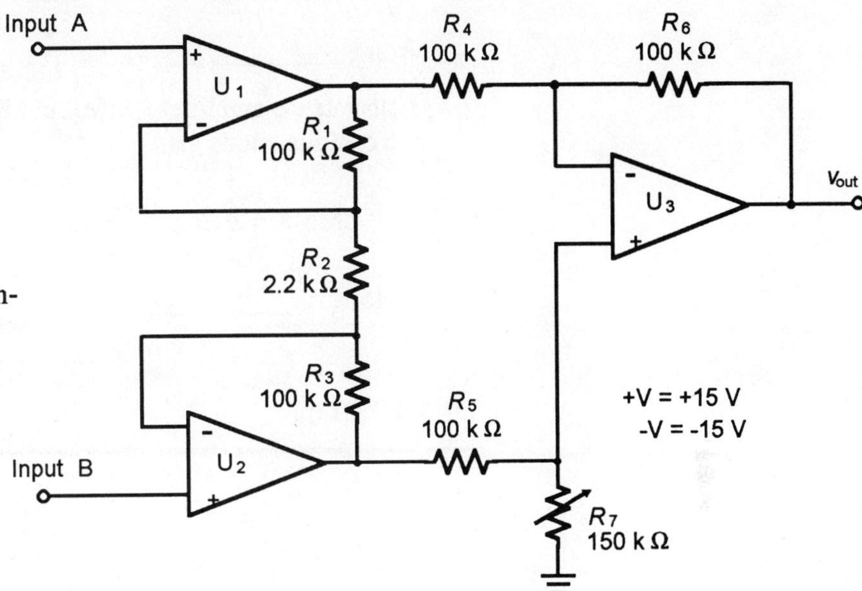

Figure 63.1

3. Measure the input voltages and set input B to 5.002 V closely as possible with input A at 5.000 V.

   Measure and record below v_{out} and its polarity.

   v_{out} = ______________

4. Reset input B to 5.000 V and adjust input A to 5.002 V. Measure and record v_{out} and its polarity.

   v_{out} = ______________

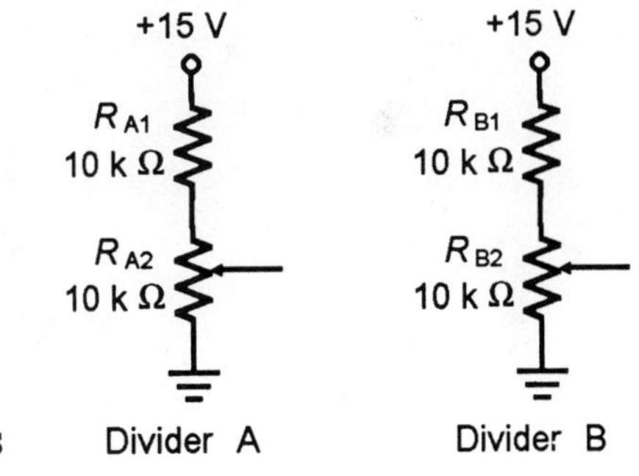

Figure 63.2

5. With a differential input of 2 mV, calculate the amplifier gain for steps 4 and 5.

   A_{A1} = ______________ A_{VZ} = ______________

   Average the gain values to obtain an average gain for the amplifier.

   A_{NA} = ______________

   Although this demonstration is not exact, notice that your instrumentation amplifier amplified a differential signal of 2 mV with a common-mode input of 5 V.

6. Disconnect the two DC input voltage dividers. Connect a function generator to supply both inputs. Adjust the function generator to provide a 60-Hz sine wave at 250 $mV_{p\text{-}p}$. Use your oscilloscope to measure both the com-

mon input signal and the peak-to-peak amplifier output voltage. Record these voltages below.

V_{in} = ______________ V_{out} = ______________

Calculate the common-mode gain.

A_{CM} = ___________

7. Calculate the amplifier CMMR in dB from the amplifier average gain and the common mode gain.

$$\text{CMMR} = 20 \log \frac{A_{VA}}{A_{VM}}$$

CMMR = ___________

DISCUSSION

Section I

1. Discuss why you feel this instrumentation amplifier is useful in detecting a small signal change.

2. Considering the amount of circuitry (three op-amps and seven resistors) and the data results you obtained, discuss your perceptions of the effectiveness of the instrumentation amplifier.

Quick Check

1. One method of increasing the common-mode rejection is to use precision resistors.

   True False

2. Instrumentation amplifiers will amplify large common-mode signals as well as small signals.

   True False

3. The output voltage is an amplified version of the difference between the two input signals.

   True False

4. Which resistor sets the gain for the instrumentation amplifier?

   ____

64

DIGITAL-TO-ANALOG CONVERSION

INTRODUCTION

Digital-to-analog (D/A) conversion is the process of changing a binary number value to a decimal value. This process has an ever-increasing application in modern electronics. It is the mechanism through which systems such as CD players and digital cellular phones operate.

The converters in such systems are sophisticated chips. This experiment will use a much simpler circuit to let you see the basics of D/A conversion and to help you understand the operation.

In this experiment you will build a basic four-bit convertor and make measurements for a variety of input conditions. Through this process you will be able to see and understand D/A conversion.

REFERENCE

Principles of Electronic Devices and Circuits - Chapter 16, Section 16.3

OBJECTIVES

In this experiment you will:

✓ Utilize the binary numbering system in a practical application

✓ Verify the conversion of a digital to an analog signal

✓ Calculate and relate the output voltages to a variety of digital inputs

✓ Relate the digital inputs to the various analog outputs

EQUIPMENT AND MATERIALS

DC power supply
Digital multimeter
Circuit protoboard
Operational amplifier, 741 or equivalent
Resistors: 300 Ω, 4.7 kΩ [2], 6.8 kΩ, 10 kΩ, 20 kΩ, 39 kΩ, 1 MΩ
Single-pole double-throw switch [4] (You may substitute jumper wires for switches.)

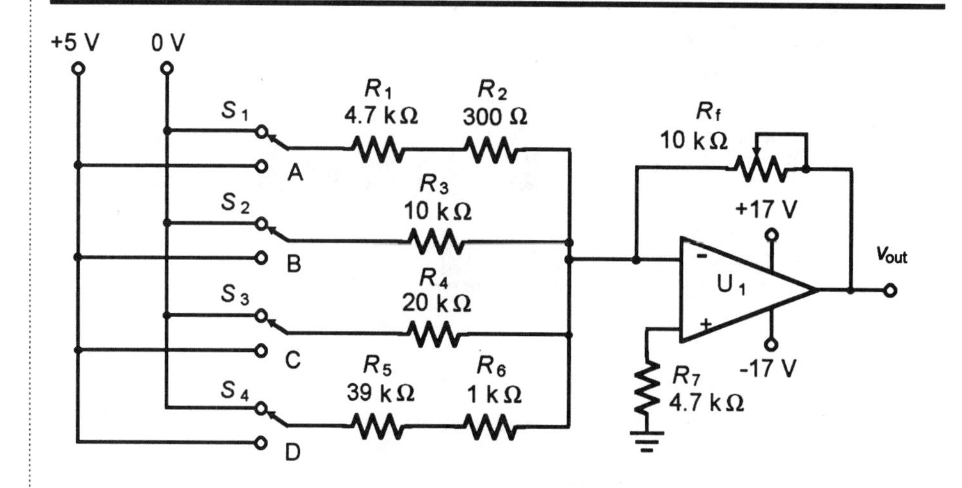

Figure 64.1

If you don't have a 20-kΩ resistor for R_4, use a 22-kΩ resistor in parallel with a 220-kΩ resistor.

1. Construct the circuit of Figure 64.1. Set potentiometer R_f to 8 kΩ.

   You can substitute a jumper wire for each switch. There must be a jumper wire for each input resistor (R_1 through R_5), and each must make a connection to either +5 V or ground.

   Note: Do not leave any input open-circuited.

   When a switch (or jumper) is connected to ground, a logic "0" is at that input. When a switch (or jumper) is connected to +5 V, a logic "1" is applied. Initially, set all switches for a logic input of 0000.

2. Apply 17 VDC to the op-amp, and +5 V to the 5-V switch buss. Measure the converter output voltage. Ideally, this should be zero. In a practical circuit without nulling, you should find an output that doesn't exceed 100 mV. For the purpose of this experiment disregard the output offset and consider it as zero.

3. Notice that your converter is just a summing amplifier. For the logic input of 0000, record the converter output in Table 64.1 in the appropriate row.

4. Apply a digital 0001, point D to +5 V, all others at 0 V. Measure and record the converter output in Table 64.1. Using your knowledge of summing amplifiers, calculate the expected converter output and record in Table 64.1.

5. Repeat the process of step 4 for each digital entry in Table 64.1.

   The resistor sizes and feedback scaling were selected so that your converter should have given you the decimal value of each 4-bit digital input. For example, for the 0111 input you should have had an output of essentially 7 V.

Binary Inputs				Output Voltages	
A S_1	B S_2	C S_3	D S_4	Measured	Calculated
0	0	0	0		
0	0	0	1		
0	0	1	0		
0	0	1	1		
0	1	0	0		
0	1	0	1		
0	1	1	0		
0	1	1	1		
1	0	0	0		
1	0	0	1		
1	0	1	0		
1	0	1	1		
1	1	0	0		
1	1	0	1		
1	1	1	0		
1	1	1	1		

Table 64.1

SECTION II TROUBLESHOOTING

Fault 1 - Feedback resistor (R_4) open

1. Turn off all circuit power. Disconnect the feedback potentiometer and replace it with a 1-MΩ resistor.

   Note: The 1-MΩ resistor does not totally simulate a complete open; however, in this circuit the simulation is sufficient.

2. Reapply the op-amp and 5 VDC power. Monitor the converter output and apply 4 to 6 different digital inputs from 0000 to 1111. Describe below the converter response to each input.

   ______________________ ______________________

   ______________________ ______________________

   ______________________ ______________________

3. Make any additional measurements that you feel appropriate. List your measurement and measured result below.

   ______________________ ______________________

   ______________________ ______________________

DISCUSSION

Section I

1. Describe how the digital input is converted to an analog output by your converter.

2. Discuss why this D/A converter form is not used in digital circuits and what the limiting factors are that preclude the use of this circuit.

Section II

Fault 1 - Feedback resistor R_f open

1. Discuss the converter response to the logic input of 0000 versus all other input combinations. What conclusions does this set of conditions justify?

2. Describe the additional measurements you made and what the measured result indicated about the circuit fault. If you made no additional measurements, describe why your conclusions of the step 2 measured data isolated the failure of R_f.

Quick Check

1. If the inputs were all high (+5 V), you would expect to get the maximum output signal.

   True False

2. With all switches at zero, you would measure half the output because these are weighted resistors.

   True False

3. The output voltage is equal to the feedback voltage.

   True False

4. The signal represented by the switches $S_1 = 0$, $S_2 = 1$, $S_3 = 0$, and $S_4 = 1$ would represent what decimal number? ___________

5. If R_f in the circuit of Figure 64.1 were 10 kΩ, what would the converter output be for the logic switches set to 1001? ___________

65

555 TIMER CIRCUITS

INTRODUCTION

The 555 timer is widely used as an IC timing device. It operates in both astable (multivibrator) and monostable (triggered) modes. This versatile IC timing circuit requires a minimum of external components to serve a range of applications, from a clock pulse generator to a pulse width modulator.

In this experiment you will build and explore the operation of the 555 in the astable mode as a pulse generator oscillator and in the monostable mode as a one-shot pulse generator.

The troubleshooting section will let you see the effect of a component failure on the circuit operation.

REFERENCE

Principles of Electronic Devices and Circuits - Chapter 17, Section 17.4

OBJECTIVES

In this experiment you will:

✓ Test the operation of a 555 timer as an astable multivibrator

✓ Test the operation of the 555 timer as a monostable multivibrator

✓ Determine the circuit parameters for a simulated fault

EQUIPMENT AND MATERIALS

DC power supply
Dual-trace oscilloscope
Circuit protoboard
555 timer
Resistors: 1 kΩ, 3.3 kΩ, 6.8 kΩ, 10 kΩ [2], 12 kΩ, 15 kΩ
Capacitors: 0.001 μF, 0.01 μF, 0.022 μF, 1 μF

Astable Operation

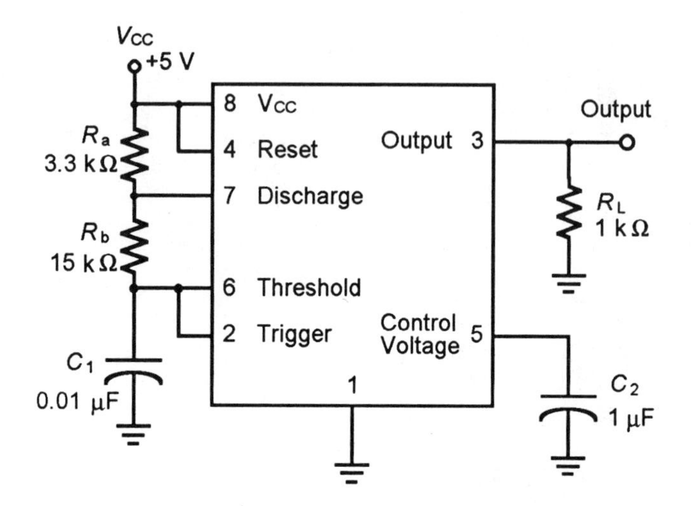

Figure 65.1

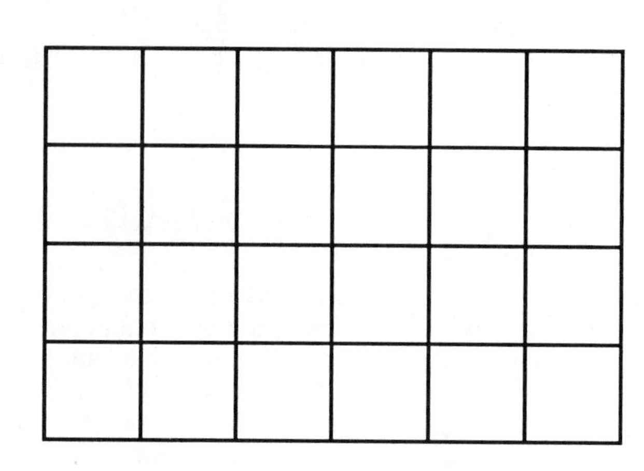

Graph 65.1

1. Connect the circuit as shown in Figure 65.1. Adjust the DC power supply to +5 V for the 555 timer.

   Note: Be sure that you don't forget to connect pin 8 for the $+V_{CC}$ and pin 1 for ground.

2. Connect the oscilloscope to the output and measure the time for one complete cycle. Use teh reult to determine the frequency of the output waveform. Record it below.

   f_o = ___________

3. With your oscilloscope set for dual-trace operation, connect one channel (DC coupled) to the junction of R_b and C_1 to monitor the waveform of C_1, and connect the other channel to the circuit output.

4. Observe the capacitor voltage waveform versus the output waveform. Sketch the oscilloscope display in Graph 65.1. Note in your diagram the minimum and maximum capacitor voltage values.

   You should have found that V_{c1} varies from approximately $1/3V_{cc}$ to $2/3V_{cc}$. What is the frequency of operation?

   f_o = ___________

5. Calculate the duty cycle of the output waveform.

   DC = ___________

6. Turn circuit power off. Replace R_a with a 6.8 kΩ resistor and R_b with a 12- kΩ resistor. The multivibrator frequency of operation will be nearly the same as before. What do you predict the duty cycle will be?

   DC = ____________ (predicted)

7. Turn on circuit power and measure the output frequency and the pulse on time. Record below the output frequency and the new duty cycle.

   f_o = ____________ DC = ____________

Monostable Operation

1. Connect the circuit as shown in Figure 65.2. Connect the signal generator to the input. Adjust the generator for 8-kHz square-wave input at 2 $V_{p\text{-}p}$. Connect channel 1 of the oscilloscope to the input and channel 2 to the output. Sketch the input versus the output in Graph 65.2. Label the timing of the pulses.

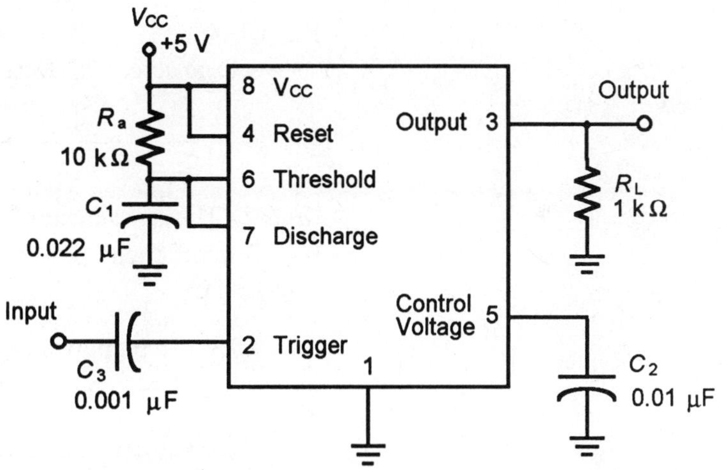

Figure 65.2

2. Observe the input versus the output of your monostable timer while you change the frequency of the input from approximately 4 kHz to 12 kHz.

   Is there any change in the output pulse width? ____________

   Does the circuit generate an output pulse for every negative-going input signal? ____________

3. Set the signal generator to 8 kHz. Connect your oscilloscope to monitor the input signal and the voltage waveform of C_1. Sketch your oscilloscope display in Graph 65.3. Note the minimum and maximum voltage values of the C_1 waveform.

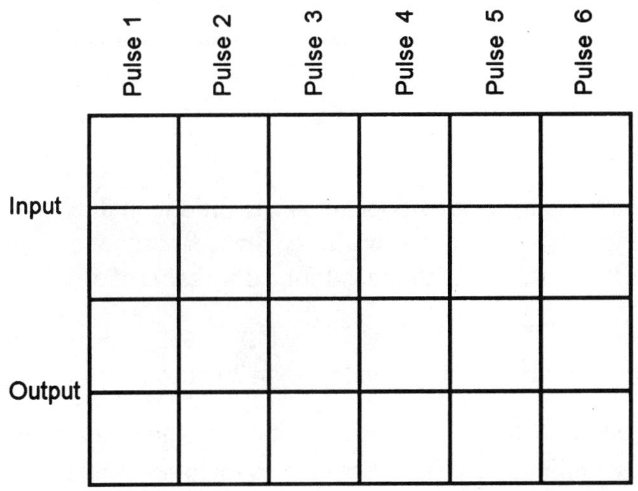

Graph 65.2

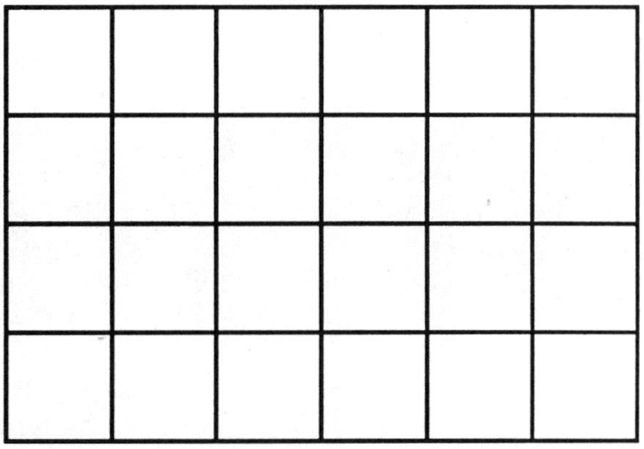

Graph 65.3

SECTION II TROUBLESHOOTING

Fault 1 - Timing resistor R_a open

1. Use the circuit of Figure 65.2. Turn off DC power. Simulate an open timing resistor (R_a) by disconnecting the end of R_a from V_{cc}.

2. Apply power to your circuit and an 8-kHz squarewave signal at 2 $V_{p\text{-}p}$. Connect your oscilloscope to monitor the trigger input and circuit output. Describe below the resulting waveforms.

   ____

3. Change the oscilloscope connections to monitor the trigger input and the voltage across C_1. Describe the waveforms below.

   ____

Fault 2 - Timing capacitor C_1 leaky

1. Turn off circuit power. Reconnect R_a. Connect a 10-kΩ resistor in parallel with C_1.

2. Apply circuit power and trigger input. Monitor the trigger input and circuit output. Describe the circuit waveforms.

   ____

3. Reconnect the oscilloscope to monitor the trigger input and the V_{c1} waveform. Describe the signal waveforms.

   ____

   What is the peak voltage of the V_{c1} waveform?

   V_{c1} (peak) = __________

DISCUSSION

Section I

1. Discuss the effect on the astable circuit operation and the circuit duty cycle when R_a is large or small with respect to R_b. Since R_a cannot be zero, what value of R_a would you recommend for a duty cycle of 50% when R_b is 15 kΩ?

   Note: R_a must limit C_1 discharge current to less than 200 mA.

2. Suppose you are working with a logic controller that (a) requires an input pulse width of 10 μs minimum and (b) is driven by a logic sensor that has negative-going pulses that vary in width from 1 μs to 15 μs. How would you use a 555 timer to ensure proper operation of the system?

Describe your solution and include a schematic diagram as part of your description.

Section II

Fault 1 - Timing resistor R_a open

Discuss the circuit waveform results of this failure. Could you be reasonably certain that the failure was the timing resistor?

Fault 2 - Timing capacitor C_1 leaky

While electrolytic capacitors can fail leaky, this fault is not likely to occur in ceramic or disc capacitors. However, describe how well your measured data indicated a capacitor failure. Would you need additional measurements to prove the capacitor failure?

Quick Check

1. Which resistors in Figure 65.1 control the timing of the 555 timer?

   ____

2. Increasing the size of R_a will cause the pulse width to increase.

   True False

3. The output is always a square wave when the 555 timer is operating in monostable operation.

   True False

4. The 555 timer can operate as a voltage-controlled oscillator.

   True False

66

SILICON-CONTROLLED RECTIFIERS

INTRODUCTION

The silicon-controlled rectifier (SCR) is one of many devices classified as a *thyristor*. It is a four-layer, three-terminal device which acts like a latch with a trigger input. These devices have many applications in commercial and industrial electronics.

In this experiment you will see the basic operation of the SCR and make measurements to observe the characteristics. Then, you will test the SCR in a basic phase-control circuit, emulating one application of an SCR.

REFERENCE

Principles of Electronic Devices and Circuits - Chapter 18, Section 18.2

OBJECTIVES

In this experiment you will:

✓ Understand the basic operational characteristics of an SCR

✓ Learn about SCR operation in a basic phase control circuit

EQUIPMENT AND MATERIALS

DC power supply
Digital multimeter [2]
Oscilloscope
Power transformer, 115 V:12.6 V
Circuit protoboard
SCR, C-106 or similar
Small-signal diode, 1N914 or similar
Resistors: 150 Ω (2 W), 1 kΩ, 1.8 kΩ (2 W), 10 kΩ, 27 kΩ
Potentiometers: 2-kΩ ten-turn trimpot, 50-kΩ ten-turn trimpot

SECTION I FUNCTIONAL EXPERIMENT

1. Construct the circuit of Figure 66.1. Set potentiometer R_2 so that the output voltage to the gate will be 0 V (wiper to ground).

2. Apply 20 VDC to your circuit. Measure and record V_{SCR} (anode to cathode) and V_{RL} in the *SCR Off* box of Table 66.1.

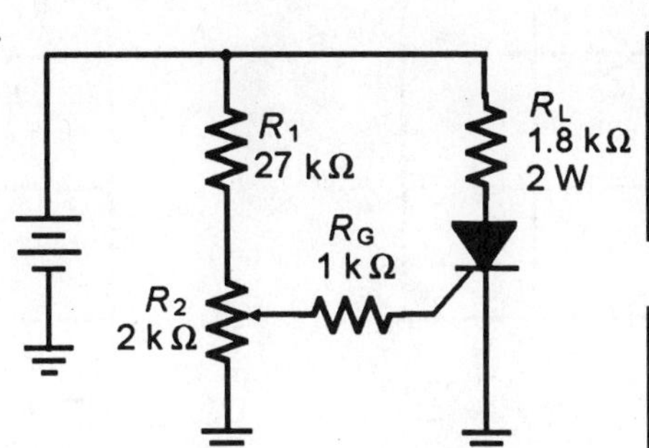

Figure 66.1

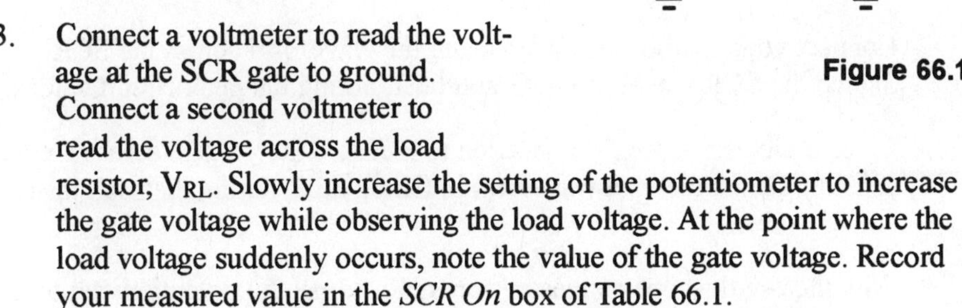

3. Connect a voltmeter to read the voltage at the SCR gate to ground. Connect a second voltmeter to read the voltage across the load resistor, V_{RL}. Slowly increase the setting of the potentiometer to increase the gate voltage while observing the load voltage. At the point where the load voltage suddenly occurs, note the value of the gate voltage. Record your measured value in the *SCR On* box of Table 66.1.

   Note: When you want to repeat this measurement, turn down the gate voltage slightly, and switch the power supply off, then back on. It might take several operations of this measurement to obtain good data.

   Once you have a good and consistent gate voltage reading, leave the SCR on and measure the voltage across the SCR. Record your measured values in the *SCR On* box of Table 66.1.

4. With the potentiometer set to the point of turning the SCR on, turn off the DC supply and connect an ammeter in series with R_L and the SCR. Switch the power supply on. Slowly decrease the power supply voltage and note the SCR current at the point where the SCR stops conducting. This is the SCR holding current I_{HX}. Record your measured value in the *SCR On* box of Table 66.1.

5. Construct the circuit of Figure 66.2. Before applying AC power, recheck your circuit connections to ensure proper connections. Set the potentiometer R_2 to approximately mid-point.

6. Apply AC power.

   Note: Since the transformer secondary is floating from ground, you may apply one (and only one) ground in your circuit.

SCR Off
$V_{SCR(off)}$ = ____________
V_{RL} = ____________

SCR On
V_{GT} = ____________
$V_{SCR(on)}$ = ____________
V_{RL} = ____________
V_{HX} = ____________

Table 66.1

Beginning from step 5, you will set up a basic phase-control circuit. The setting of R_2 will determine the point in the AC sine wave where the SCR will begin to conduct.

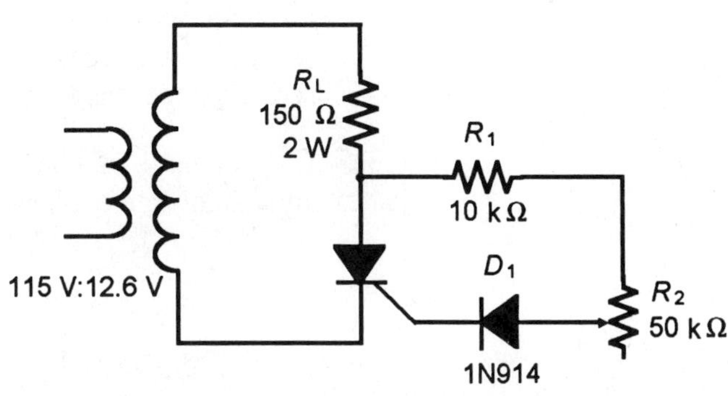

Figure 66.2

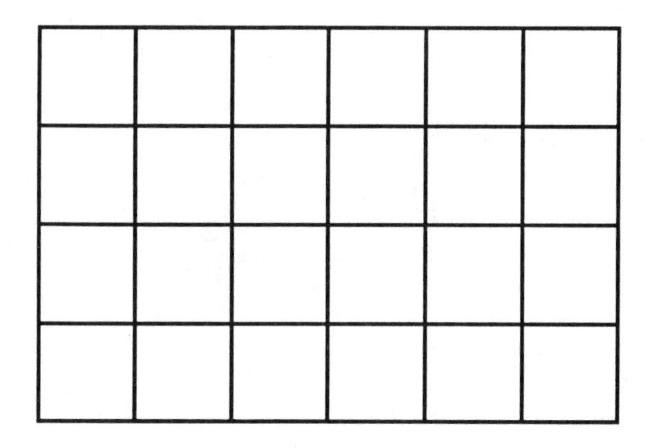

Graph 66.1

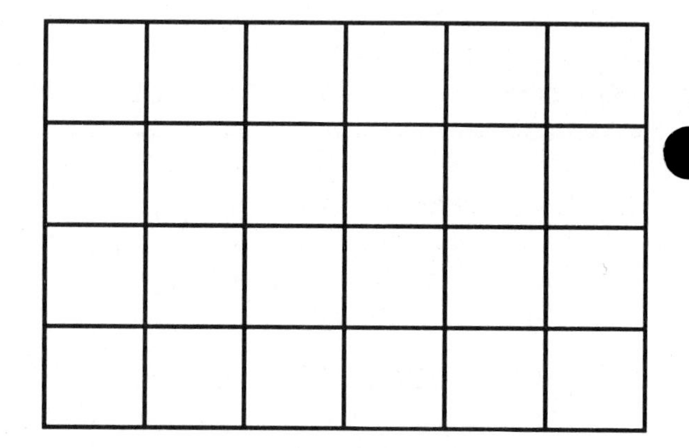

Graph 66.2

Connect your oscilloscope to measure the waveform across the SCR. Sketch the SCR waveform in Graph 66.1, noting the peak circuit values.

7. Change the oscilloscope connection to measure the voltage across the load resistor. Sketch the load waveform in Graph 66.2, noting the peak load voltage values.

8. With the oscilloscope connected across R_L adjust the potentiometer to increase the resistance in the gate circuit and observe the load waveform. Then adjust the potentiometer to decrease the resistance in the gate circuit and observe the new waveform.

You should find that with more resistance, the waveform approximates the diagram of Figure 66.3 (a) and with less resistance in the gate circuit, the load waveform approximates that of Figure 66.3 (b).

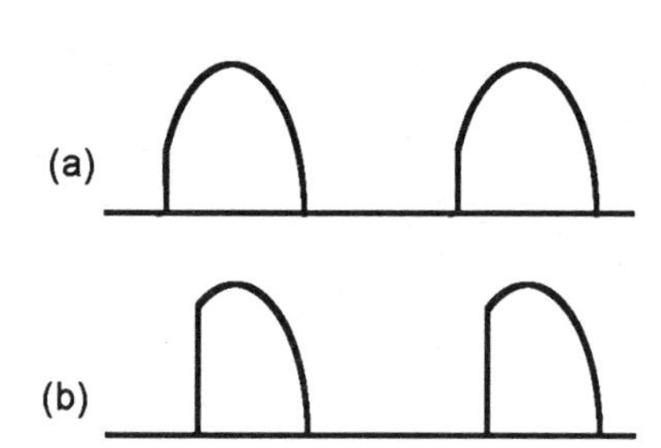

Figure 66.3

As the SCR conducts for more of the positive half cycle more power is delivered to the load. And as the SCR conducts for less of the positive half cycle there is less power to the load.

DISCUSSION

1. In step 8 or your procedure, it is stated that as the SCR conducts for more of the positive half-cycle, more power is delivered to the load. Expand on this and discuss why this should be the case.

2. Figures 66.3 (a) and (b) give an approximate waveform for the load for different SCR turn-on times (phase angles). For each of these figures, sketch the waveform you would expect to see across the SCR and describe the reasons for your waveform sketches.

3. Can you think of a household application for the SCR? Discuss this application and explain how it would operate.

Quick Check

1. The term *SCR* stands for _____.

   (a) Semi-Controlled Regulator
   (b) Silicon-Controlled Regulator
   (c) Silicon-Controlled Rectifier
   (d) Solid-Control Rectifier

2. SCRs are often used in industry.

   True
   False

3. SCRs cannot operate in AC applications.

   True
   False

4. The SCR will stop conducting (turn off) when the SCR current is less than I_{HX}.

   True
   False

67

TRIACS

INTRODUCTION

The triac acts like two SCRs in parallel. It allows current to pass in both directions and is therefore bidirectional. A forward bias is required to trigger the triac, and like the SCR, the triac is often used in industrial applications where control of high current demand loads is common.

In this experiment you will construct a triac control circuit in which the triggering time is determine by an RC time constant. The troubleshooting section will let you see the effect on a triac control circuit that has a timing control failure.

OBJECTIVES

In this experiment you will:

✓ Understand the characteristics of the triac

✓ See how a RC time constant is used to trigger a triac

REFERENCE

Principles of Electronic Devices and Circuits - Chapter 18, Section 18.3

EQUIPMENT AND MATERIALS

Dual-trace oscilloscope
Isolation transformer
Circuit protoboard
SPST switch
Incandescent lamp, 60 W at 120 V
Triac, SC141D or similar
Resistors: 5.6 kΩ, 10 kΩ, 1 MΩ
Potentiometer, 100-kΩ ten-turn trimpot
Capacitor, 0.22 μF [2]

SECTION I FUNCTIONAL EXPERIMENT

> **IMPORTANT!**
>
> Measurements in this experiment require grounding one side of the AC line. Therefore the circuit must be supplied through an isolation transformer to avoid shorting the source.

1. Construct the circuit in Figure 67.1. Make sure S_1 is open.

2. Set the potentiometer R_1 to its maximum setting.

3. Calculate the RC time constant for R = 100 kΩ.

4. Connect your oscilloscope channel 1 to the triac MT_2, and connect channel 2 to the junction of C_1 and the triac gate.

   Note: The voltage level of channel 1 will be 120 V_{RMS}. You might need a 10:1 scope probe to make this measurement.

5. Apply AC power. Close switch S_1. Since the oscilloscope is measuring the voltage across the triac which will be a 120-V sine wave when the triac is not conducting, the firing point is identified as the point in the waveform where the triac voltage drops suddenly to approximately 2 V. See the waveform diagrams in Figure 67.2 to identify the firing and conduction angles.

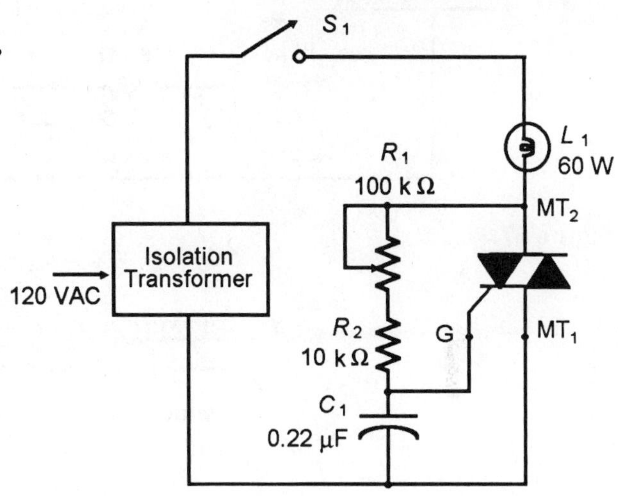

Figure 67.1

6. Slowly adjust R_1 to decrease the resistance while observing the triac waveform until you observe the triac firing. Measure and record below the peak voltage across capacitor C_1. This is the gate voltage required to trigger the triac.

   V_{C1} (peak) = ___________

Note: The RC trigger may cause the lamp to delay lighting, then suddenly turn on brightly. In this case readjust R_1 to make the lamp dim to its minimum value.

7. Adjust R_1 until the lamp turns on and stays lit, even though dim. Measure and record below the conduction angle.

   Conduction angle = ___________

   Turn off the circuit power. Disconnect R_1 and measure its resistance. Using this resistance and the fixed resistance in the circuit, calculate the timing circuit RC time constant.

   RC = ___________

8. Reconnect R_1. Apply AC power. Adjust R_1 to its minimum value. Because of the fixed resistance in the timing circuit you won't obtain full lamp brightness. Measure and record below the circuit conduction angle.

   Conduction angle = ___________

   Calculate the timing circuit RC time constant.

   RC = ___________

   Observe and describe the capacitor C_1 waveform.

   ____

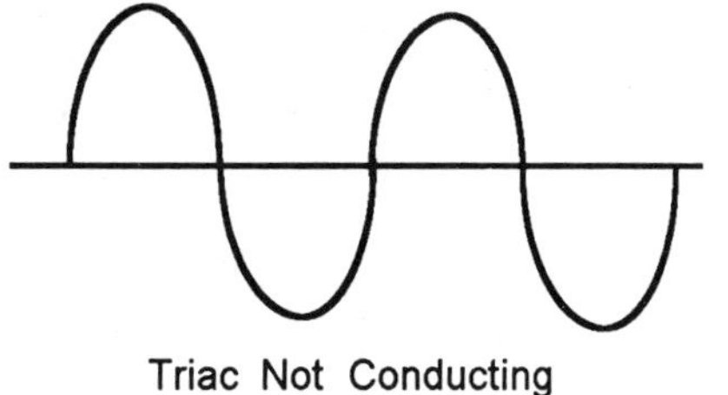

Triac Not Conducting

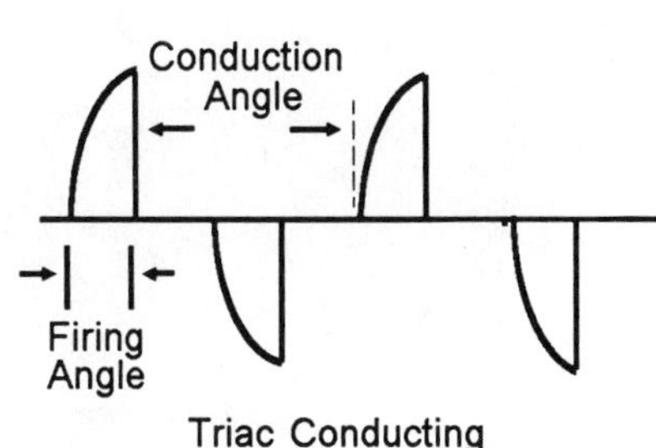

Triac Conducting

Figure 67.2

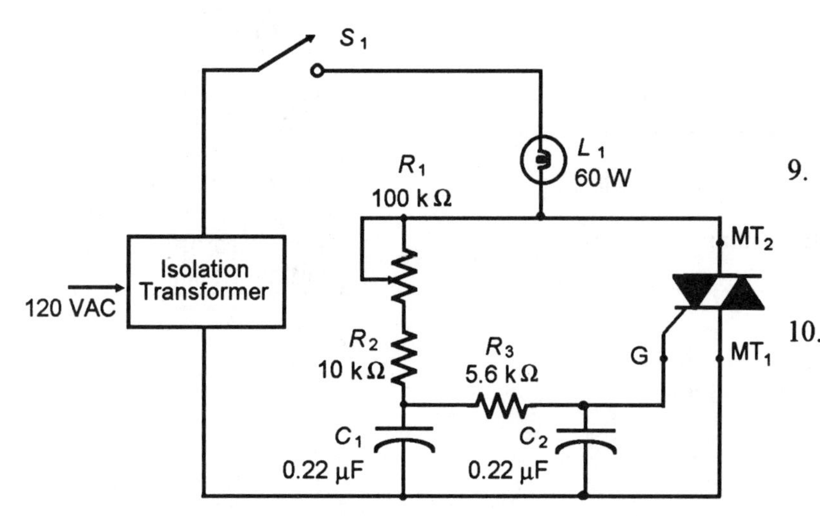

Figure 67.3

The circuit of Figure 67.3 has an added RC network to give greater range of the conduction angle.

9. Turn off AC power and add this additional network to your circuit. Set R_1 to its mid-point value. Open switch S_1. Apply AC power to your circuit.

10. Adjust R_1 while observing the lamp. Does your circuit operate better in controlling the lamp brightness? ___________

SECTION II TROUBLESHOOTING

Fault 1 - Timing resistor open

1. Turn off the AC power. Remove R_2, the 10-kΩ resistor, and replace with a 1-MΩ resistor. Open S_1 and connect your oscilloscope to monitor the triac gate and triac waveform.

2. Turn on AC power. Close S_1. Try adjusting R_1 to turn on the lamp. Can you get the lamp to light? ___________

3. Measure the peak capacitor voltage and record below.

   $V_{c1(pk)}$ = ___________

   Observe and describe the triac voltage waveform.

   ____

DISCUSSION

Section I

1. In your own words, describe the circuit operation of Figure 67.1. How does it operate? Formulate your description as though you were the technician writing the service manual for the field technician.

2. At what point did the triac begin to conduct? ___________

   Did the trigger voltage match the specs in the data book? ___________

Section II

Fault 1 - Timing resistor open

1. What does the triac waveform tell you about the triac? Could the triac be partially shorted, for example?

2. What does the capacitor voltage waveform tell you about C_1? Can it be open? Shorted?

Quick Check

1. A triac acts like two SCRs in parallel.

   True False

2. Trigger voltages are listed on data sheets for the triac.

   True False

3. In the circuit in Figure 67.1, the triac is triggered after the _____.

   (a) capacitor charge reaches the trigger voltage
   (b) resistor charge reaches the trigger voltage
   (c) load reaches maximum current
   (d) load reaches minimum current

4. The triac is a bidirectional device.

   True False

68

FULL-WAVE PHASE CONTROL

INTRODUCTION

The diac is a thrysistor device that, like the triac, can latch on and conduct current in either direction. Unlike the triac, the diac has no gate. The diac switches on when the applied voltage reaches the breakover potential. One of its main applications is to serve as a trigger diode for the triac or SCR.

In this experiment you will see the diac as a trigger diode for a triac in a phase-control circuit. The diac trigger provides a current pulse to the triac gate, giving more consistent triggering than the sine wave of an RC trigger. The section on troubleshooting will let you see the effect on a phase-control circuit when the triac fails. You will make circuit measurements to determine the fault effect.

OBJECTIVES

In this experiment you will:

- ✓ Become more familiar with the characteristics of a triac phase-control circuit
- ✓ Observe the characteristics of the diac in a phase-control circuit
- ✓ Through measurements determine the effect of a triac failure on a phase-control circuit

REFERENCE

Principles of Electronic Devices and Circuits - Chapter 18, Section 18.3

EQUIPMENT AND MATERIALS

Digital multimeter
Dual-trace oscilloscope
Circuit protoboard
Isolation transformer
Potentiometer, 100-kΩ ten-turn trimpot
Capacitor, 0.33 μF
Triac, SC141D or similar
Diac, ST-2 or similar
Incandescent lamp: 60 W,120 V

SECTION I FUNCTIONAL EXPERIMENT

CAUTION 

Do not operate the circuit without an isolation transformer. Your test equipment will ground one of the AC lines in your circuit. The isolation transformer will prevent the possibility of shorting the AC source.

1. Build the circuit of Figure 68.1. Make sure switch S_1 is open. Set potentiometer R_1 to its maximum resistance. **Do not apply AC power at this time.**

2. Connect one channel of your oscilloscope to the MT_2 terminal of the triac to measure the voltage across the triac.

   Note: The voltage will at times be 120 V_{RMS}. You may need to use a 10:1 probe for your oscilloscope.

   Connect the other channel of your oscilloscope to the junction of C_1 and the diac to observe the voltage waveform of the capacitor C_1.

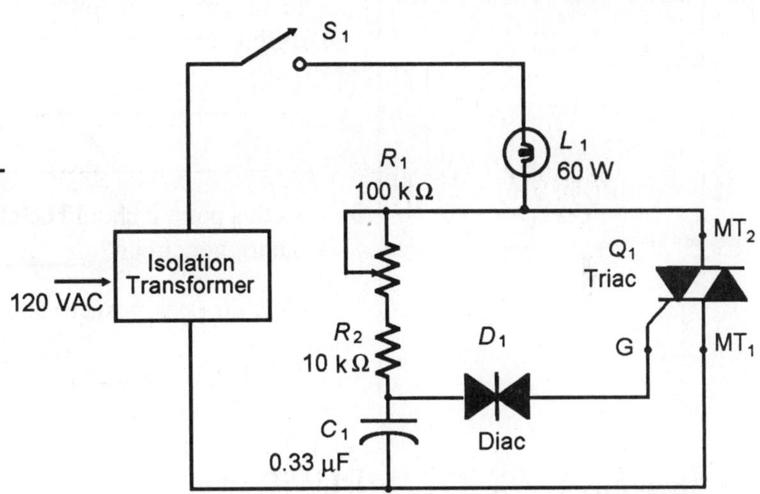

Figure 68.1

3. Apply AC power to your circuit. Observe the voltage waveform of the triac and capacitor C_1. Note below the peak positive and negative values of C_1 voltage.

   $V_{C1}(+pk)$ = ___________

   $V_{C1}(-pk)$ = ___________

4. Adjust R_1 to decrease the resistance in the timing circuit until the lamp just begins to light. If the lamp flashes on, wait a few seconds and adjust R_1 for minimum brightness.

   Sketch the oscilloscope display in Graph 68.1, noting the peak voltage values of the waveforms.

Graph 68.1

5. From your triac waveform of step 4, calculate the circuit conduction angle, and record below.

   Conduction angle = ___________

6. Adjust R_1 for maximum lamp brightness. From the oscilloscope display calculate the circuit conduction angle. Record this angle below.

   Conduction angle = ___________

SECTION II TROUBLESHOOTING

Fault 1 - Triac shorted

1. Turn off the AC power. Set S_1 to the off position. Simulate a shorted triac by installing a jumper wire in parallel with the triac. Connect your oscilloscope to monitor the triac voltage and capacitor C_1 waveforms.

2. Apply AC power. Switch S_1 on. Your lamp should light with full brightness. Observe the voltage waveform of the triac. Compare this waveform to the waveform observed in Section I, steps 4 and 6. Note the differences.

   ____

   ____

3. At this point it should be clear that the fault is the triac. Can you rule out the trigger circuit? __________

   Note the peak positive and negative values of the capacitor waveform.

   $V_{C1}(+pk)$ = __________ $V_{C1}(-pk)$ = __________

DISCUSSION

Section I

1. From your measurements of step 4, determine the diac breakover voltage. Assume that the triac gate potential is 2 V at the instant of trigger application.

   V_{BO} = __________

   Discuss your observation of the capacitor C_1 voltage waveform. Was there a large difference in the positive and negative peak values at the point of triac triggering over several cycles of operation?

2. Describe another application for the full-wave phase-control circuit. Use a functional block diagram and give sufficient information in your description to indicate a full understanding of the phase-control circuit.

Section II

Fault 1 - Triac shorted

1. Discuss the troubleshooting steps you followed in testing the phase-control circuit for the simulated failure. Was step 3 a necessary test to help confirm that the failure was the triac?

2. Describe the test steps you would make to troubleshoot a triac gate failure.

Quick Check

1. A diac is a bidirectional device.

True False

2. Breakover voltage is the voltage required for the diac to begin to conduct.

True False

3. In the circuit in Figure 68.1, the triac is triggered after the _____.

(a) capacitor charge reaches the trigger voltage
(b) capacitor reaches diac breakdown
(c) load reaches maximum current
(d) load reaches minimum current

4. The triac in this circuit acts as a bidirectional device.

True False

69

SERIES-PASS REGULATORS

INTRODUCTION

A series-pass regulator is a circuit where the load voltage, and consequently the load current, is controlled by a transistor in series with the load. The series-pass regulator is the basis of many of the currently available integrated-circuit voltage regulators. This basic circuit provides simple and effective load and line regulation, with nominal power consumption from the source.

In this experiment you will construct a discrete device series-pass regulator. Your measurements of this circuit will let you see the basic operation of a voltage regulator.

OBJECTIVES

Through this experiment you will:

✓ Understand the operation of a series-pass regulator

✓ Determine through measurement the regulation of a series-pass regulator

REFERENCE

Principles of Electronic Devices and Circuits - Chapter 19, Section 19.2

EQUIPMENT AND MATERIALS

DC Power Supply
NPN transistor [2], 2N3904 or equivalent
Small-signal diode, 1N914 or similar
Resistors: 150 Ω, 200 Ω, 470 Ω, 1 kΩ, 3.3 kΩ
Potentiometer, 1-kΩ ten-turn trimpot

SECTION I FUNCTIONAL EXPERIMENT

The circuit used in this experiment is nearly identical to Figure 19.7 in your text. This experiment, however, uses a forward-biased diode in place of the zener diode shown in your text.

1. Construct the circuit of Figure 69.1. Set the potentiometer R_3 to approximate mid-range.

2. Apply 10.0 VDC to your circuit input. Measure the regulator output, and adjust R_3 to obtain an output of 5.00 V as closely as you can. Record the DC input and output of your regulator in Table 69.1 for the load resistance of 1 kΩ.

3. Switch off the DC supply and install the next-listed load resistance of Table 69.1. Turn on the DC input and maintain this voltage at the same value set in step 2. Record in Table 69.1 the value of the input and output voltages.

4. Repeat the procedure of step 3 for each load resistance value of Table 69.1. *No load* is simply the regulator operating with R_L removed.

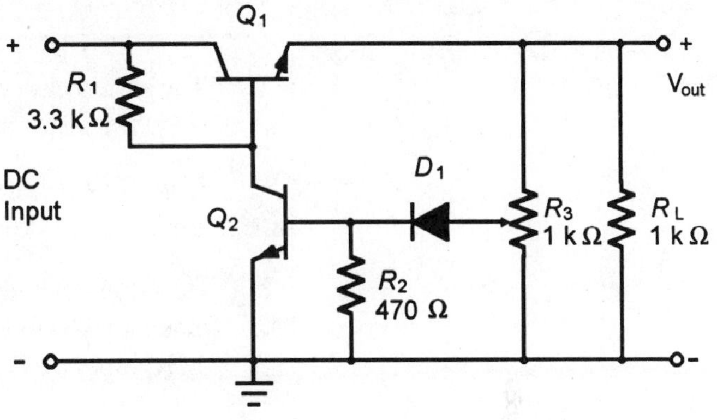

Figure 69.1

Note: Since your regulator is a low-power circuit, turn off the DC input as soon as you have completed a measurement.

V_{in}	V_{out}	R
________	________	1 kΩ
________	________	470 Ω
________	________	220 Ω
________	________	150 Ω
________	________	No load

Table 69.1

5. Using the regulator output with a 1-kΩ resistor as nominal, calculate the difference between the nominal and the output readings having the greatest amount of difference. This is the output variation for load changes. Record the variation below.

   Variation = ____________

6. Calculate the regulation percentage using the formula below.

$$\text{Regulation } \% = \frac{\text{Variation}}{\text{Nominal Output}} \times 100$$

   The Nominal Output is the regulator voltage measured with a 1-kΩ load and Variation is the value calculated in step 5.

7. Turn the DC power off. Install the 1-kΩ load. Apply the DC input. Measure and record the DC input level.

   DC input = ____________

   Measure the regulator output and the base voltage of Q_1 the series-pass transistor, and record below.

   V_{out} = ______________ V_B (Q1) = __________

8. To see the adjustment range of your regulator, adjust potentiometer R_3 to one extreme end and measure the output voltage. Then set R_3 to the other extreme end and measure the output voltage. Notice that the control transistor Q_2 does not have sufficient gain to cut off or saturate the pass transistor, Q_1.

DISCUSSION

1. The series-pass regulator of your experiment is just a DC coupled two-stage transistor amplifier. Describe the configuration of each stage.

   Hint: If necessary, sketch the pass transistor in a vertical position to aid in identifying the configuration.

2. Describe what changes you would make to the circuit of Figure 69.1 to make a fixed 9-V regulator capable of providing 0.5-A load current with a DC input of 12.5 V.

Quick Check

1. A series-pass regulator puts the regulating device in series with the load.

   True False

2. Regulators are designed to deliver a constant current to changing loads.

   True False

3. In a series-pass regulator like that of Figure 69.1 the pass transistor configuration is _____.

   (a) base-biased common emitter (b) DC-coupled common base
   (c) DC-coupled common collector (d) base-biased common base

4. In a series-pass regulator like that of Figure 69.1, the output voltage should be approximately 0.7 V less than the base voltage of the pass transistor.

   True False

70

IC REGULATORS

INTRODUCTION

There are many integrated circuit (IC) forms of regulators available. These regulators make possible very simple regulated power supplies with good line and load regulation. One popular regulator is the 7805. It is part of a family of fixed voltage regulators, intended primarily to provide an output of +5 V.

In this experiment you will construct a circuit using the 7805 regulator and, through measurement, evaluate the effects of varying line input and circuit loads.

REFERENCE

Principles of Electronic Devices and Circuits - Chapter 19, Section 19.6

OBJECTIVES

In this experiment you will:

✓ Observe the characteristics of the IC voltage regulator

✓ Make measurements to verify line and load regulation of the 7805 regulator

EQUIPMENT AND MATERIALS

DC power supply
Digital multimeter
Circuit protoboard
Regulator, MC7805
Resistors: 25 Ω (1 W), 47 Ω (2 W), 100 Ω (1 W), 220 Ω
Capacitors: 0.33 μF, 1 μF

CAUTION 

Since your regulator is not mounted on a heat sink, its power dissipation is limited to 2 W. Switch off the input DC voltage as soon as your output measurement is complete to avoid excess heating of the regulator.

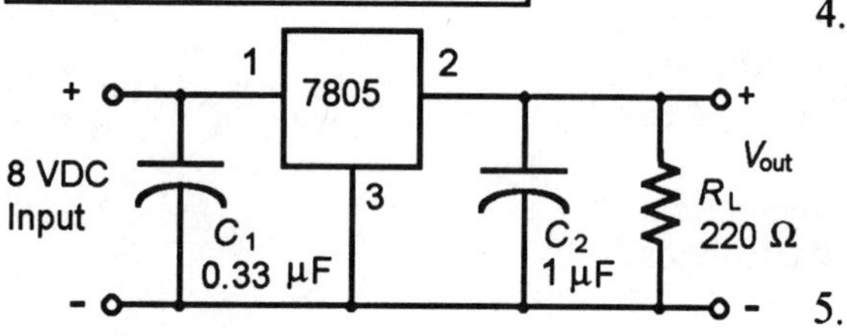

Figure 70.1

Load Regulation	
R_L	V_{out}
220 Ω	________
100 Ω	________
50 Ω	________
25 Ω	________
Line Regulation	
V_{in}	V_{out}
8.0 V	________
8.5 V	________
7.5 V	________
7.0 V	________

Table 70.1

SECTION I FUNCTIONAL EXPERIMENT

1. Construct the circuit in Figure 70.1.
2. Apply 8 VDC power. Measure the voltage across the R_L and record it in Table 70.1.
3. Switch off the DC power, and replace R_L with the next value listed in the *Load Regulation* part of Table 70.1. Apply DC power; then measure and ensure that the input voltage is 8.0 V. Measure and record in Table 70.1 the regulated output. Repeat this process for each load resistance value in Table 70.1.
4. From your load regulation data of Table 70.1, what is the largest variation of the output voltage (ΔV_{out}) from the nominal 5 V measured for a 220-Ω load?

   ΔV_{out} = ____________

5. Calculate the load regulation as a percentage, using the formula below and the variation (ΔV_{out}) from step 4.

$$\% \text{ Regulation} = \frac{\Delta V_{out}}{5 \text{ V}} \times 100$$

   % Regulation = ____________

6. Make sure the DC power is off. Change the load resistor to the 100-Ω resistor. Reapply DC power, and ensure that the input is 8.0 V. Measure and record the actual value in the *Line Regulation* part of Table 70.1.
7. Change the input voltage to each of the values in Table 70.1. Measure and record the regulated output voltage in the table.
8. Although the input voltage variations exceeded that of a ripple of 1 $V_{p\text{-}p}$, the output variation should have been less than 50 mV. How well did your regulator maintain the output?

   Output variation (max) = ________________ V

DISCUSSION

1. Describe how you would use the 7805 regulator with a full-wave bridge rectifier to make a regulated 5-V power supply. Include in your description a schematic showing your complete power supply.
2. Discuss the advantages of using a regulator such as the 7805 in a power supply. List as many advantages as you can.

Quick Check

1. Most three-terminal regulators come in TO-3 or TO-220 packages.

   True False

2. The 7805 regulator is often used for TTL ICs.

   True False

3. The voltage regulator is designed to deliver a constant voltage for varying loads.

   True False

4. The data sheet indicates that the 7805 regulator can deliver a constant 5-V level for input voltages between 7 V and 12 V.

   True False

71

SIGNAL MODULATION AND DEMODULATION

INTRODUCTION

Modern electronic devices such as radio and television are made possible through the process of impressing an intelligence signal upon a carrier. The high-frequency signal is called the *carrier*, the low-frequency signal is called the *modulation* or *intelligence*, and the resultant is called the *modulated signal.*

In this experiment you will construct an amplitude modulator and detector. Through measurement of the input and output signals of each, you will see the process of modulation and the process of demodulation, extracting the intelligence from the carrier.

REFERENCE

Principles of Electronic Devices and Circuits - Chapter 20, Section 20.2

OBJECTIVES

Through this experiment you will:

✓ Understand the characteristics of amplitude modulation

✓ Understand the characteristics of amplitude modulation detection (demodulation)

EQUIPMENT AND MATERIALS

DC power supply
Dual-trace oscilloscope
Function generator [2]
Circuit protoboard
NPN transistor, 2N3904 or equivalent
Small-signal diode, 1N914 or similar
Resistors: 2.2 kΩ, 27 kΩ, 100 kΩ
Capacitors: 680 pF, 0.01 μF, 0.022 μF

SECTION I FUNCTIONAL EXPERIMENT

1. Construct the circuit in Figure 71.1. Apply DC power.

2. Inject a 120-kHz carrier signal of approximately 200 $mV_{p\text{-}p}$ to the base of your modulator.

3. With your oscilloscope, observe the modulator output signal, and adjust the generator frequency for a maximum amplitude output. When the generator is tuned to the resonant frequency, set the generator output level to obtain a modulator output of approximately 18 $V_{p\text{-}p}$.

4. Connect a second generator to provide the modulating (intelligence) signal to the emitter of your modulator. Set the generator to its minimum output voltage level with a 1-kHz sinewave output.

5. Vary the modulation input signal level while observing the modulator output. You should find that you can vary the modulated signal from a very low modulation level to overmodulation (a distorted, nonsinusoidal modulation envelope). Set the modulation generator to provide 100% modulation. See the modulation waveform in Figure 71.2.

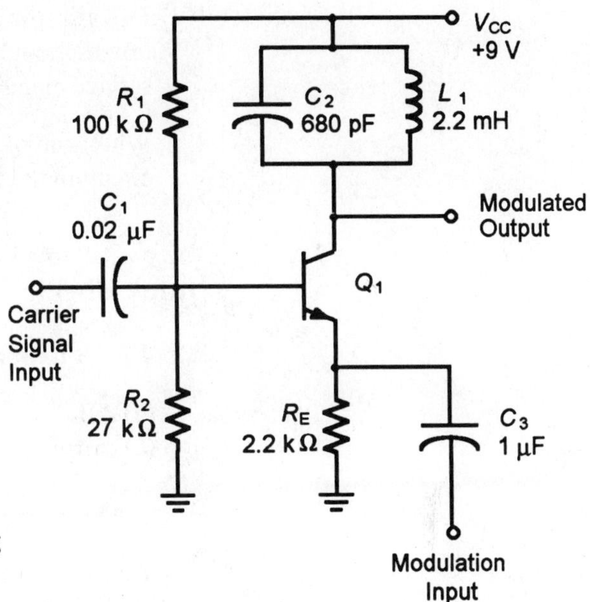

Figure 71.1

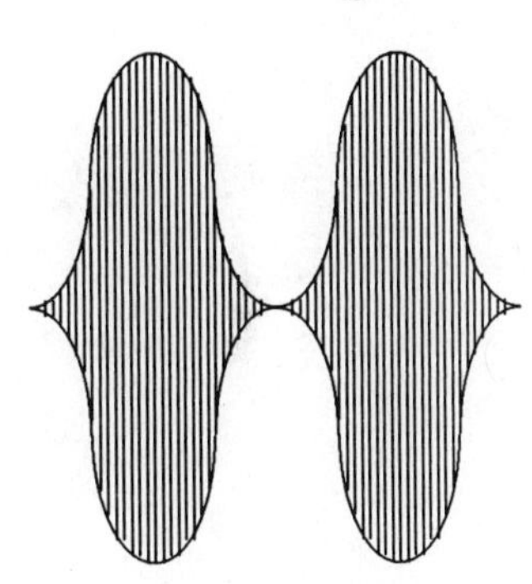

Figure 71.2

6. When you varied the modulation generator output level, what part of the modulated output waveform changed?

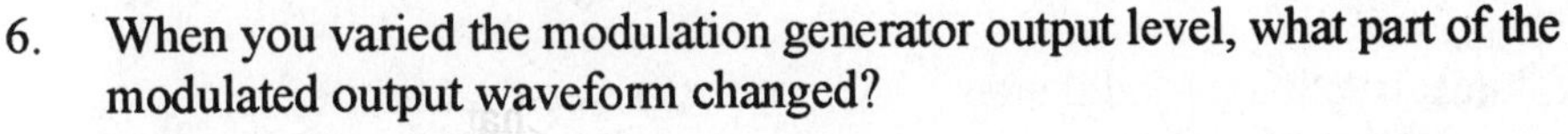

Vary the modulation generator frequency while observing the modulator output. What part of the modulated signal changed?

____

7. Construct the circuit in Figure 71.3.

8. Connect the detector circuit to the output of the modulator.

9. Note the signal at the input of the detector circuit. It should look like the modulated wave you observed in step 5.

10. Observe the output signal of the detector. Does it look like the intelligence part of the modulated waveform you observed in step 5?

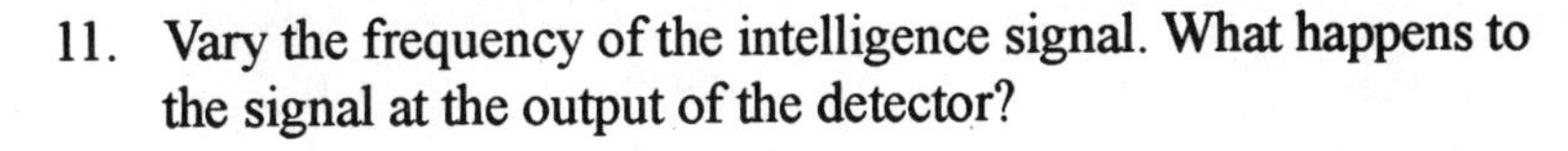

11. Vary the frequency of the intelligence signal. What happens to the signal at the output of the detector?

____

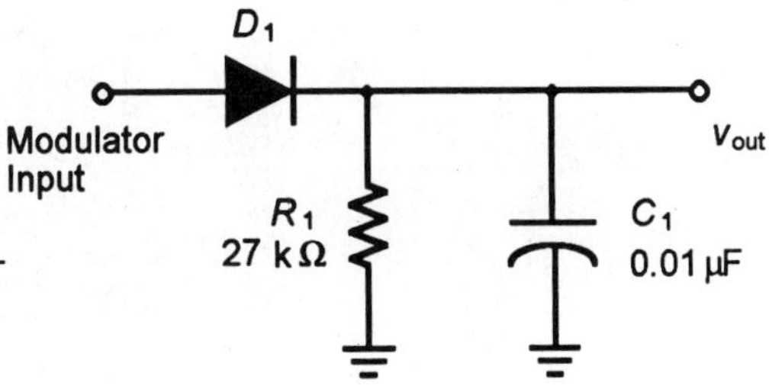

Figure 71.3

DISCUSSION

1. Describe the operation of the modulator circuit. Formulate your description as though you were the engineering technician assigned to write the service manual for the field technicians.

2. What kind of circuit is the detector circuit? (*Hint:* You saw this type of circuit in the beginning chapters.)

Quick Check

1. The term for mixing two signals is _____.

   (a) AB (b) heterodyning
   (c) carrying (d) AM

2. The high-frequency signal is called the carrier.

   True False

3. The modulating signal is also called the intelligence.

   True False

4. The term *AM* stands for _____.

   (a) Amplitude Modulation (b) Always Mobile
   (c) After Modulation (d) Alternate Modulation

72

PHASE-LOCKED LOOPS

INTRODUCTION

The phase-locked loop (PLL) has several important applications in communications and industrial electronics. It can provide voltage-to-frequency conversion, frequency modulation, and frequency demodulation, and it is the central element in a frequency synthesis system.

In this experiment you will explore the phase-lock characteristics of a phase locked loop and the effect of input frequency changes on the DC output of the chip.

REFERENCE

Principles of Electronic Devices and Circuits - Chapter 20, Section 20.5

OBJECTIVES

In this experiment you will:

✓ Gain an understanding of the operation of the phase locked loop

✓ Understand, by experiment, the relationship between input frequency shifts and the DC output voltage

✓ Measure the lock range of a phase-locked loop

EQUIPMENT AND MATERIALS

Dual DC power supply
Digital multimeter
Dual-trace oscilloscope
Function generator
Circuit protoboard
PLL, 565 or equivalent
Resistor, 7 kΩ
Potentiometer, 50-kΩ ten-turn trimpot
Capacitors: 0.001 μF, 0.01 μF, 0.1 μF

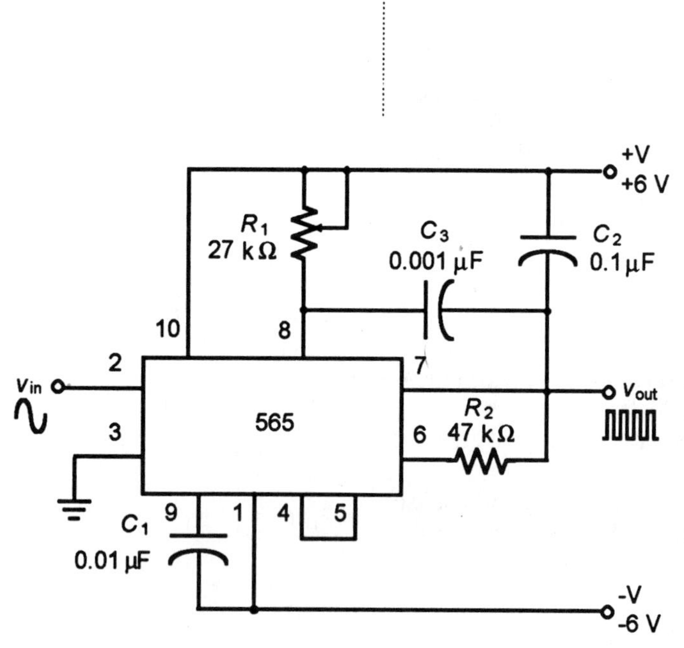

Figure 72.1

Phase F_{in} vs. f_o	Frequency	Demodulated Output Voltage V_d
0	Upper lock	
90		
180	Lower lock	

Table 72.1

Demodulator Equations for Figure 72.1

$$f_o = \frac{1}{3.7R_1C_1}$$

$$f_h = \pm\frac{8\ f_o}{V_{CC}}$$

1. Preset potentiometer R_1 for 20 kΩ. Carefully construct the circuit shown in Figure 72.1. Apply power.

2. Measure the frequency (f_o) of the PLL's internal VCO (pins 4 and 5). Adjust R_1 to set this frequency accurately to 10 kHz.

   Important: Set up the function generator as described in the fol-lowign step *prior to connecting the generator* to the input of th ecir-cuit.

3. Set the function generator for a V_{in} of 2 $V_{p\text{-}p}$,10-kHz square wave.

4. Apply V_{in} to the circuit. Measure the phase relationship of $f_o = f_{in}$ with the oscilloscope. Record the value in Table 72.1. With a DVM, measure and record the demodulated voltage (V_d) at pin 7.

5. Slowly increase and decrease the frequency f_{in}. Observe that as f_{in} de-creases, f_o decreases, the phase relationship between the two changes, and V_d amplitude changes–the loop is locked.

Steps 6 through 8 provide data necessary to analyze the linear-ity of the frequency-to-voltage conversion, the lock range, and the capture range of the circuit.

6. Slowly decrease f_{in} below 10 kHz.

   As the input frequency decreases, does f_o become more out of phase or nearer in phase with f_{in}? ___________

   As f_{in} decreases, what happens to the output voltage at pin 7?

   ____

7. The internal VCO will track input frequency changes until the phase relationship of f_{in} vs. f_o becomes in phase or 180° out of phase. Beyond this point, the PLL cannot remain locked on f_{in}. Decrease f_{in} and measure as accurately as possible the frequency and demodulated voltage at the lower lock range limit.

   Lower lock range = ___________

8. After you have gone below the lower lock range limit, slowly increase f_{in}. Record the frequency at which the PLL locks with f_{in}. This is the lower capture range limit. The capture ranges are functions of the low-pass filter (C_2) and loop gain control (R_2).

   Lower capture range = ___________

9. Repeat steps 7 and 8 to analyze the tracking response of the PLL for frequencies above f_o. Record the upper lock range and upper capture range.

   Upper lock range = ___________

   Upper capture range = ___________

10. Incrementally change the input frequency throughout the lock range. Measure and record the input frequency, the amplitude of V_{out}, and the phase relationships of f_{in} vs. f_o in Table 72.1.

11. Plot in Graph 72.1, the input frequency vs. demodulated voltage from measurements logged in Table 72.1. Label the lock and capture ranges.

V_d f_{in}

Graph 72.1

12. Return f_{in} to 10 kHz. Remove the low-pass filter, C_2. Briefly describe the duty cycle, frequency, and DC reference of the unfiltered squarewave V_{out}.

13. Sweep f_{in} throughout the lock range. Briefly describe the effects that a change in f_{in} has on the duty cycle (and hence the average voltage) of V_{out}.

    Note: In many applications where minimum demodulated ripple, fast lock-up time, and broad capture range are of primary concern, quite higher-order, low-pass filtering of V_d is necessary.

DISCUSSION

Section I

1. From a practical viewpoint, without technical detail, explain why you think the capture frequency for the PLL is less (smaller) than the lock frequency.

2. Describe how you would use the 565 PLL as a voltage-to-frequency convertor where the output frequency is proportional to the input DC voltage.

3. Describe, using the PLL block diagram, how you would use the PLL as a frequency modulator.

Quick Check

1. The phase lock range for the LM565 PLL is 90°.

   True False

2. If a PLL has an output frequency of 26 kHz at 1.8-V input and an output frequency of 32 kHz at 3.2-V input, what is the frequency-to-voltage conversion of the VCO?

   (a) 3.15 kHz/V (b) 4.29 kHz/V
   (c) 5.01 kHz/V (d) 6.23 kHz/V

3. A PLL can be used to generate a frequency-modulated signal.

   True False

4. A PLL has two states–capture and lock.

   True False

5. When a PLL is used as a frequency demodulator, the output is taken from the _____.

   (a) voltage-controlled oscillator output
   (b) phase comparator input
   (c) filter output
   (d) reference signal input

73

VARACTOR DIODES

INTRODUCTION

The varactor (voltage-variable-capacitor) is a special purpose-silicon diode used in communications equipment such as television and FM receivers. The capacitance value of the varactor is controlled by a voltage.

Because the capacitance is variable, the varactor can replace a mechanically variable capacitor. In this experiment you will construct a circuit to see the ability of the varactor diode to modify the circuit operation in response to a control voltage.

OBJECTIVES

Through this experiment you will:

✓ Understand the characteristics of the varactor

✓ Observe a practical application of the varactor in a circuit controlling the frequency of operation

REFERENCE

Principles of Electronic Devices and Circuits - Chapter 20, Section 20.4

EQUIPMENT AND MATERIALS

DC power supply
Dual-trace oscilloscope
Circuit protoboard
Inductors: 2.2 mH, 300 mH
Capacitors: 680 pF, 0.1 μF
Varactor diode, 1N5148 or equivalent
NPN transistor, 2N3904 or equivalent

SECTION I FUNCTIONAL EXPERIMENT

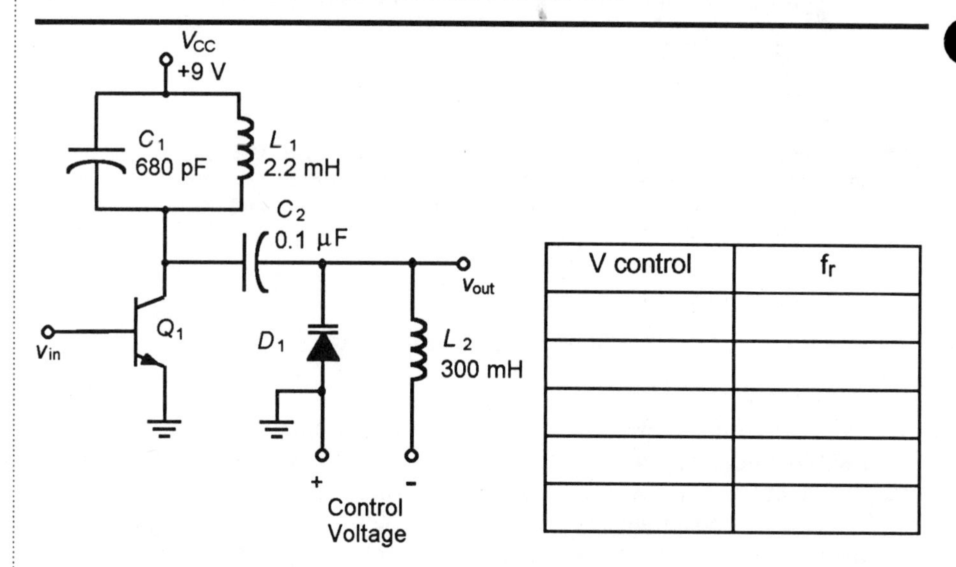

Figure 73.1

V control	f_r

Table 73.1

1. Construct the circuit in Figure 73.1.

2. Apply +9 VDC to your circuit. Connect varactor control voltage source and adjust it to the supply 25 VDC.

3. Calculate the expected frequency of operation of the circuit tank.

   f_o = ____________

4. Connect your signal generator to the circuit input, and use the oscilloscope to adjust the generator for a sinewave signal of 2 V_{p-p} at the frequency you calculated in step 3. Make sure there is no DC offset in the signal generator output. Then connect your oscilloscope to observe the circuit output.

5. Adjust the signal generator frequency to obtain the maximum output signal from your circuit. This is the circuit resonant frequency with a minimum capacity of the varactor diode.

6. Measure the circuit resonant frequency using your oscilloscope and record below.

   f_r = ____________

You should have found that the resonant frequency shifted down approximately 5% to 8%, and that you can control, within the resonant frequency range of steps 6 and 7, the circuit operating frequency with the varactor control voltage.

7. Decrease the varactor diode control voltage to 1 V. Adjust the signal generator to the circuit resonant frequency. Measure the frequency of the generator and record below.

   f_r = ____________

8. Increase the varactor control voltage in 5-V steps and track the tank resonant frequency by tuning the signal generator to each new resonant frequency.

WARNING

Do not exceed a 25-V varactor control voltage in steps 8 and 9.

Record the control voltage and circuit resonant frequency in Table 73.1.

9. Is there a linear change in resonant frequency shift versus the control voltage? ___________

DISCUSSION

1. Recall that reverse bias of a diode widens the depletion area of the diode, effectively increasing the distance between the conductors (plates) of the diode appearing as a capacitor. Discuss the operation of your circuit and its response to an increasing and decreasing bias of the varactor diode.

2. Suppose that the Q of the circuit of Figure 73.1 was relatively low, approximately 10, and that the circuit was supplied with a fixed input frequency at or very near resonance. Describe the circuit output if the varactor control voltage was a sinewave clamped with the positive peak at 0 V (a negative sinusoidal waveform).

   What would you call this circuit in this case?

Quick Check

1. The varactor is a special-purpose diode operated in the reverse bias mode.

   True False

2. The varactor reacts like a standard diode at high frequencies.

   True False

3. The varactor reacts like a capacitor at high frequencies.

   True False

4. Varactors are often used in FM receivers to control VCOs.

   True False

APPENDIX A

COMPOSITE EQUIPMENT AND MATERIALS LIST

Laboratory Equipment

DC power supply [2]
Dual-trace oscilloscope
VOM
Digital multimeter [2]
Function generator [2]
Frequency counter
Circuit protoboard

Fixed Carbon Resistors

All are 1/4 or 1/2 W unless stated otherwise

25 Ω, 1W
47 Ω
47 Ω, 2W
82 Ω
100 Ω [2]
100 Ω, 1W
150 Ω
150 Ω, 2W
180 Ω
200 Ω
200 Ω, 2W
220 Ω
270 Ω
300 Ω
330 Ω [2]
470 Ω
470 Ω, 1/2W
500 Ω
560 Ω
680 Ω
820 Ω
1 kΩ [2]
1 kΩ 1/2W
1.2 kΩ
1.5 kΩ
1.8 kΩ [2]
1.8 kΩ 2W
2 kΩ
2.2 kΩ [2]
2.7 kΩ [2]
3 kΩ
3.3 kΩ [2]
3.9 kΩ [2]
4.6 kΩ
4.7 kΩ [2]
5 kΩ [2]
5.1 kΩ
5.6 kΩ
6.8 kΩ [3]
7 kΩ
8.2 kΩ
9.1 kΩ
10 kΩ [3]
12 kΩ [4]
15 kΩ
18 kΩ
20 kΩ
22 kΩ [2]
27 kΩ
30 kΩ
33 kΩ [2]
36 kΩ
39 kΩ
40 kΩ
47 kΩ
50 kΩ [2]
51 kΩ
56 kΩ
68 kΩ
82 kΩ
100 kΩ [5]
120 kΩ
150 kΩ
220 kΩ
240 kΩ
270 kΩ
300 kΩ
470 kΩ
500 kΩ
510 kΩ
680 kΩ
910 kΩ
1 M [2]
2.2 M
4.7 M

Potentiometers

Any available type
100 Ω
25 kΩ
200 kΩ

Ten-turn trimpots
1 kΩ
2 kΩ
5 k Ω [2]
10 kΩ [2]
50 kΩ
2 MΩ or 5 MΩ

Capacitors

Working voltage rating of 25 V or better

0.1 mF
0.22 mF [2]
0.33 mF
0.47 mF [2]
1 mF [3]
4.7 mF
10 mF [2]
22 mF
47 mF
100 mF
200 mF
470 mF (50 V) [3]

20 pF
47 pF
68 pF
100 pF [2]
470 pF [2]
1000 pF [3]
0.001 mF
0.0033 mF
0.0047 mF [2]
0.01 mF [2]
0.022 mF [2]
0.033 mF
0.047 mF

Semiconductor Devices

Small-signal diode, 1N914 or similar
Recifier diode [4], 1N4001 or similar
9.1-V zener diode, 1N5239B or equivalent
Varactor diode, 1N5148 or equivalent
Red LED, TIL221 or similar
UJT, 2N2646 or equivalent
NPN transistor, 2N3440 or equivalent
NPN transistor [3], 2N3904 or equivalent*
PNP transistor, 2N3906 or equivalent
N-channel JFET, 2N3819 or MPF102 or equivalent*
N-channel JFET, 2N5459 or equivalent
N-channel D MOSFET, 2N3796 or equivalent
SCR, C-106 or similar
Diac, ST-2 or similar
Triac, SC141D or similar
555 timer IC*
Regulator IC, MC7805
PLL IC, 656 or equivalent
Operational amplifier IC [3], 741 or equivalent*

*Data sheet available in Appendix B

Inductors and Transformers

15 μH inductor
1 mH inductor
10 mH inductor
15 mH inductor

Power transformer, 115 V:12.6 V CT
Isolation transformer

Miscellaneous Items

SPST toggle switch [4]
1-0-1 mA meter movement
Incandescent lamp, 60 W at 120 V
Heat gun or freeze spray

APPENDIX B

DATA SHEETS

2N3903
2N3904★

CASE 29-04, STYLE 1
TO-92 (TO-226AA)

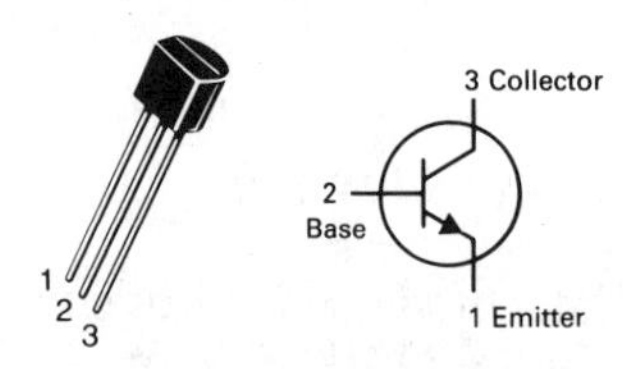

GENERAL PURPOSE TRANSISTORS

NPN SILICON

★This is a Motorola designated preferred device.

MAXIMUM RATINGS

Rating	Symbol	Value	Unit
Collector-Emitter Voltage	V_{CEO}	40	Vdc
Collector-Base Voltge	V_{CBO}	60	Vdc
Emitter-Base Voltage	V_{EBO}	6.0	Vdc
Collector Current — Continuous	I_C	200	mAdc
Total Device Dissipation @ $T_A = 25°C$ Derate above 25°C	P_D	625 5.0	mW mW/°C
*Total Device Dissipation @ $T_C = 25°C$ Derate above 25°C	P_D	1.5 12	Watts mW/°C
Operating and Storage Junction Temperature Range	T_J, T_{stg}	−55 to +150	°C

***THERMAL CHARACTERISTICS**

Characteristic	Symbol	Max	Unit
Thermal Resistance, Junction to Ambient	$R_{\theta JA}$	200	°C/W
Thermal Resistance, Junction to Case	$R_{\theta JC}$	83.3	°C/W

*Indicates Data in addition to JEDEC Requirements.

ELECTRICAL CHARACTERISTICS ($T_A = 25°C$ unless otherwise noted.)

Characteristic		Symbol	Min	Max	Unit
OFF CHARACTERISTICS					
Collector-Emitter Breakdown Voltage(1) ($I_C = 1.0$ mAdc, $I_B = 0$)		$V_{(BR)CEO}$	40	—	Vdc
Collector-Base Breakdown Voltage ($I_C = 10$ μAdc, $I_E = 0$)		$V_{(BR)CBO}$	60	—	Vdc
Emitter-Base Breakdown Voltage ($I_E = 10$ μAdc, $I_C = 0$)		$V_{(BR)EBO}$	6.0	—	Vdc
Base Cutoff Current ($V_{CE} = 30$ Vdc, $V_{EB} = 3.0$ Vdc)		I_{BL}	—	50	nAdc
Collector Cutoff Current ($V_{CE} = 30$ Vdc, $V_{EB} = 3.0$ Vdc)		I_{CEX}	—	50	nAdc
ON CHARACTERISTICS					
DC Current Gain(1)		h_{FE}			—
($I_C = 0.1$ mAdc, $V_{CE} = 1.0$ Vdc)	2N3903 2N3904		20 40	— —	
($I_C = 1.0$ mAdc, $V_{CE} = 1.0$ Vdc)	2N3903 2N3904		35 70	— —	
($I_C = 10$ mAdc, $V_{CE} = 1.0$ Vdc)	2N3903 2N3904		50 100	150 300	
($I_C = 50$ mAdc, $V_{CE} = 1.0$ Vdc)	2N3903 2N3904		30 60	— —	
($I_C = 100$ mAdc, $V_{CE} = 1.0$ Vdc)	2N3903 2N3904		15 30	— —	
Collector-Emitter Saturation Voltage(1) ($I_C = 10$ mAdc, $I_B = 1.0$ mAdc) ($I_C = 50$ mAdc, $I_B = 5.0$ mAdc)		$V_{CE(sat)}$	— —	0.2 0.3	Vdc
Base-Emitter Saturation Voltage(1) ($I_C = 10$ mAdc, $I_B = 1.0$ mAdc) ($I_C = 50$ mAdc, $I_B = 5.0$ mAdc)		$V_{BE(sat)}$	0.65 —	0.85 0.95	Vdc
SMALL-SIGNAL CHARACTERISTICS					
Current-Gain — Bandwidth Product ($I_C = 10$ mAdc, $V_{CE} = 20$ Vdc, f = 100 MHz)	2N3903 2N3904	f_T	250 300	— —	MHz

2N3903, 2N3904

ELECTRICAL CHARACTERISTICS (continued) (T_A = 25°C unless otherwise noted.)

Characteristic		Symbol	Min	Max	Unit
Output Capacitance (V_{CB} = 5.0 Vdc, I_E = 0, f = 1.0 MHz)		C_{obo}	—	4.0	pF
Input Capacitance (V_{EB} = 0.5 Vdc, I_C = 0, f = 1.0 MHz)		C_{ibo}	—	8.0	pF
Input Impedance (I_C = 1.0 mAdc, V_{CE} = 10 Vdc, f = 1.0 kHz)	2N3903	h_{ie}	1.0	8.0	k ohms
	2N3904		1.0	10	
Voltage Feedback Ratio (I_C = 1.0 mAdc, V_{CE} = 10 Vdc, f = 1.0 kHz)	2N3903	h_{re}	0.1	5.0	$X 10^{-4}$
	2N3904		0.5	8.0	
Small-Signal Current Gain (I_C = 1.0 mAdc, V_{CE} = 10 Vdc, f = 1.0 kHz)	2N3903	h_{fe}	50	200	—
	2N3904		100	400	
Output Admittance (I_C = 1.0 mAdc, V_{CE} = 10 Vdc, f = 1.0 kHz)		h_{oe}	1.0	40	μmhos
Noise Figure (I_C = 100 μAdc, V_{CE} = 5.0 Vdc, R_S = 1.0 k ohms, f = 1.0 kHz)	2N3903	NF	—	6.0	dB
	2N3904		—	5.0	

SWITCHING CHARACTERISTICS

Characteristic	Conditions		Symbol	Min	Max	Unit
Delay Time	(V_{CC} = 3.0 Vdc, V_{BE} = 0.5 Vdc, I_C = 10 mAdc, I_{B1} = 1.0 mAdc)		t_d	—	35	ns
Rise Time			t_r	—	35	ns
Storage Time	(V_{CC} = 3.0 Vdc, I_C = 10 mAdc, I_{B1} = I_{B2} = 1.0 mAdc)	2N3903	t_s	—	175	ns
		2N3904		—	200	
Fall Time			t_f	—	50	ns

(1) Pulse Test: Pulse Width ≤ 300 μs, Duty Cycle ≤ 2.0%.

FIGURE 1 – DELAY AND RISE TIME EQUIVALENT TEST CIRCUIT

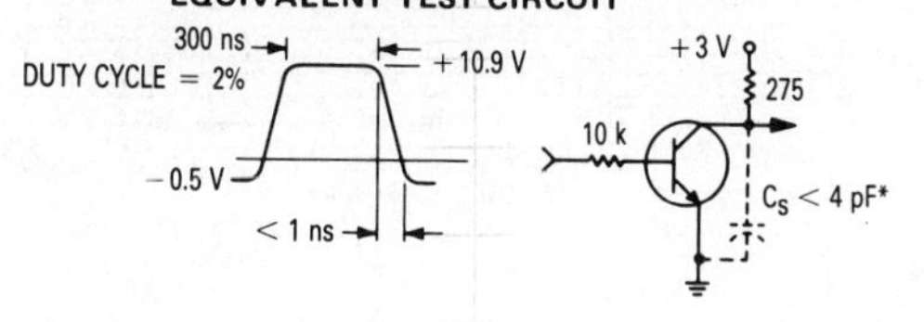

FIGURE 2 – STORAGE AND FALL TIME EQUIVALENT TEST CIRCUIT

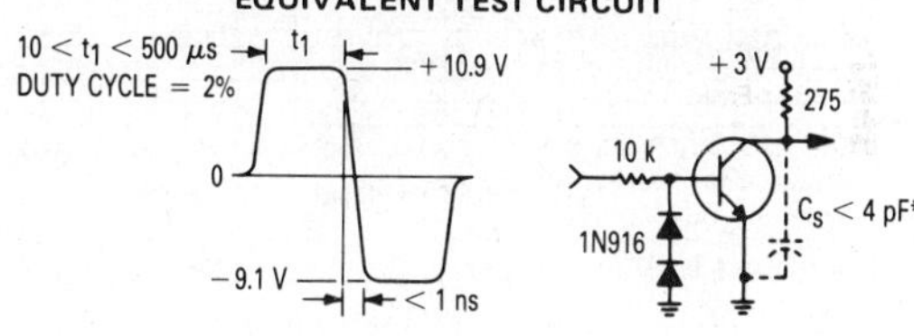

*Total shunt capacitance of test jig and connectors

TYPICAL TRANSIENT CHARACTERISTICS

— T_J = 25°C - - - T_J = 125°C

FIGURE 3 – CAPACITANCE

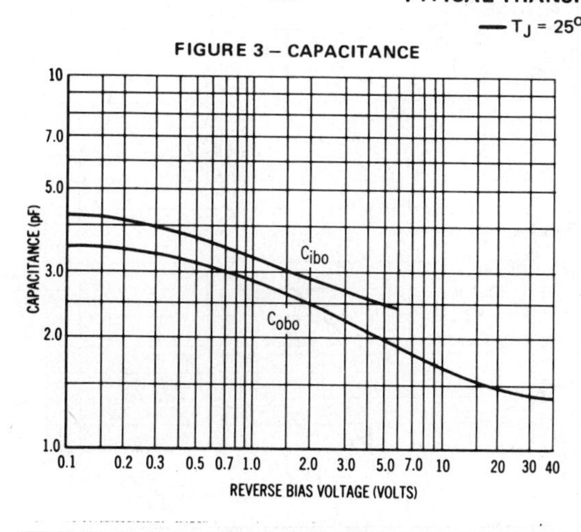

FIGURE 4 – CHARGE DATA

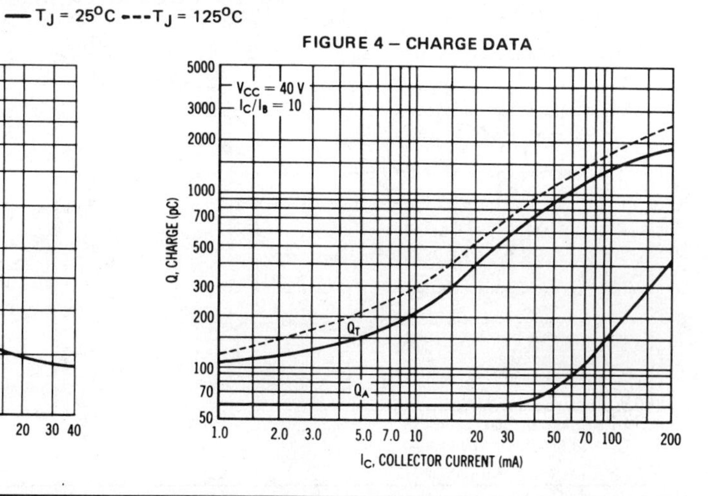

FIGURE 5 – TURN-ON TIME

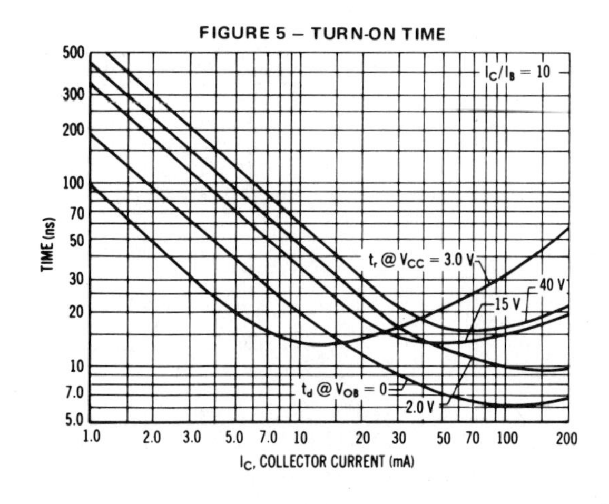

FIGURE 6 – RISE TIME

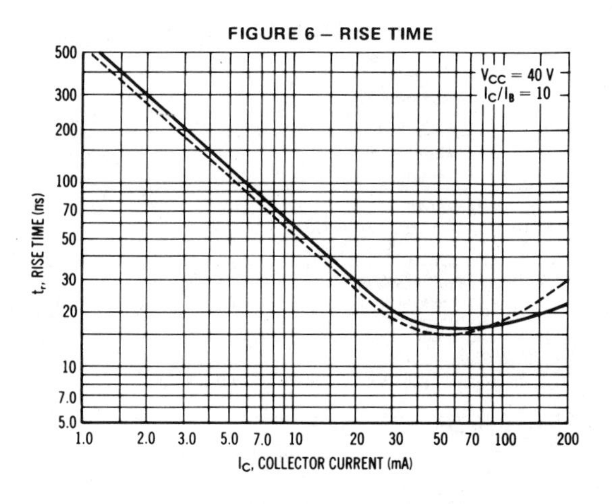

FIGURE 7 – STORAGE TIME

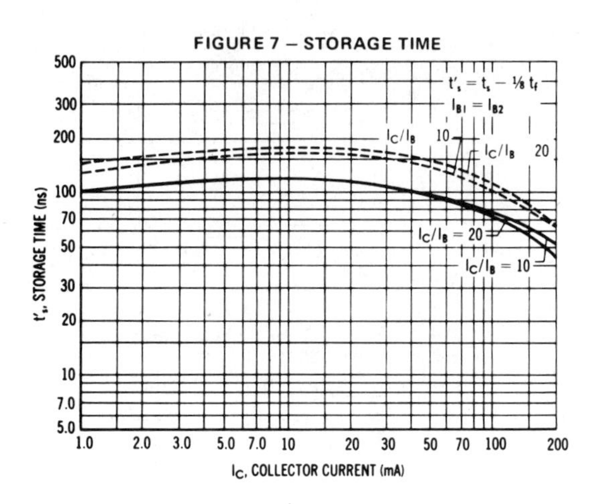

FIGURE 8 – FALL TIME

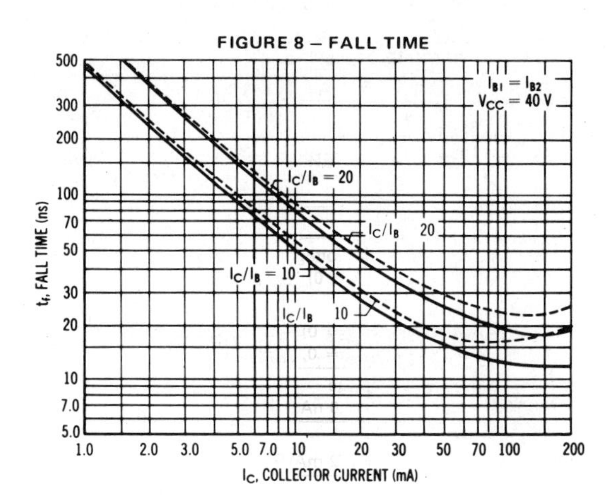

TYPICAL AUDIO SMALL-SIGNAL CHARACTERISTICS

NOISE FIGURE VARIATIONS

V_{CE} = 5.0 Vdc, T_A = 25°C,
Bandwidth = 1.0 Hz

FIGURE 9

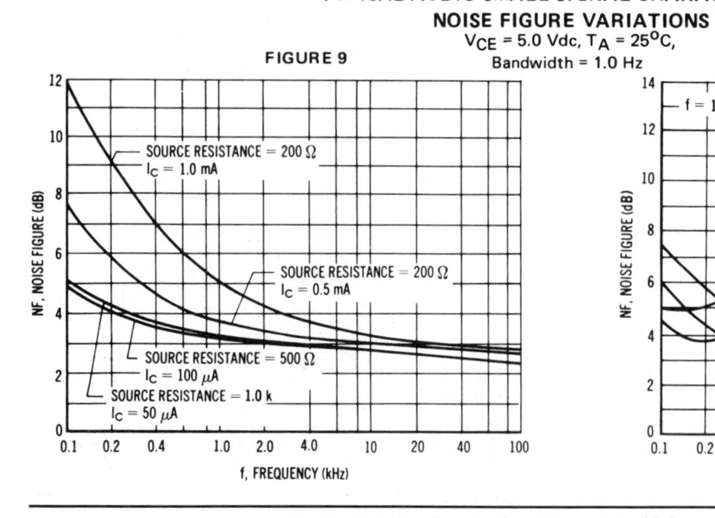

FIGURE 10

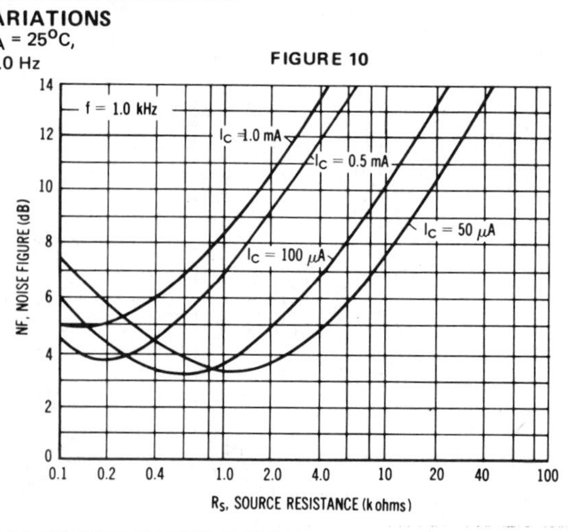

MPF102

CASE 29-04, STYLE 5
TO-92 (TO-226AA)

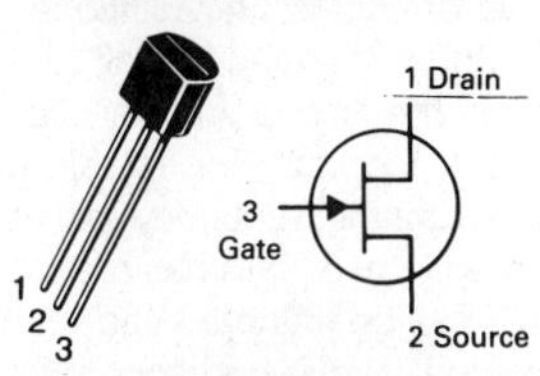

JFET
VHF AMPLIFIER

N-CHANNEL — DEPLETION

Refer to 2N5484 for graphs.

MAXIMUM RATINGS

Rating	Symbol	Value	Unit
Drain-Source Voltage	V_{DS}	25	Vdc
Drain-Gate Voltage	V_{DG}	25	Vdc
Gate-Source Voltage	V_{GS}	−25	Vdc
Gate Current	I_G	10	mAdc
Total Device Dissipation @ T_A = 25°C Derate above 25°C	P_D	350 2.8	mW mW/°C
Junction Temperature Range	T_J	125	°C
Storage Temperature Range	T_{stg}	−65 to +150	°C

ELECTRICAL CHARACTERISTICS (T_A = 25°C unless otherwise noted.)

Characteristic	Symbol	Min	Max	Unit
OFF CHARACTERISTICS				
Gate-Source Breakdown Voltage (I_G = −10 μAdc, V_{DS} = 0)	$V_{(BR)GSS}$	−25	—	Vdc
Gate Reverse Current (V_{GS} = −15 Vdc, V_{DS} = 0) (V_{GS} = −15 Vdc, V_{DS} = 0, T_A = 100°C)	I_{GSS}	 — —	 −2.0 −2.0	 nAdc μAdc
Gate Source Cutoff Voltage (V_{DS} = 15 Vdc, I_D = 2.0 nAdc)	$V_{GS(off)}$	—	−8.0	Vdc
Gate Source Voltage (V_{DS} = 15 Vdc, I_D = 0.2 mAdc)	V_{GS}	−0.5	−7.5	Vdc
ON CHARACTERISTICS				
Zero-Gate-Voltage Drain Current* (V_{DS} = 15 Vdc, V_{GS} = 0 Vdc)	I_{DSS}	2.0	20	mAdc
SMALL-SIGNAL CHARACTERISTICS				
Forward Transfer Admittance* (V_{DS} = 15 Vdc, V_{GS} = 0, f = 1.0 kHz) (V_{DS} = 15 Vdc, V_{GS} = 0, f = 100 MHz)	$\|y_{fs}\|$	 2000 1600	 7500 —	μmhos
Input Admittance (V_{DS} = 15 Vdc, V_{GS} = 0, f = 100 MHz)	$Re(y_{is})$	—	800	μmhos
Output Conductance (V_{DS} = 15 Vdc, V_{GS} = 0, f = 100 MHz)	$Re(y_{os})$	—	200	μmhos
Input Capacitance (V_{DS} = 15 Vdc, V_{GS} = 0, f = 1.0 MHz)	C_{iss}	—	7.0	pF
Reverse Transfer Capacitance (V_{DS} = 15 Vdc, V_{GS} = 0, f = 1.0 MHz)	C_{rss}	—	3.0	pF

*Pulse Test: Pulse Width ≤ 630 ms; Duty Cycle ≤ 10%.

LM555/LM555C Timer

General Description

The LM555 is a highly stable device for generating accurate time delays or oscillation. Additional terminals are provided for triggering or resetting if desired. In the time delay mode of operation, the time is precisely controlled by one external resistor and capacitor. For astable operation as an oscillator, the free running frequency and duty cycle are accurately controlled with two external resistors and one capacitor. The circuit may be triggered and reset on falling waveforms, and the output circuit can source or sink up to 200 mA or drive TTL circuits.

Features

- Direct replacement for SE555/NE555
- Timing from microseconds through hours
- Operates in both astable and monostable modes
- Adjustable duty cycle
- Output can source or sink 200 mA
- Output and supply TTL compatible
- Temperature stability better than 0.005% per °C
- Normally on and normally off output

Applications

- Precision timing
- Pulse generation
- Sequential timing
- Time delay generation
- Pulse width modulation
- Pulse position modulation
- Linear ramp generator

Schematic Diagram

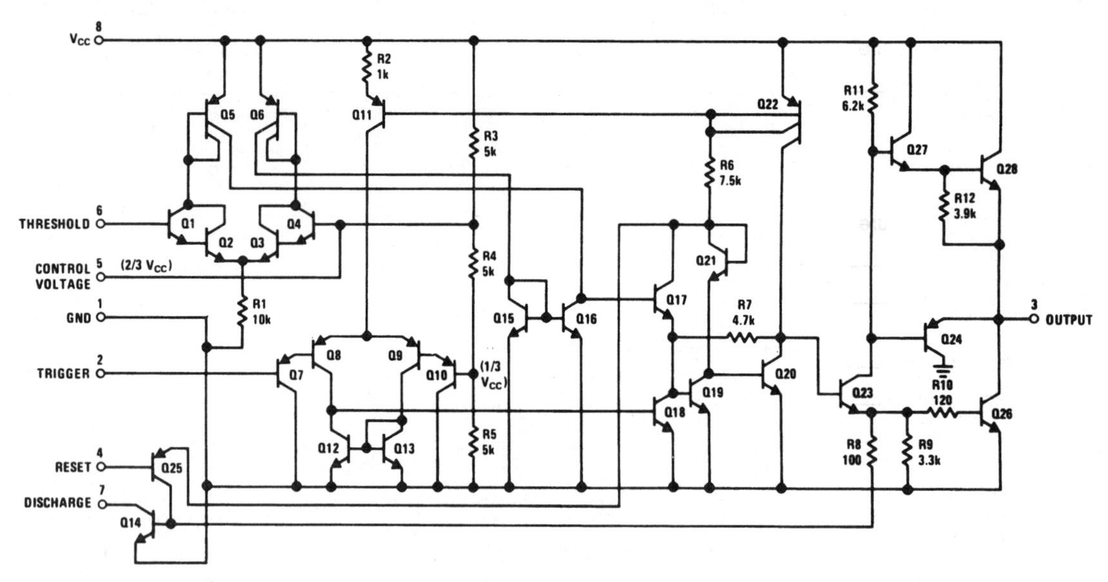

TL/H/7851-1

Absolute Maximum Ratings

If Military/Aerospace specified devices are required, please contact the National Semiconductor Sales Office/Distributors for availability and specifications.

Supply Voltage	+18V
Power Dissipation (Note 1)	
LM555H, LM555CH	760 mW
LM555, LM555CN	1180 mW
Operating Temperature Ranges	
LM555C	0°C to +70°C
LM555	−55°C to +125°C
Storage Temperature Range	−65°C to +150°C
Soldering Information	
Dual-In-Line Package	
Soldering (10 Seconds)	260°C
Small Outline Package	
Vapor Phase (60 Seconds)	215°C
Infrared (15 Seconds)	220°C

See AN-450 "Surface Mounting Methods and Their Effect on Product Reliability" for other methods of soldering surface mount devices.

Electrical Characteristics (T_A = 25°C, V_{CC} = +5V to +15V, unless othewise specified)

Parameter	Conditions	Limits						Units
		LM555			LM555C			
		Min	Typ	Max	Min	Typ	Max	
Supply Voltage		4.5		18	4.5		16	V
Supply Current	V_{CC} = 5V, R_L = ∞		3	5		3	6	mA
	V_{CC} = 15V, R_L = ∞ (Low State) (Note 2)		10	12		10	15	mA
Timing Error, Monostable								
Initial Accuracy			0.5			1		%
Drift with Temperature	R_A = 1k to 100 kΩ, C = 0.1 μF, (Note 3)		30			50		ppm/°C
Accuracy over Temperature			1.5			1.5		%
Drift with Supply			0.05			0.1		%/V
Timing Error, Astable								
Initial Accuracy			1.5			2.25		%
Drift with Temperature	R_A, R_B = 1k to 100 kΩ, C = 0.1 μF, (Note 3)		90			150		ppm/°C
Accuracy over Temperature			2.5			3.0		%
Drift with Supply			0.15			0.30		%/V
Threshold Voltage			0.667			0.667		x V_{CC}
Trigger Voltage	V_{CC} = 15V	4.8	5	5.2		5		V
	V_{CC} = 5V	1.45	1.67	1.9		1.67		V
Trigger Current			0.01	0.5		0.5	0.9	μA
Reset Voltage		0.4	0.5	1	0.4	0.5	1	V
Reset Current			0.1	0.4		0.1	0.4	mA
Threshold Current	(Note 4)		0.1	0.25		0.1	0.25	μA
Control Voltage Level	V_{CC} = 15V	9.6	10	10.4	9	10	11	V
	V_{CC} = 5V	2.9	3.33	3.8	2.6	3.33	4	V
Pin 7 Leakage Output High			1	100		1	100	nA
Pin 7 Sat (Note 5)								
Output Low	V_{CC} = 15V, I_7 = 15 mA		150			180		mV
Output Low	V_{CC} = 4.5V, I_7 = 4.5 mA		70	100		80	200	mV

Electrical Characteristics $T_A = 25°C$, $V_{CC} = +5V$ to $+15V$, (unless othewise specified) (Continued)

Parameter	Conditions	Limits						Units
		LM555			LM555C			
		Min	Typ	Max	Min	Typ	Max	
Output Voltage Drop (Low)	$V_{CC} = 15V$							
	$I_{SINK} = 10$ mA		0.1	0.15		0.1	0.25	V
	$I_{SINK} = 50$ mA		0.4	0.5		0.4	0.75	V
	$I_{SINK} = 100$ mA		2	2.2		2	2.5	V
	$I_{SINK} = 200$ mA		2.5			2.5		V
	$V_{CC} = 5V$							
	$I_{SINK} = 8$ mA		0.1	0.25				V
	$I_{SINK} = 5$ mA					0.25	0.35	V
Output Voltage Drop (High)	$I_{SOURCE} = 200$ mA, $V_{CC} = 15V$		12.5			12.5		V
	$I_{SOURCE} = 100$ mA, $V_{CC} = 15V$	13	13.3		12.75	13.3		V
	$V_{CC} = 5V$	3	3.3		2.75	3.3		V
Rise Time of Output			100			100		ns
Fall Time of Output			100			100		ns

Note 1: For operating at elevated temperatures the device must be derated above 25°C based on a +150°C maximum junction temperature and a thermal resistance of 164°c/w (T0-5), 106°c/w (DIP) and 170°c/w (S0-8) junction to ambient.

Note 2: Supply current when output high typically 1 mA less at $V_{CC} = 5V$.

Note 3: Tested at $V_{CC} = 5V$ and $V_{CC} = 15V$.

Note 4: This will determine the maximum value of $R_A + R_B$ for 15V operation. The maximum total $(R_A + R_B)$ is 20 MΩ.

Note 5: No protection against excessive pin 7 current is necessary providing the package dissipation rating will not be exceeded.

Note 6: Refer to RETS555X drawing of military LM555H and LM555J versions for specifications.

Connection Diagrams

Metal Can Package

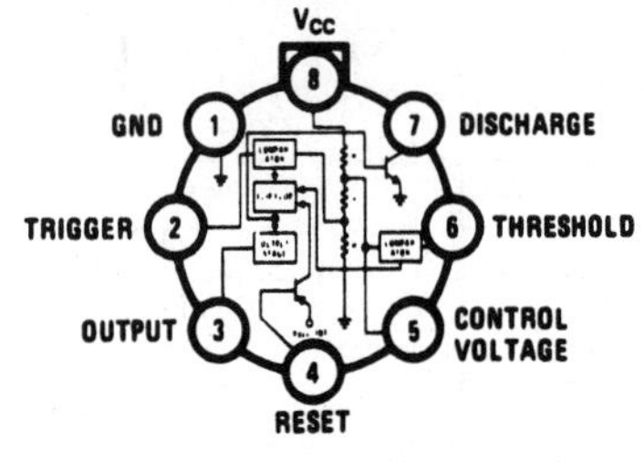

TL/H/7851-2

Top View

Order Number LM555H or LM555CH
See NS Package Number H08C

Dual-In-Line and Small Outline Packages

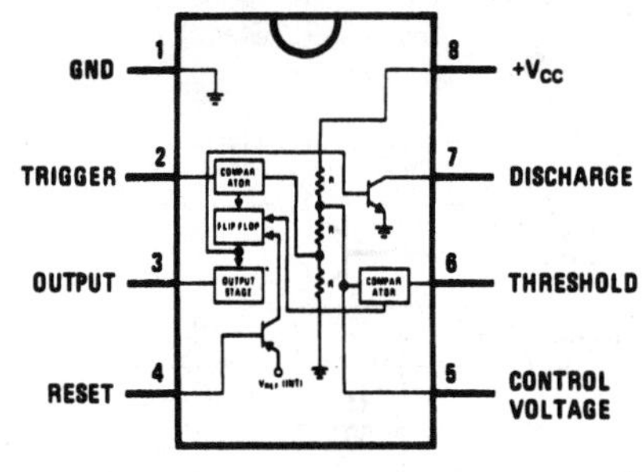

TL/H/7851-3

Top View

Order Number LM555J, LM555CJ, LM555CM or LM555CN
See NS Package Number J08A, M08A or N08E

National Semiconductor

LM741 Operational Amplifier

General Description

The LM741 series are general purpose operational amplifiers which feature improved performance over industry standards like the LM709. They are direct, plug-in replacements for the 709C, LM201, MC1439 and 748 in most applications.

The amplifiers offer many features which make their application nearly foolproof: overload protection on the input and output, no latch-up when the common mode range is exceeded, as well as freedom from oscillations.

The LM741C/LM741E are identical to the LM741/LM741A except that the LM741C/LM741E have their performance guaranteed over a 0°C to +70°C temperature range, instead of −55°C to +125°C.

Schematic Diagram

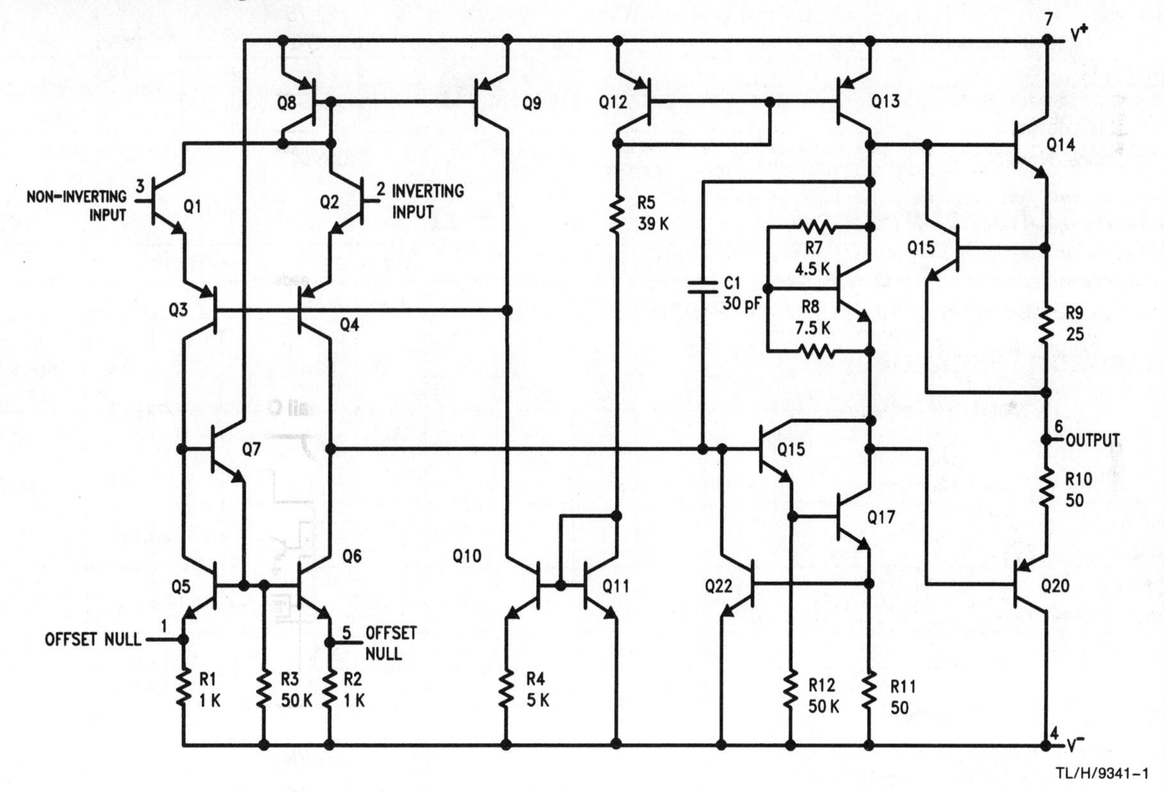

TL/H/9341-1

Offset Nulling Circuit

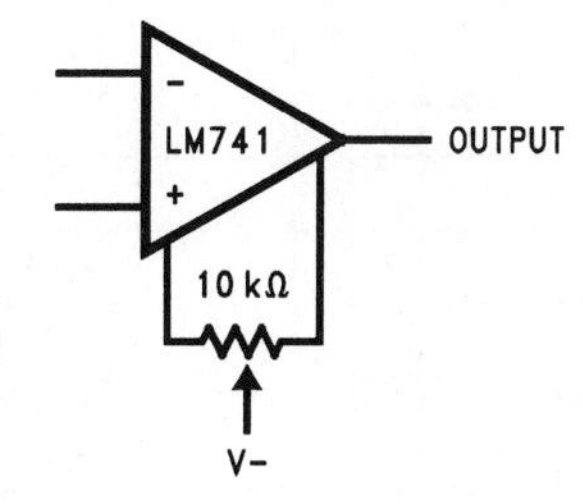

TL/H/9341-7

Absolute Maximum Ratings

If Military/Aerospace specified devices are required, please contact the National Semiconductor Sales Office/ Distributors for availability and specifications.
(Note 5)

	LM741A	LM741E	LM741	LM741C
Supply Voltage	±22V	±22V	±22V	±18V
Power Dissipation (Note 1)	500 mW	500 mW	500 mW	500 mW
Differential Input Voltage	±30V	±30V	±30V	±30V
Input Voltage (Note 2)	±15V	±15V	±15V	±15V
Output Short Circuit Duration	Continuous	Continuous	Continuous	Continuous
Operating Temperature Range	−55°C to +125°C	0°C to +70°C	−55°C to +125°C	0°C to +70°C
Storage Temperature Range	−65°C to +150°C	−65°C to +150°C	−65°C to +150°C	−65°C to +150°C
Junction Temperature	150°C	100°C	150°C	100°C
Soldering Information				
N-Package (10 seconds)	260°C	260°C	260°C	260°C
J- or H-Package (10 seconds)	300°C	300°C	300°C	300°C
M-Package				
Vapor Phase (60 seconds)	215°C	215°C	215°C	215°C
Infrared (15 seconds)	215°C	215°C	215°C	215°C

See AN-450 "Surface Mounting Methods and Their Effect on Product Reliability" for other methods of soldering surface mount devices.

	LM741A	LM741E	LM741	LM741C
ESD Tolerance (Note 6)	400V	400V	400V	400V

Electrical Characteristics (Note 3)

Parameter	Conditions	LM741A/LM741E			LM741			LM741C			Units
		Min	Typ	Max	Min	Typ	Max	Min	Typ	Max	
Input Offset Voltage	$T_A = 25°C$										
	$R_S \leq 10\ k\Omega$					1.0	5.0		2.0	6.0	mV
	$R_S \leq 50\Omega$		0.8	3.0							mV
	$T_{AMIN} \leq T_A \leq T_{AMAX}$										
	$R_S \leq 50\Omega$			4.0							mV
	$R_S \leq 10\ k\Omega$						6.0			7.5	mV
Average Input Offset Voltage Drift				15							μV/°C
Input Offset Voltage Adjustment Range	$T_A = 25°C, V_S = \pm 20V$	±10				±15			±15		mV
Input Offset Current	$T_A = 25°C$		3.0	30		20	200		20	200	nA
	$T_{AMIN} \leq T_A \leq T_{AMAX}$			70		85	500			300	nA
Average Input Offset Current Drift				0.5							nA/°C
Input Bias Current	$T_A = 25°C$		30	80		80	500		80	500	nA
	$T_{AMIN} \leq T_A \leq T_{AMAX}$			0.210			1.5			0.8	μA
Input Resistance	$T_A = 25°C, V_S = \pm 20V$	1.0	6.0		0.3	2.0		0.3	2.0		MΩ
	$T_{AMIN} \leq T_A \leq T_{AMAX}$, $V_S = \pm 20V$	0.5									MΩ
Input Voltage Range	$T_A = 25°C$							±12	±13		V
	$T_{AMIN} \leq T_A \leq T_{AMAX}$				±12	±13					V
Large Signal Voltage Gain	$T_A = 25°C, R_L \geq 2\ k\Omega$										
	$V_S = \pm 20V, V_O = \pm 15V$	50									V/mV
	$V_S = \pm 15V, V_O = \pm 10V$				50	200		20	200		V/mV
	$T_{AMIN} \leq T_A \leq T_{AMAX}$, $R_L \geq 2\ k\Omega$,										
	$V_S = \pm 20V, V_O = \pm 15V$	32									V/mV
	$V_S = \pm 15V, V_O = \pm 10V$				25			15			V/mV
	$V_S = \pm 5V, V_O = \pm 2V$	10									V/mV

Electrical Characteristics (Note 3) (Continued)

Parameter	Conditions	LM741A/LM741E			LM741			LM741C			Units
		Min	Typ	Max	Min	Typ	Max	Min	Typ	Max	
Output Voltage Swing	$V_S = \pm 20V$										
	$R_L \geq 10\ k\Omega$	±16									V
	$R_L \geq 2\ k\Omega$	±15									V
	$V_S = \pm 15V$										
	$R_L \geq 10\ k\Omega$				±12	±14		±12	±14		V
	$R_L \geq 2\ k\Omega$				±10	±13		±10	±13		V
Output Short Circuit Current	$T_A = 25°C$	10	25	35		25			25		mA
	$T_{AMIN} \leq T_A \leq T_{AMAX}$	10		40							mA
Common-Mode Rejection Ratio	$T_{AMIN} \leq T_A \leq T_{AMAX}$										
	$R_S \leq 10\ k\Omega$, $V_{CM} = \pm 12V$				70	90		70	90		dB
	$R_S \leq 50\Omega$, $V_{CM} = \pm 12V$	80	95								dB
Supply Voltage Rejection Ratio	$T_{AMIN} \leq T_A \leq T_{AMAX}$, $V_S = \pm 20V$ to $V_S = \pm 5V$										
	$R_S \leq 50\Omega$	86	96								dB
	$R_S \leq 10\ k\Omega$				77	96		77	96		dB
Transient Response	$T_A = 25°C$, Unity Gain										
Rise Time			0.25	0.8		0.3			0.3		µs
Overshoot			6.0	20		5			5		%
Bandwidth (Note 4)	$T_A = 25°C$	0.437	1.5								MHz
Slew Rate	$T_A = 25°C$, Unity Gain	0.3	0.7			0.5			0.5		V/µs
Supply Current	$T_A = 25°C$					1.7	2.8		1.7	2.8	mA
Power Consumption	$T_A = 25°C$										
	$V_S = \pm 20V$		80	150							mW
	$V_S = \pm 15V$					50	85		50	85	mW
LM741A	$V_S = \pm 20V$										
	$T_A = T_{AMIN}$			165							mW
	$T_A = T_{AMAX}$			135							mW
LM741E	$V_S = \pm 20V$										
	$T_A = T_{AMIN}$			150							mW
	$T_A = T_{AMAX}$			150							mW
LM741	$V_S = \pm 15V$										
	$T_A = T_{AMIN}$					60	100				mW
	$T_A = T_{AMAX}$					45	75				mW

Note 1: For operation at elevated temperatures, these devices must be derated based on thermal resistance, and T_j max. (listed under "Absolute Maximum Ratings"). $T_j = T_A + (\theta_{jA} P_D)$.

Thermal Resistance	Cerdip (J)	DIP (N)	HO8 (H)	SO-8 (M)
θ_{jA} (Junction to Ambient)	100°C/W	100°C/W	170°C/W	195°C/W
θ_{jC} (Junction to Case)	N/A	N/A	25°C/W	N/A

Note 2: For supply voltages less than ±15V, the absolute maximum input voltage is equal to the supply voltage.

Note 3: Unless otherwise specified, these specifications apply for $V_S = \pm 15V$, $-55°C \leq T_A \leq +125°C$ (LM741/LM741A). For the LM741C/LM741E, these specifications are limited to $0°C \leq T_A \leq +70°C$.

Note 4: Calculated value from: BW (MHz) = 0.35/Rise Time(µs).

Note 5: For military specifications see RETS741X for LM741 and RETS741AX for LM741A.

Note 6: Human body model, 1.5 kΩ in series with 100 pF.

Connection Diagrams

Metal Can Package

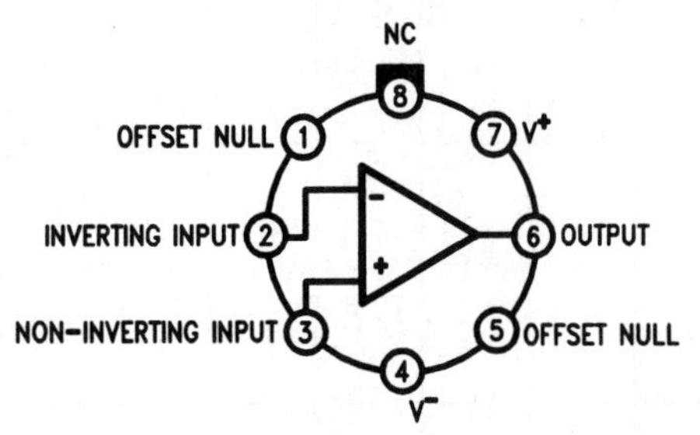

TL/H/9341-2

Order Number LM741H, LM741H/883*, LM741AH/883
LM741CH or LM741EH
See NS Package Number H08C

Ceramic Dual-In-Line Package

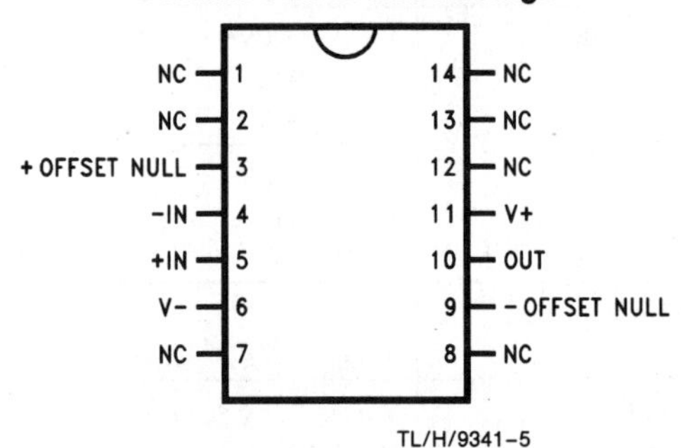

TL/H/9341-5

Order Number LM741J-14/883*, LM741AJ-14/883**
See NS Package Number J14A

*also available per JM38510/10101

**also available per JM38510/10102

Dual-In-Line or S.O. Package

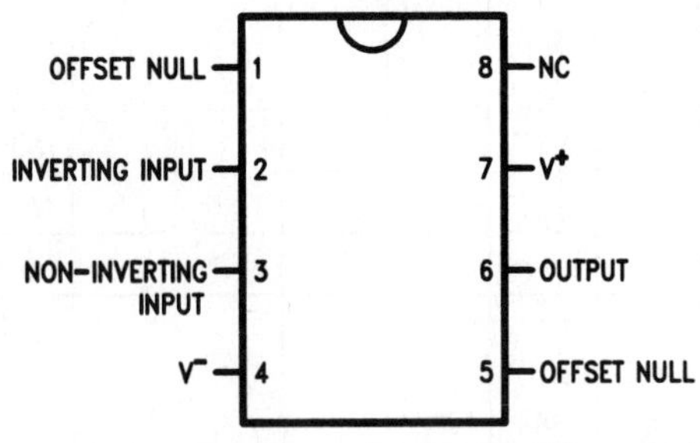

TL/H/9341-3

Order Number LM741J, LM741J/883, LM741CJ,
LM741CM, LM741CN or LM741EN
See NS Package Number J08A, M08A or N08E

Ceramic Flatpak

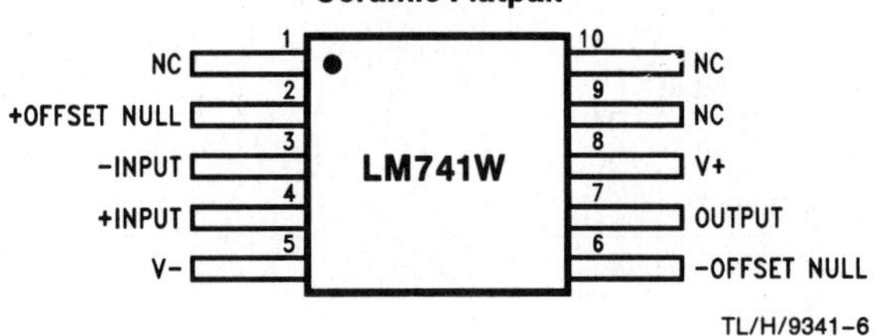

TL/H/9341-6

Order Number LM741W/883
See NS Package Number W10A

*LM741H is available per JM38510/10101